MW00861881

50 EFFECTIVE KNIFE TECHNIQUES

MICHAEL J. McGREAL

AMERICAN TECHNICAL PUBLISHERS
Orland Park, Illinois

Cover Photo: Knives provided by Canada Cutlery Inc.

American Technical Publishers, Inc., Editorial Staff

Editor in Chief:
Jonathan F. Gosse
Vice President—Editorial:
Peter A. Zurlis
Director of Product Development:
Cathy A. Scruggs
Assistant Production Manager:
Nicole D. Bigos
Technical Editor:
Sara M. Marconi
Supervising Copy Editor:
Catherine A. Mini
Copy Editor:
Talia J. Lambarki

Cover Design:
Bethany J. Fisher
Art Supervisor:
Sarah E. Kaducak
Illustration/Layout:
Bethany J. Fisher
Digital Media Coordinator:
Adam T. Schuldt
Digital Resources:
Carl R. Hansen
Cory S. Butler
James R. Hein

The author and publisher would like to thank the following companies for providing images.

Browne Foodservice
Canada Cutlery Inc.
Carlisle FoodService Products
Dexter-Russell, Inc.
Edlund Co.
Kyocera Advanced Ceramics
Mercer Cutlery
Messermeister
Paderno World Cuisine

1 2 3 4 5 6 7 8 9 – 17 – 9 8 7 6 5 4 3 2 1

Printed in the United States of America

ISBN 978-0-8269-4241-8

This book is printed on recycled paper.

ACKNOWLEDGMENTS

The foundation of an impeccable dish starts with cutting foods in a safe and efficient manner that promotes even cooking and enhances presentation. *50 Effective Knife Techniques* provides culinary and hospitality students as well as novice cooks with the foundational knife techniques necessary for culinary success.

Author, Chef Michael J. McGreal, M.Ed., CEC, CCE, CHE, FMP, CHA, MCFE, has over 35 years of experience in the foodservice industry, holding chef positions at some of Chicago's premier restaurants and hotels. Chef McGreal joined the prestigious culinary arts program at Joliet Junior College as an instructor 20 years ago and currently serves as department chair. In addition to *50 Effective Knife Techniques*, Chef McGreal is the author of *Culinary Math Principles and Applications* and *Culinary Arts Principles and Applications*.

As a member of the Chef's Move to Schools National Advisory Committee White House Initiative and a chef consultant for the United States Department of Agriculture's Institute of Child Nutrition, Chef McGreal's contributions to education reach beyond the classroom. This commitment to education is also reflected in the many honors he has received, including the 2011 American Culinary Federation (ACF) Presidential Medallion and the 2010 National Institute for Staff and Organizational Development's (NISOD) Excellence in Teaching Award.

The author and publisher would like to acknowledge Chef Timothy Bucci, CEC, CCE, CHE, FMP, MCFE, CCJ, for his exceptional contributions to this book and the companion videos. After working in restaurants and hotels for 14 years, Chef Bucci taught at the Cooking and Hospitality Institute of Chicago for 10 years prior to coming to Joliet Junior College, where he currently teaches students his passion for cooking. As a member of the 2012 Culinary Team USA, Chef Bucci earned two silver medals at the World Culinary Olympics, took second place overall among 37 countries, and won two gold medals at the 2010 World Culinary Cup. As a mentor, Chef Bucci has coached many culinary students, including the 2013 and 2011 ACF National Champions and the 2013 and 2011 ACF National Student Culinarian of the Year. In 2009, Chef Bucci was named the ACF National Educator of the Year.

The author and publisher also appreciate the technical information and assistance provided by the following:

Chef R. Andrew McColley CEC, CCE, CE
Chef Instructor
Evergreen School District, Vancouver, WA

Joliet Junior College, Joliet, IL
Culinary Arts Department
Students and staff

CONTENTS

CONTENTS

LEARNER RESOURCES

50 Effective Knife Techniques includes access to learner resources that enhance and reinforce the content of the textbook.

- Quick Quizzes®
- Interactive Glossary
- Flash Cards
- Study Questions
- Technique Video Preview
- ATPeResources.com

These online resources can be accessed using either of the following methods.

- Key ATPeResources.com/QuickLinks™ into a web browser and then enter QuickLink™ code 583241.
- Use a Quick Response (QR) reader app on a mobile device to scan the QR code located on the opening page of each section.

Note to Educators: A separate instructor access code is available for the *50 Effective Knife Techniques Instructor Resources*. For more information, visit atplearning.com/knifetechniques.

INTRODUCTION

50 Effective Knife Techniques explains and demonstrates the foundational knife skills necessary to create successful dishes. All 50 knife techniques include step-by-step instructions to reinforce proper cutting techniques for foods ranging from fruits and vegetables to poultry, seafood, and meats. Each instructional step is enhanced with a corresponding image to clearly depict the technique being described. This exceptional book also includes the opportunity to view videos for each technique where a professional chef demonstrates safe and accurate cuts that minimize waste, promote even cooking, and enhance presentations.

Section I: Knife Basics, explains how to select, maintain, and use knives for safe and optimal performance. Preparing a safe work station, sharpening and honing knives, holding knives properly, and correctly positioning foods for cutting are essential skills used daily in the professional kitchen.

Section II: Basic Knife Cuts, demonstrates the foundational knife techniques needed to produce consistent cuts. The ability to safely execute basic knife cuts, such as slicing, dicing, chopping, and mincing, helps build the skills needed to efficiently cut a wide variety of foods.

Section III: Cutting Produce, showcases the various ways in which vegetables and fruits can be cut effectively for the desired application. Cutting produce appropriately enhances a finished dish with complimentary textures and appealing presentations.

Section IV: Cutting and Deboning Poultry, demonstrates the essential techniques required to safely cut and break down whole poultry. Cutting poultry into various pieces and removing bones when desired can decrease food costs, provide portion control, and increase consistency of cooking time.

Section V: Cutting and Trimming Seafood, features the knife techniques necessary to cut a wide range of seafood, including crustaceans, cephalopods, roundfish, flatfish, and mollusks. The ability to cut different types of seafood is cost effective, helps ensure that the product is fresh, and enables recipe collections to expand.

Section VI: Fabricating Meats, highlights the fundamental knife techniques needed to cut and trim raw meats. Properly fabricating meats limits waste and increases the tenderness of the plated item.

Section VII: Carving and Cutting Cooked Proteins, shows the importance of using proper knife techniques to carve and cut cooked poultry, seafood, and meats. Knowing how to properly carve and cut cooked proteins takes an average dish to an exceptional dish.

Knife Techniques Videos DVD
Each knife technique is further reinforced with a high-quality video that showcases the cuts being used in finished dishes, provides step-by-step audio narratives, and allows the user to pause or replay each step as often as needed while refining their knife skills. The *Knife Techniques Videos* DVD can be purchased with the book (Item 4245-6) or separately (Item 4242-5).

SECTION I
KNIFE BASICS

- KNIFE PARTS
- KNIFE TYPES
- KNIFE CONSTRUCTION
- PREPARING SAFE WORK STATIONS
- CARING FOR KNIVES
- HANDLING KNIVES SAFELY
- THE KNIFE HAND
- THE GUIDING HAND
- POSITIONING FOOD ITEMS ON CUTTING BOARDS
- POSITIONING KNIVES FOR CUTTING
- CUTTING METHODS

LEARNER RESOURCES
ATPeResources.com/QuickLinks™
Access Code: 583241

KNIFE BASICS

Knives are one of the most fundamental tools in a kitchen, and knowing how to use knives properly is an indispensable skill. Proper knife techniques create uniform cuts that help reduce waste, promote even cooking, and add visual interest to the way food is presented.

Mastering proper knife techniques starts by selecting the most appropriate knife (or cutting tool) for the intended application. It also includes the ability to handle knives safely and maintain knives correctly. It is equally important to remember that producing knife cuts with precision and speed requires not only the proper technique, but also practice and attention to detail.

KNIFE PARTS

An understanding of knife parts enables the user to choose a well-constructed knife that is suitable for the intended task. Well-constructed knives are comfortable and balanced in the hand. Each part of the knife—the blade, handle, bolster, tang, and rivets—has a specific function.

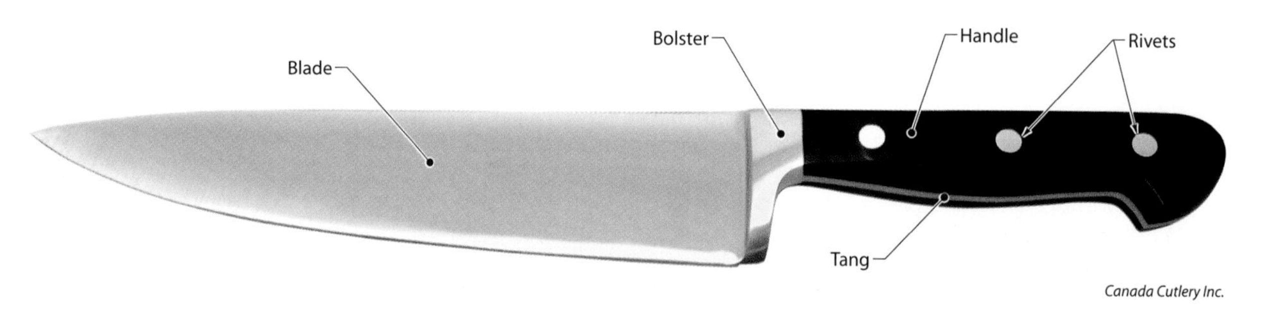

Canada Cutlery Inc.

BLADES

The *blade* is the sharp, flat portion of a knife that is used for cutting food items. Regardless of the blade's flexibility, length, and weight, knife blades have five parts: the heel, tip, point, edge, and spine.

- The *heel* is the rear portion of a knife blade that is most often used to cut thick food items in which more force is required.
- The *tip* is the front quarter of a knife blade. Most cutting is accomplished with the section of the blade between the tip and the heel.
- The *point* is the foremost section of a knife tip that can be used as a piercing tool.

- The *edge* is the sharpened part of a knife blade that extends from the heel to the tip.
- The *spine* is the unsharpened top part of a knife blade that is opposite the edge.

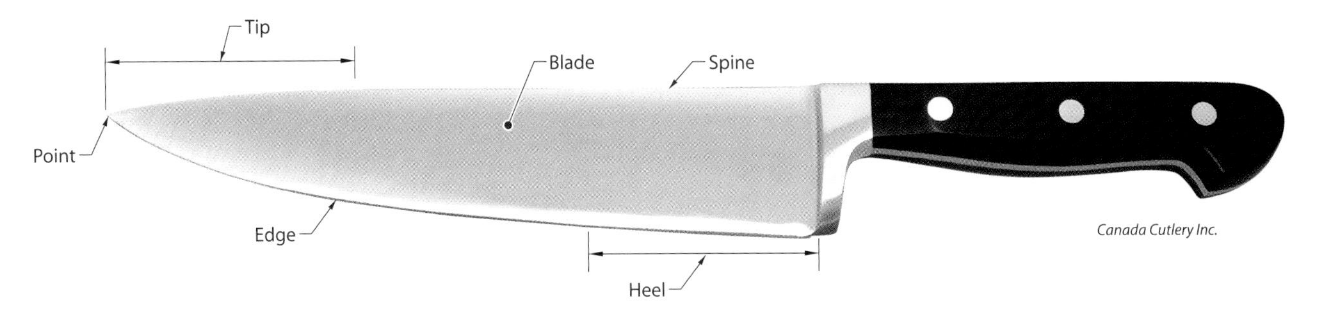

HANDLES

The *handle* is the area of a knife designed to be held in the hand. The knife handle should be comfortable and properly fit the hand. The knife handle should not feel slippery or cause the user to grip excessively hard.

BOLSTERS

The *bolster* is a thick band of metal located where a knife blade joins the handle. The purpose of the bolster is to provide strength to the knife blade and prevent food from entering the seam between the blade and the handle. The bolster also helps balance the knife and helps prevent the user's fingers from slipping during the cutting process. Knives considered of lesser quality often lack a bolster.

TANGS AND RIVETS

The *tang* is the unsharpened tail of a knife blade that extends into the handle. The tang is often secured to the handle with rivets. A *rivet* is a metal fastener used to securely attach the tang of a knife to the handle. High-quality knives typically have a full tang and rivets that are flush with the surface of the handle.

A *full tang* is the tail of a knife blade that extends to the end of the handle and typically contains several rivets. A *partial tang* is a shorter tail of a knife blade that has fewer rivets than a full tang. Partial-tang knives are less durable than full-tang knives but may be used for infrequent or light use. A *rat-tail tang* is a narrow rod of metal that runs the length of a knife handle but is narrower than the handle. Rat-tail tangs are fully enclosed in the handle and are less durable than full or partial tangs.

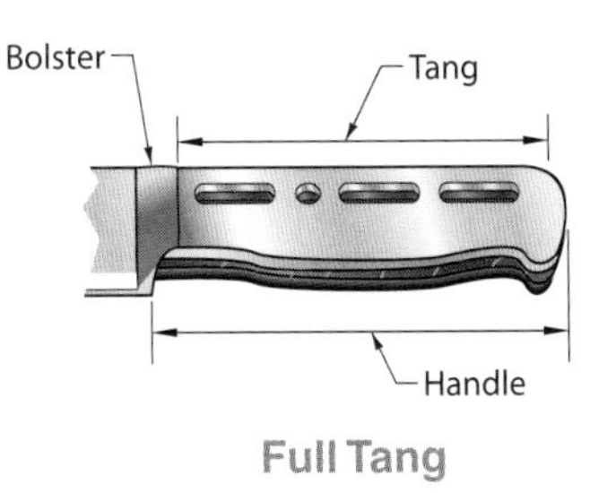

Full Tang

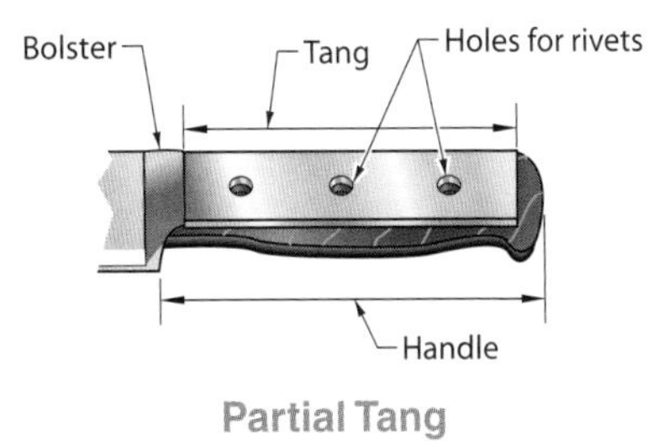

Partial Tang

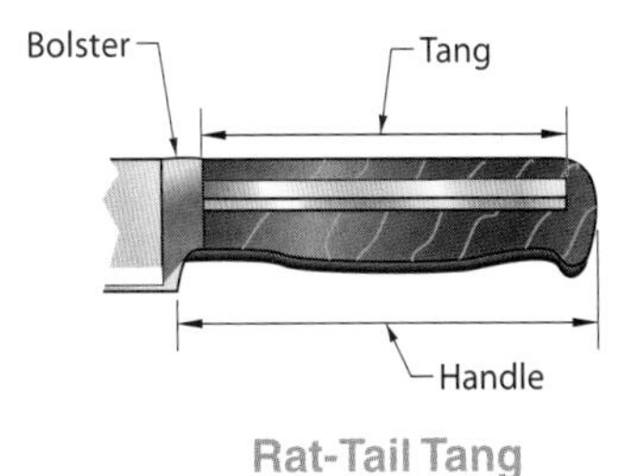

Rat-Tail Tang

KNIFE TYPES

Many different types of knives and special cutting tools are used in the kitchen. Knowing which knife or special cutting tool to use for a given application makes using them safer and more efficient.

LARGE KNIVES

Large knives are used to perform a wide variety of tasks, such as chopping produce, deboning meats, filleting fish, and carving cooked proteins. Large knives commonly used in a professional kitchen include chef's knives, santoku knives, utility knives, boning knives, fillet knives, scimitars, butcher's knives, cleavers, carving knives, slicers, and bread knives.

Chef's Knives. A *chef's knife,* also known as a French knife, is a versatile knife typically with an 8-, 10-, or 12-inch tapering blade that is used to slice, dice, chop, and mince foods. The heel of a chef's knife is wide and tapers to a point.

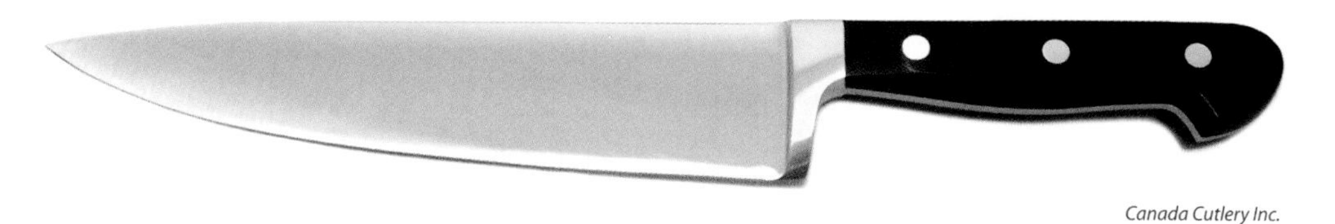

Canada Cutlery Inc.

QUICK TIP

The weight of a chef's knife should be evenly balanced between the blade and the handle to prevent hand and wrist fatigue.

Santoku Knives. A *santoku knife* is a knife with a broad blade and a razor-sharp edge that is less tapered than a chef's knife. A santoku knife blade is typically 5–8 inches in length and commonly has a granton edge (an edge with hollowed out grooves).

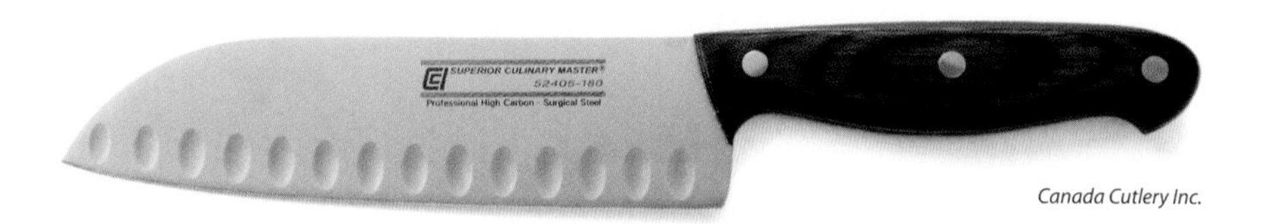

Canada Cutlery Inc.

Utility Knives. A *utility knife* is a multipurpose knife with a stiff blade that is 6–10 inches long and is similar in shape to a chef's knife but much narrower at the heel. The edge of a utility knife may be straight or serrated (have saw-like teeth).

Canada Cutlery Inc.

Boning Knives. A *boning knife* is a knife with a thin, pointed blade that is 5–6 inches long and is used to separate flesh from bones with minimal waste. The blade of a boning knife may be flexible or stiff. Flexible boning knives are often used to fillet fish. Stiff boning knives have a straight or curved blade and are commonly used with meat and poultry to cut between and around bones, joints, and muscle groups.

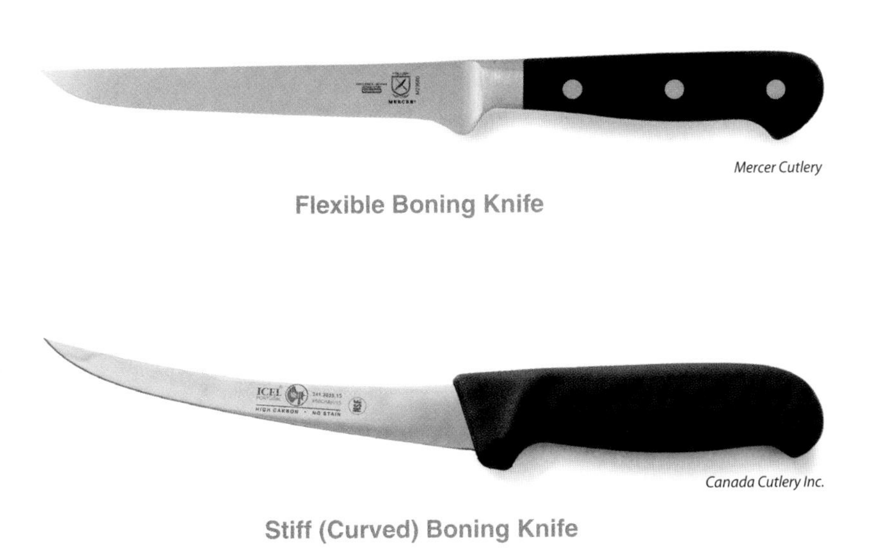
Mercer Cutlery

Flexible Boning Knife

Canada Cutlery Inc.

Stiff (Curved) Boning Knife

Fillet Knives. A *fillet knife* is a knife with a flexible blade and a fine point that is 6–9 inches long and is used to cut delicate flesh. Compared to boning knives, fillet knives have a longer, more flexible blade that is ideal for maneuvering around the skin and bones of fish.

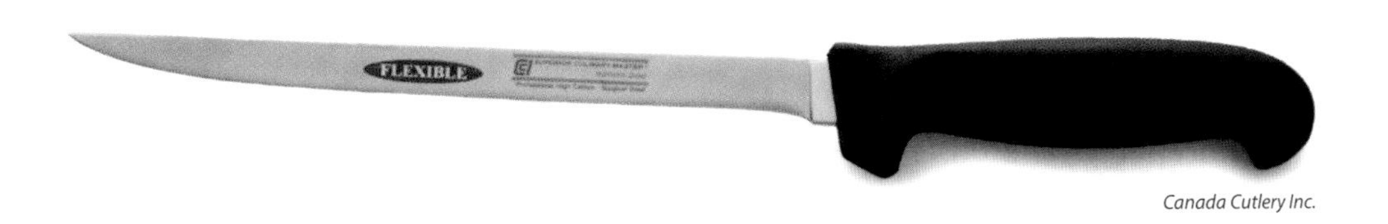

Canada Cutlery Inc.

Scimitars. A *scimitar* is a long knife with an upward-curved tip that is used to cut steaks and primal cuts of meat. The shape of the scimitar blade resembles a stiff (curved) boning knife, yet it is much larger.

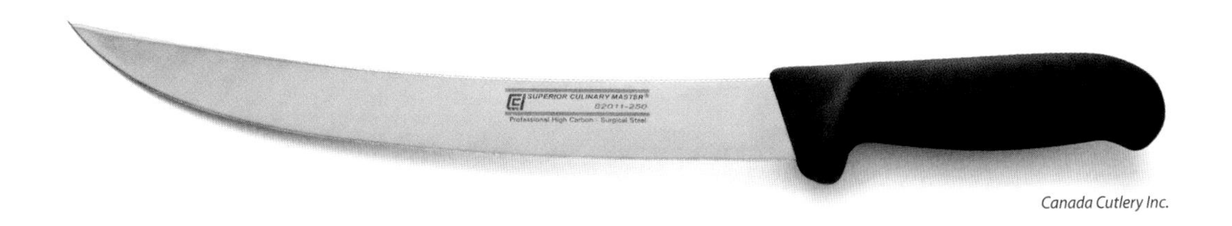
Canada Cutlery Inc.

Butcher's Knives. A *butcher's knife* is a heavy knife with a curved tip and a blade that is 7–14 inches long. The curved tip of a butcher's knife angles upward approximately 25°. A butcher's knife is typically used to cut, section, and portion raw meats.

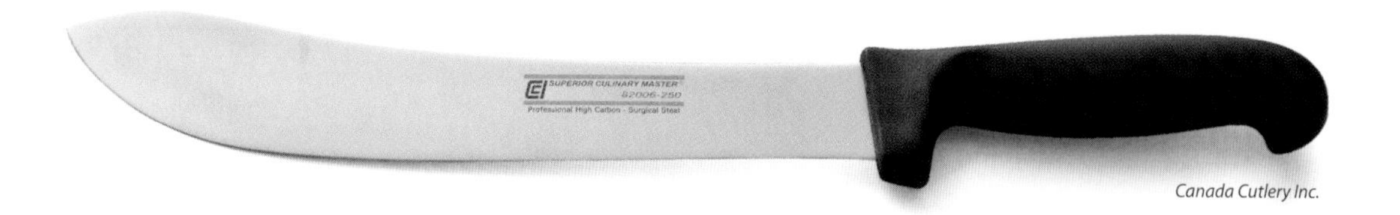

Canada Cutlery Inc.

Cleavers. A *cleaver* is a heavy knife with a rectangular blade that is typically used to cut through large bones. A cleaver is also used to cut through joints where two bones connect.

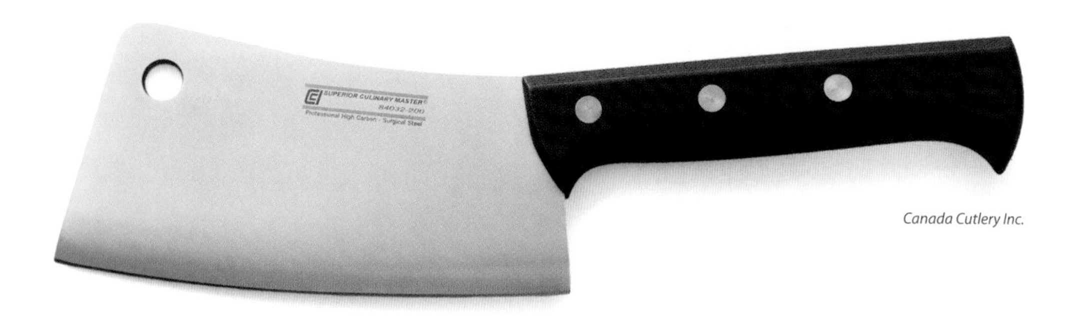

Canada Cutlery Inc.

Carving Knives. A *carving knife* is a knife with a tapered blade and pointed tip that is 8–10 inches long and used to slice cooked meats and poultry. Carving knives are narrower than chef's knives to allow for thinner, more precise slices.

Canada Cutlery Inc.

Slicers. A *slicer* is a knife with a rounded tip and a 10–14 inch long blade that is used to slice cooked meats, poultry, and fish. Slicers may have a straight, serrated, or granton edge. A slicer functions similarly to a carving knife but generally has a longer blade. Slicers are often used at carving stations to slice roasts and hams. They are also used effectively for slicing cakes into layers.

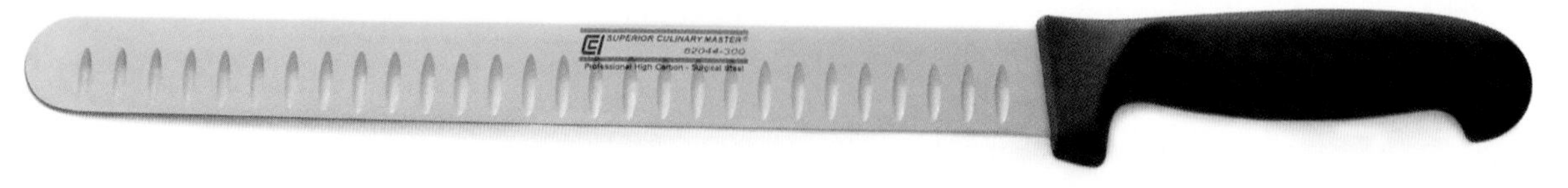

Canada Cutlery Inc.

Bread Knives. A *bread knife* is a knife with a serrated blade that is 8–12 inches long and used to cut through bread without crushing the soft interior.

Canada Cutlery Inc.

SMALL KNIVES

Small knives offer the user the ability to make precise cuts in small areas or to open food items, such as shellfish. Small knives commonly used in the kitchen include paring knives, tourné knives, oyster knives, and clam knives.

Paring Knives. A *paring knife* is a short knife with a stiff blade that is 2–5 inches long and typically used to trim or peel fruits and vegetables.

Canada Cutlery Inc.

Tourné Knives. A *tourné knife,* also known as a bird's beak knife, is a short knife with a curved blade that is primarily used to carve vegetables into a seven-sided football shape with flat ends called a tourné.

Canada Cutlery Inc.

Oyster Knives. An *oyster knife* is a short knife with a dull edge and tapered tip that is used to open oysters. The tapered tip is inserted into the hinge at the back of the oyster and used to pry the shell open.

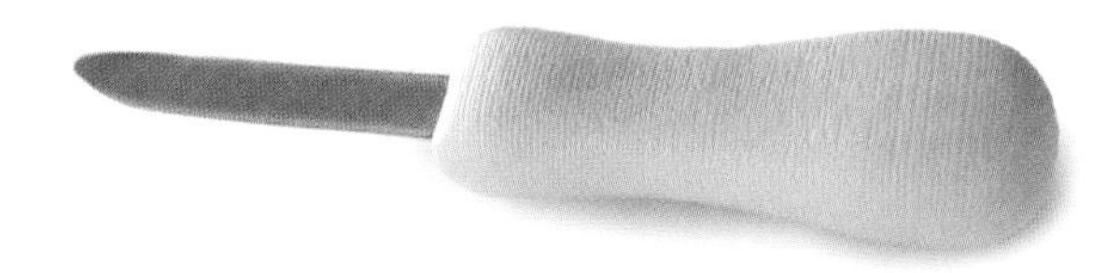

Clam Knives. A *clam knife* is a short knife with a sharp edge and rounded tip that is used to open clams. The sharp edge opens the clam by slicing between the top and bottom shell.

CUTTING TOOLS

In addition to knives, cutting tools are used to cut food items for specific applications. Although there are many special cutting tools, those commonly used in the kitchen include mandolines, peelers, zesters, kitchen shears, channel knives, and parisienne scoops.

Mandolines. A *mandoline* is a cutting tool with adjustable steel blades that is used to cut food into consistently sized pieces. A mandoline can be used to cut foods paper thin as well as produce stick cuts and waffle cuts, also known as gaufrettes.

Paderno World Cuisine

QUICK TIP

A hand guard needs to be in place when using a mandoline because it has a sharp, exposed blade.

Peelers. A *peeler* is a cutting tool with a swiveling, double-edged blade attached to a handle that is used to remove the skin or peel from fruits and vegetables.

Messermeister

Zesters. A *zester* is a cutting tool with tiny blades inside small holes that are attached to a handle. Zesters are typically used on the outer peel of citrus fruits, such as lemons and limes, to yield "zest" that can be used as a natural flavoring.

Dexter-Russell, Inc.

Kitchen Shears. *Kitchen shears* is a cutting tool that operates like heavy-duty scissors and is used to cut through the skin, bones, joints, and ligaments of poultry, the fins of fish, and the shells of some crustaceans.

Canada Cutlery Inc.

Channel Knives. A *channel knife* is a cutting tool with a thin metal blade within a raised channel that is used to remove large strings from the surface of a food item. A channel knife leaves a decorative pattern on the surface of a food item, such as a cucumber.

Dexter-Russell, Inc.

Parisienne Scoops. A *parisienne scoop,* also known as a melon baller, is a cutting tool with either one or two sharp-edged scoops attached to a handle that is used to cut fruits and vegetables into uniform spheres. The scoops on each end of a melon baller range from ½ to 1½ inches in size.

Paderno World Cuisine

KNIFE CONSTRUCTION

Knife construction affects the sharpness and durability of a knife and how a knife will be used and cared for. Key factors to consider when evaluating the construction of a knife blade include the materials used to make the blade, whether the blade is stamped or forged, the type of edge on the blade, and the grind used to produce the edge.

The construction of the knife handle must be considered as well. The knife handle should be comfortable in the hand and easy to keep clean and sanitized. There are also two main knife styles to consider: European/American-style knives and Asian-style knives.

KNIFE BLADE MATERIALS

The most common knife blade materials include ceramic, carbon steel, stainless steel, and high-carbon stainless steel. Each type of material has qualities that affect the performance of the knife.

Ceramic Blades. Ceramic knife blades provide the sharpest edge, stay sharp longer than any other blade materials, do not oxidize (discolor), and are easy to keep clean. However, ceramic blades are made from a rigid material (zirconium oxide) that may cause the blade to chip if it strikes hard objects, such as large bones. Ceramic knives must also be professionally sharpened on diamond wheels or specialty sharpening stones.

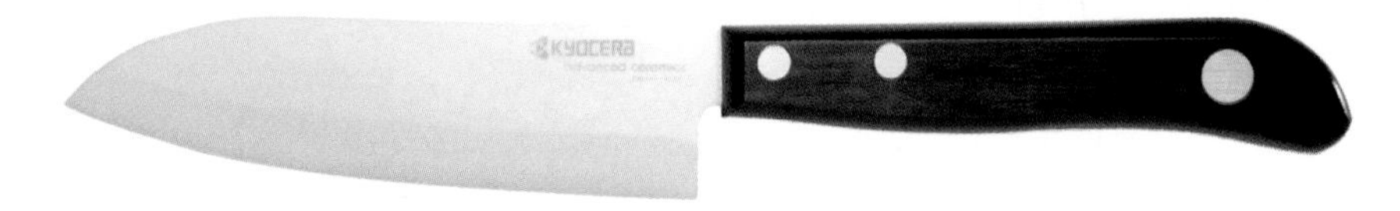

Kyocera Advanced Ceramics

Carbon Steel Blades. Carbon steel is a soft metal that makes knives easy to sharpen. However, the soft metal makes it hard to maintain a sharp edge for extended periods of use. Carbon steel knives also oxidize over time, which can cause some foods to turn brown when cut. To reduce the risk of oxidization, carbon steel knives must be cleaned and dried immediately after use.

Stainless Steel Blades. Stainless steel is a hard metal that produces a strong, durable knife blade. Stainless steel blades keep a sharp edge longer than knives with a carbon steel blade, yet the hardness of stainless steel makes the blade more difficult to sharpen. Knives made from stainless steel are also resistant to oxidation due to chromium, which is an element contained in stainless steel.

High-Carbon Stainless Steel Blades. Most knives used in professional kitchens are made of high-carbon stainless steel. High-carbon stainless steel combines the best qualities of carbon steel and stainless steel. High-carbon stainless steel produces a blade that is easy to keep sharp and does not oxidize.

Canada Cutlery Inc.

QUICK TIP

Damascus-style knives have a unique looking blade. They are produced by folding together two or more types of steel to produce thin layers of varying steel. The steel is then acid-washed, resulting in a knife blade with color variations and a wavy design.

STAMPED AND FORGED KNIFE BLADES

Steel knife blades are either stamped or forged. Stamped knife blades are thinner, lighter blades cut from a flat sheet of steel and then ground to form a sharp edge. Forged knife blades are thicker, heavier blades formed from red-hot steel that is hammered into shape and then ground to create a sharp edge.

High-quality knives are typically forged. Forged knife blades are stronger and keep a sharp edge longer than stamped knife blades. The forging process also allows manufacturers to make a bolster for each knife. A bolster protects the knife hand by allowing the fingers to be positioned safely when cutting and also helps to balance the weight of the knife. Stamped knives do not contain a bolster.

KNIFE BLADE TYPES

Knife blades are constructed with different types of edges to accommodate different cutting tasks. The three basic types of knife blade edges include straight-edge blades, serrated-edge blades, and granton-edge blades.

Straight-Edge Blades. A *straight-edge blade* is a knife blade with a smooth edge from the tip of the knife to the heel. A straight-edge blade is the most common type of knife blade and is designed to perform a variety of cutting tasks. Straight-edge blades cut neatly through foods and allow for greater control than blades with a serrated edge.

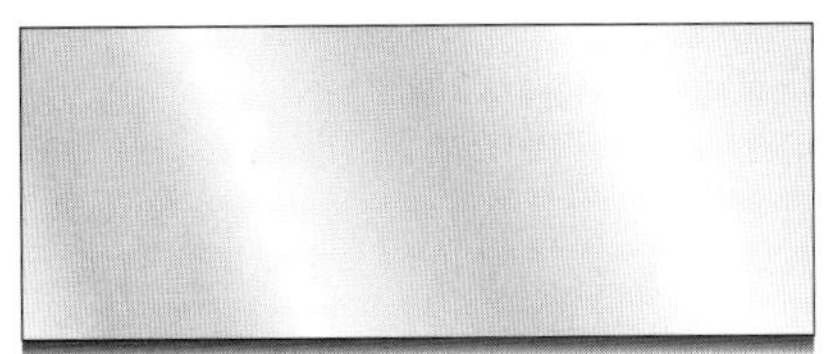

Serrated-Edge Blades. A *serrated-edge blade* is a knife blade with a scallop-shaped or sawlike edge. A serrated-edge blade is designed to neatly cut through foods that have a firmer outer texture and a softer inner texture, such as breads and tomatoes. Because of the jagged edge, serrated knife blades are difficult to sharpen.

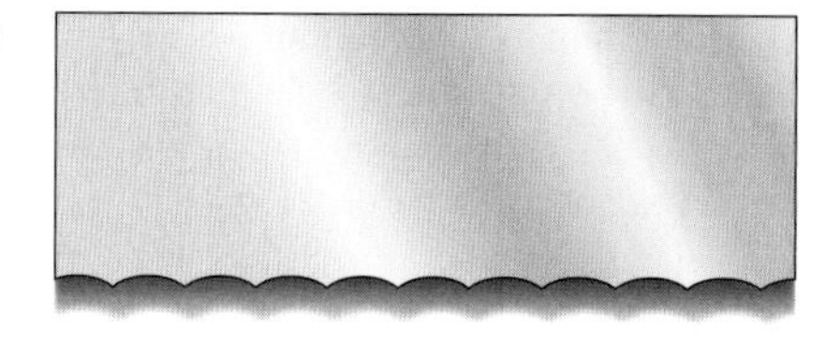

Granton-Edge Blades. A *granton-edge blade* is a knife blade with hollowed out grooves along both sides of the edge. The hollowed grooves create air pockets that prevent foods from sticking to the blade. The grooves also reduce friction on the edge to help produce a smooth cut. Granton-edge blades are often used to thinly slice foods.

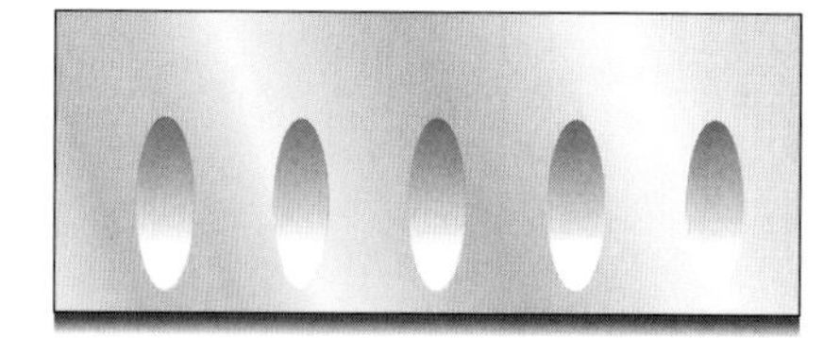

KNIFE EDGE GRINDS

Knife blades are ground to produce the cutting edge. The shape of the cutting edge is determined by the angle of the bevel. A *bevel* is the angled region of a knife blade that has been ground to form the cutting edge. A knife edge with a double bevel is ground on both sides of the blade. A knife edge with a single bevel is ground on one side of the blade.

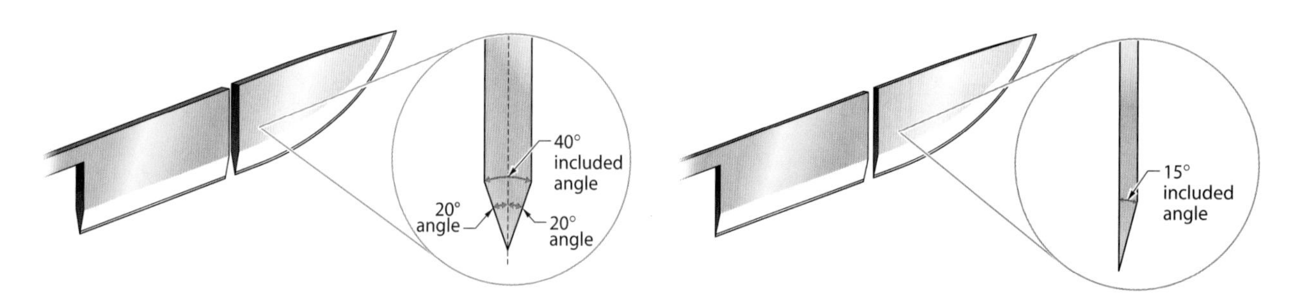

Double-Bevel Knife Edge

Single-Bevel Knife Edge

A knife with a double-beveled edge typically has a larger included angle than a knife with a single-beveled edge. An *included angle* is the sum of the angles on both sides of a knife blade. For example, if a knife has a double-beveled edge that has been ground to a 20° angle on both sides of the blade, then the included angle is 40° (20° + 20° = 40°). Knife edges with a 40° included angle are considered moderately sharp and durable. Knife edges with a 15° included angle are extremely sharp but less durable. The most common knife edge grinds produce V-ground edges, compound-ground edges, convex-ground edges, hollow-ground edges, and chisel-ground edges.

V-Ground Edges. A *V-ground edge,* also known as a taper-ground edge, is a symmetrical knife blade that is ground to form a V-shaped cutting edge. A knife blade with a V-ground edge is symmetrical because the blade is ground at the same angle on both sides of the knife. General-purpose knives, such as chef's knives, utility knives, and paring knives, commonly have a V-ground edge due to its strength and ability to cut a wide range of foods.

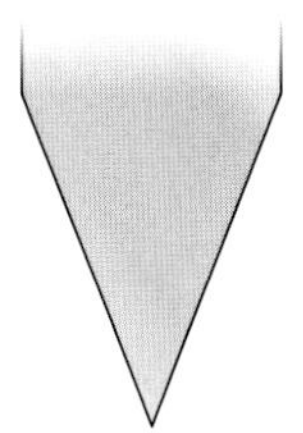

Compound-Ground Edges. A compound-ground edge is a variation of the V-ground edge. A *compound-ground edge* is a symmetrical knife blade that is ground to start forming a V-shape, but before reaching the edge, the blade tapers in again to form a smaller V-shape at the cutting edge. A compound-ground edge is stronger and more durable that a V-ground edge but not as sharp.

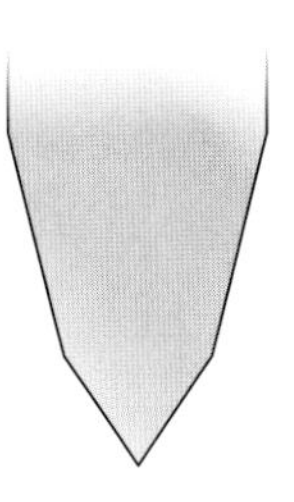

Convex-Ground Edges. A *convex-ground edge* is a symmetrical knife blade that is ground with arced (convex) sides that angle smoothly downward to meet at the cutting edge. Knives with a convex-ground edge are sharp and durable but difficult to sharpen. Due to its strength, a convex-ground edge is commonly found on cleavers.

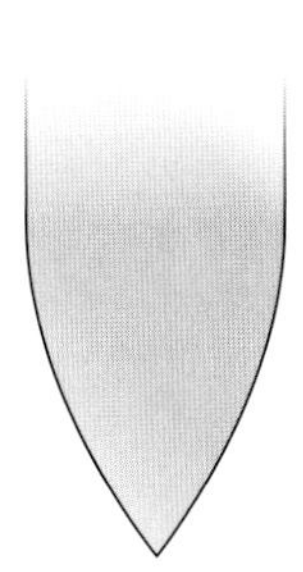

Hollow-Ground Edges. A *hollow-ground edge* is a symmetrical knife blade that is ground with sides that arch inward (concave) to meet at the cutting edge. Hollow-ground knives are sharp but are typically made from low-quality steel that is easily dulled. Eventually, hollow-ground knives become difficult to sharpen manually and require professional machining. Hollow-ground knives are commonly used for tasks that require precision, such as skinning fish and peeling fruits.

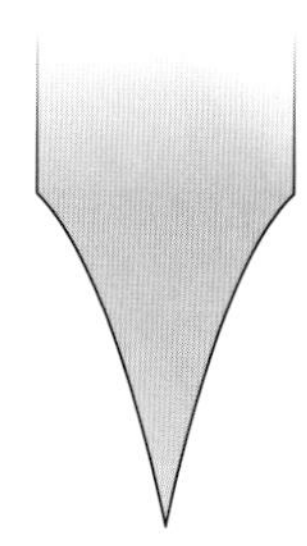

Chisel-Ground Edges. A *chisel-ground edge* is an asymmetrical knife blade that is ground with one angled side that slants toward the cutting edge. A chisel-ground edge is asymmetrical because the blade is ground on just one side of the knife. The asymmetry of a chisel-ground edge results in a small angle, making chisel-edged knives exceptionally sharp but less durable. A chisel-ground edge is commonly found on Asian-style knives that are used to make smooth cuts through delicate foods.

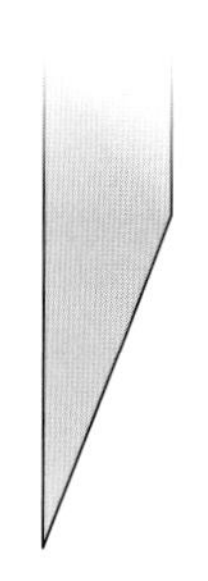

KNIFE HANDLE MATERIALS

Materials used in knife handle construction affect the design and safety of a knife. Handle materials also help determine how comfortable the knife feels in the hand. Knife handles are typically constructed out of plastic, wood, or stainless steel.

Plastic Handles. Plastic knife handles are commonly made from polypropylene plastic or materials similar to plastic, such as engineered polyoxymethylene (POM). Handles made from high-quality plastic or plasticlike materials are popular because they are durable, simple to clean and sanitize, and provide a stable grip.

Canada Cutlery Inc.

Wood Handles. Wood handles are less commonly used due to their lack of durability and the ease at which they trap bacteria. Because wood is porous, wood handles can harbor bacteria that cause foodborne illness. As a result, many local health departments prohibit the use of wood-handled knives in professional kitchens.

Some knife handles are made from a material containing wood that has been treated with a plastic resin (a substance used to produce plastic). The resin treatment reduces the sanitation concerns linked to wood-handled knives while providing the appearance of a wood handle.

Canada Cutlery Inc.

Stainless Steel Handles. Stainless steel handles tend to be heavy, durable, and sanitary but can be slippery, especially when wet. To provide a more secure grip, some knife manufacturers produce a textured steel handle.

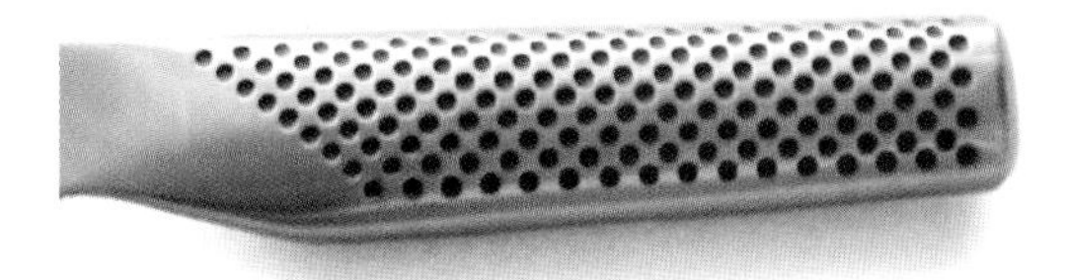

EUROPEAN/AMERICAN-STYLE KNIFE CONSTRUCTION

Knives have historically been designed to efficiently prepare foods that are common to a local culture. For example, a wide variety of fibrous meats and poultry along with hardy vegetables have been considered common staples of the European/American diet. As a result, traditional European/American-style knives are constructed with a heavy, thick blade that is often ground to a 20° angle on each side of the knife.

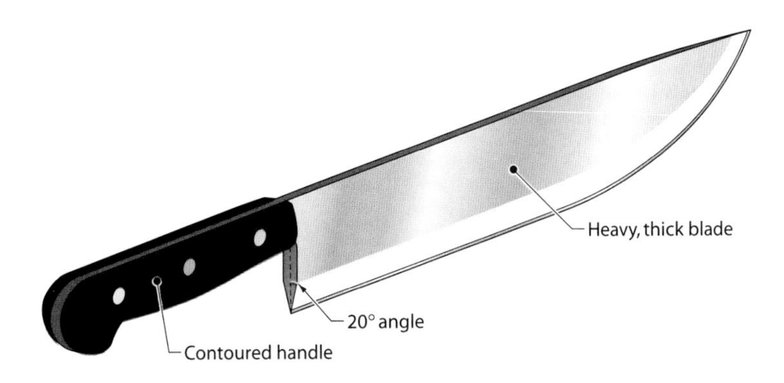

European/American-style knife handles are designed to help balance the weight of the knife in order to prevent fatigue over prolonged periods of use. The handles on European/American-style knives are also contoured, or designed ergonomically, to fit the shape of the hand and provide a comfortable grip.

ASIAN-STYLE KNIFE CONSTRUCTION

Asian-style knives are generally lighter and slimmer than traditional European/American-style knives. The difference is due to the foods indigenous to Asia and the emphasis on precise knife cuts. For example, Asian cuisine has long been associated with softer foods such as fish, shellfish, rice, and tofu. As a result, the slim design of Asian-style knives is ideal for cutting through delicate foods, such as raw fish and sushi rolls, without crushing the cell walls of the food item.

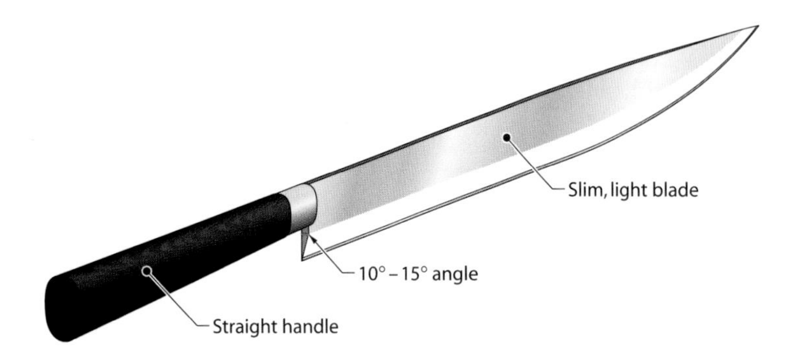

The blades on Asian-style knives are typically ground to a 10° to 15° angle on each side of the knife. Some blades are angled only on one side and are, therefore, specifically designed for the right or left hand. The handles on Asian-style knives are typically straighter and less contoured to the shape of the hand than European/American style knives.

PREPARING SAFE WORK STATIONS

A safe work station reduces the risk of knife injuries and cross-contamination. The first step in preparing a safe work station is to make sure it is clean, sanitized, and free of clutter. Then, the appropriate cutting board is secured in place, knives and equipment are properly assembled, and protective clothing is worn as needed.

It is also important to safely assemble the foods that will be cut and prepared. Produce should always be washed appropriately to reduce the risk of contaminants. Foods must also be kept at a safe temperature.

QUICK TIP

Disposable gloves are typically worn in the foodservice industry, especially when preparing ready-to-eat foods such as raw fruits and vegetables.

CUTTING BOARDS

All cutting boards should be clean, sanitized, and dry before and immediately after use. It is equally important to secure the cutting board to the work surface so that it will not slip during use. Placing a texturized rubber shelf liner under the cutting board is an effective way to anchor the board to the work surface.

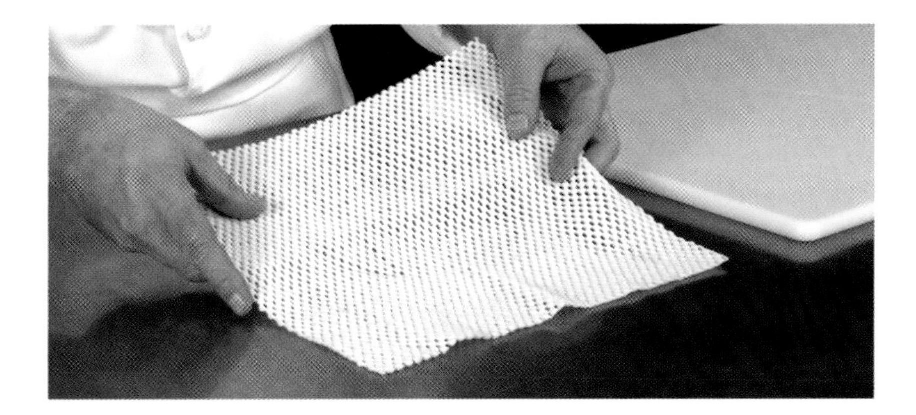

Cutting boards are made from various materials. Metal, glass, and ceramic cutting boards are typically not recommended because they can dull the knife and increase the risk of knife slippage. Most cutting boards used in the professional kitchen are made from plastic, but wood and bamboo cutting boards are also available.

Plastic Cutting Boards. Plastic cutting boards are typically made from polyethylene plastic, which is a nonporous material that reduces the risk of cross-contamination. Cutting boards made of polyethylene are also relatively inexpensive, easy to clean and sanitize, do not dull knives as quickly as other materials, and are NSF certified. Using color-coded nonporous cutting boards to designate specific types of foods may also reduce the risk of cross-contamination.

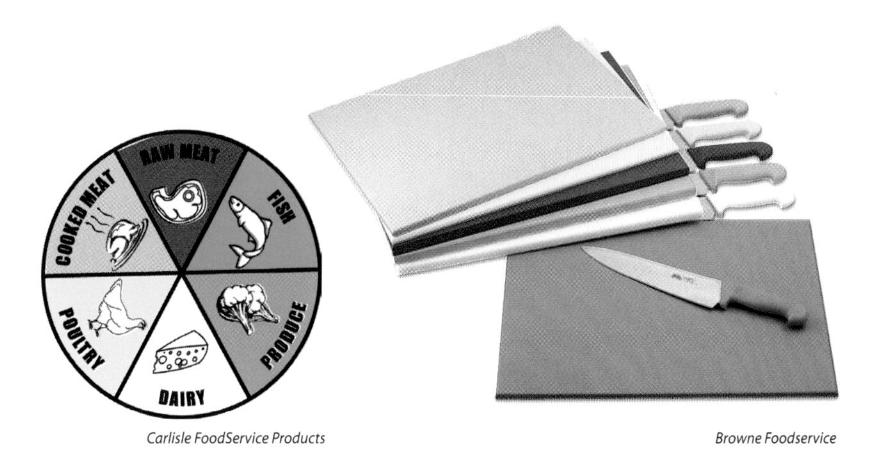

Carlisle FoodService Products *Browne Foodservice*

Wood Cutting Boards. Wood cutting boards are made of hardwoods, such as walnut or maple. The porous nature of wood cutting boards increases the risk of bacterial contamination. Wood cutting boards also tend to show scarring from knife cuts. Any cutting board that has deep scarring should be replaced as it can harbor bacteria, food, and odors.

Bamboo Cutting Boards. Bamboo cutting boards look similar to wood cutting boards but are more durable because bamboo is a stronger, less porous material. This means that bamboo cutting boards resist scratching from knives and absorb little moisture, making them more resistant to bacteria than wood cutting boards.

ASSEMBLING KNIVES AND EQUIPMENT

After selecting the appropriate knives or cutting tools for a task, it is important to assemble the knives in a manner that keeps them sharp and sanitary. This is done by setting the knives on a folded, clean kitchen towel that has been placed next to the cutting board. The towel provides a cushion between the knives and the work surface that helps prevent the knives from dulling. This practice also allows knives to remain clean and sanitary prior to use. Once a knife is used, it is placed back on the towel only after it has been properly cleaned and sanitized.

Any equipment necessary to complete the task should also be assembled. This includes items such as bowls, pots, pans, and measuring tools.

PROTECTIVE CLOTHING

Protective clothing, such as aprons and cut-resistant gloves, are commonly worn when an extensive amount of knife work is required. Aprons are made of a thick material designed to add a layer of safety over the body. Cut-resistant gloves are made of a strong woven material and are often worn to protect the hands when fabricating meats, shucking oysters, and using cutting tools, such as mandolines.

> **QUICK TIP**
> *Although cuts can easily occur in the kitchen, the frequency of cuts can be reduced by using proper knife skills and by wearing protective clothing. If a cut to the hand or fingers occurs in the professional kitchen, it must be immediately covered with a waterproof covering such as a disposable glove.*

CARING FOR KNIVES

Knives that are properly cared for are safer to use, cut more efficiently, and last longer. Proper knife care starts by checking the cutting edge to ensure that it is sharp and properly maintained. If the edge is dull, more pressure must be applied to the knife when cutting. Placing extra pressure on the knife can cause the blade or the food to slip and cause injuries.

Keeping a knife sharp and properly maintained requires both sharpening and honing the knife as needed. Having knives professionally sharpened once a year can be a good practice to follow. This helps ensure that the cutting edge is evenly sharpened and maintains the proper angle.

In addition to keeping knives sharpened and honed, knives must also be cleaned and stored properly. Knives that are cleaned and stored properly are sanitary, hold a sharp edge longer, and are less likely to warp or crack.

SHARPENING KNIVES

Although handheld or electric sharpeners are available, a whetstone is typically used to sharpen professional knives. A *whetstone* is a stone that is used to grind the edge of a blade to the proper angle for sharpness. A three-sided whetstone has a coarse-grit, a medium-grit, and a fine-grit side. A two-sided whetstone has a medium or slightly coarse-grit side and a fine-grit side. A very coarse stone removes chips from the blade to restore the cutting edge. A medium to slightly coarse stone restores the cutting edge when the knife is dull but not damaged. A fine-grit stone aligns and polishes the cutting edge.

To sharpen a knife, the knife blade is held at a specific angle to the stone. To achieve this angle, the knife blade is held at a 90° angle straight above the whetstone as if it were cutting the stone in half. Then, the knife is tilted halfway toward the stone to form a 45° angle, and then halfway again to form an angle between 20° and 25°. The knife is often tilted halfway again to find the ideal sharpening angle between 10° and 18°. The reason for the small angle is that most knives manufactured today are constructed with sharper blades.

1. Hold the knife blade straight above the whetstone as if cutting the stone in half.
2. Tilt the blade halfway toward the stone to form a 45° angle.
3. Tilt the blade halfway again to form an angle of 20°–25°.
4. Tilt the blade halfway again to form a sharpening angle of 10°–18°.

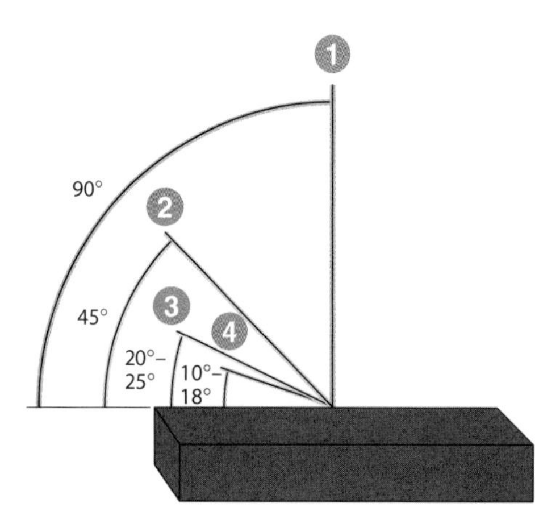

After the proper angle is achieved, the knife blade is then slowly dragged across the stone from heel to tip while applying light pressure. The knife is flipped over and the process is repeated on the other side of the blade to create a sharp, even edge.

VIDEO: KNIFE BASICS

SHARPENING KNIVES

1. Depending on the whetstone type, rub oil or water over the stone to keep it lubricated while sharpening. Use the coarse-grit side of a whetstone and place the heel of the knife near the end of the stone, tilting the spine to form a 10° to 18° angle.

2. While maintaining the 10° to 18° angle, draw the knife blade across the length of the whetstone in a continuous arclike motion, starting with the knife heel and ending with the knife tip. Repeat this step 8–10 times.

   Note: Apply approximately 2 lb of pressure on the knife while drawing it across the stone.

SHARPENING KNIVES (CONTINUED)

3. Flip the knife over and repeat Steps 1–2 on the other side of the knife blade.

4. Turn the whetstone over to the fine-grit side and repeat the sharpening procedure.

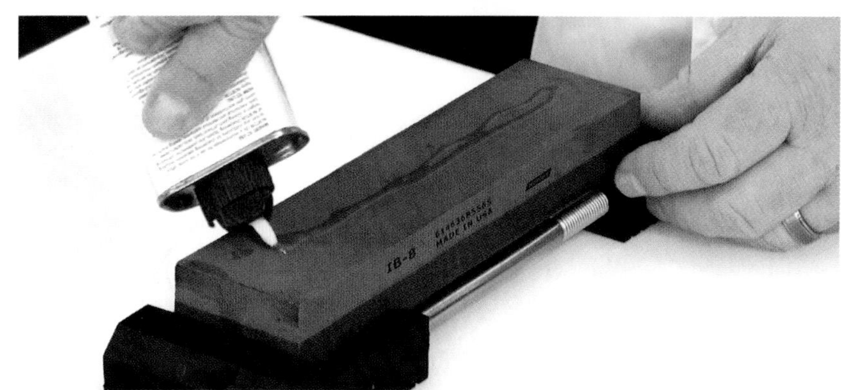

5. With the blade facing away from the body, use a clean, folded towel to wipe residue from the knife blade.

   Note: Clean and sanitize the knife prior to using.

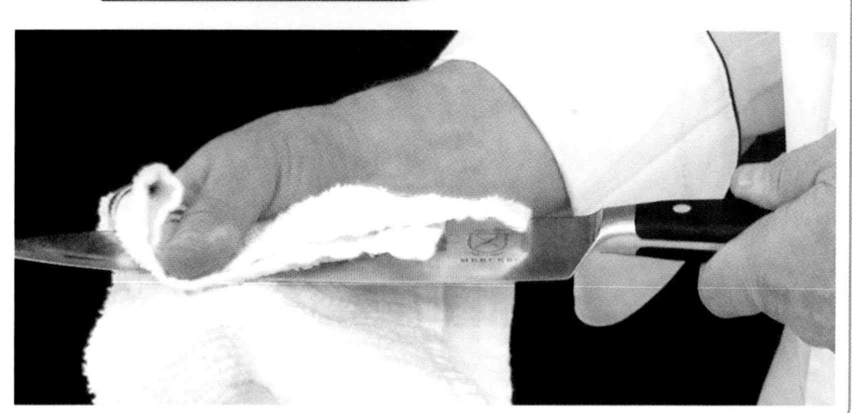

HONING KNIVES

Honing, also known as truing, is the process of aligning a knife blade's edge and removing any burrs or rough spots on the blade. A steel is used to hone professional knives. A *steel,* also known as a butcher's steel, is a steel rod approximately 18 inches long that is attached to a handle and used to align the edge of knife blades.

A steel should be used to hone the knife blade after sharpening and between sharpenings to maintain a sharp, smooth edge. The steel has a magnetic tip that catches the metal fragments as they are removed from the blade. After each use, the steel should be wiped from the handle to the tip with a clean towel to remove any metal fragments.

The 10° to 18° angle used to sharpen knives is also used to hone knives. To achieve this angle, hold the steel perpendicular or pointed toward the cutting board and hold the knife blade at a 10° to 18° angle in relation to the steel.

VIDEO: KNIFE BASICS

POSITIONING KNIVES FOR HONING

1. Hold the knife blade at a 90° angle to the steel.

2. Adjust the knife blade to half that angle (45°).

POSITIONING KNIVES FOR HONING (CONTINUED)

3. Adjust the knife blade to half that angle (20° to 25°).

4. Adjust the blade about halfway again to reach the correct 10° to 18° angle.

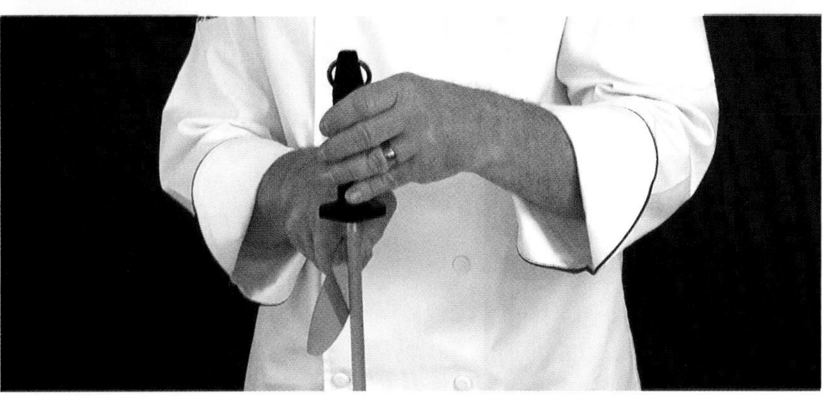

Panderno World Cuisine

HONING KNIVES

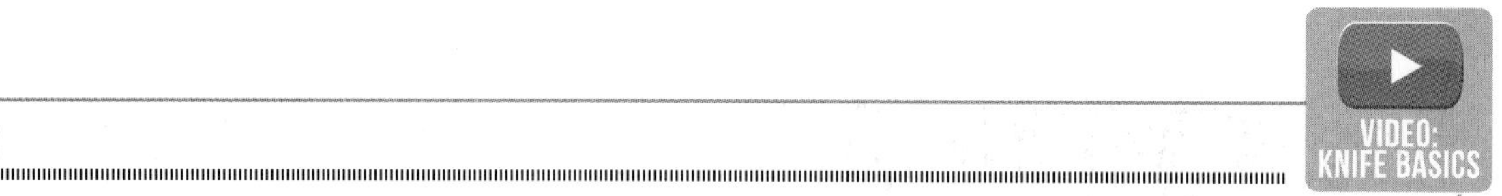

1. Grasp the handle of a steel and place the tip on the cutting board so that the steel is perpendicular to the board. Place the heel of a knife at a 10° to 18° angle along one side of the steel near the handle.

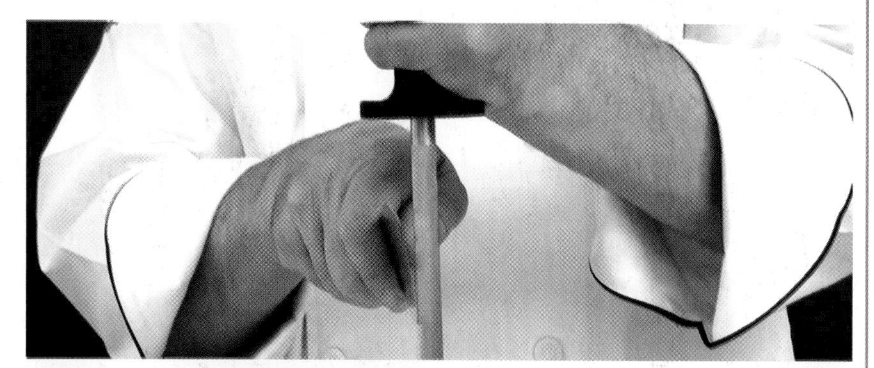

2. With gentle pressure, slide the knife down the steel, moving the blade in an arc along the steel. Finish the stroke with the tip of the knife at the bottom of the steel.

3. Place the heel of the knife at a 10° to 18° angle along the other side of the steel near the handle and repeat Step 2.

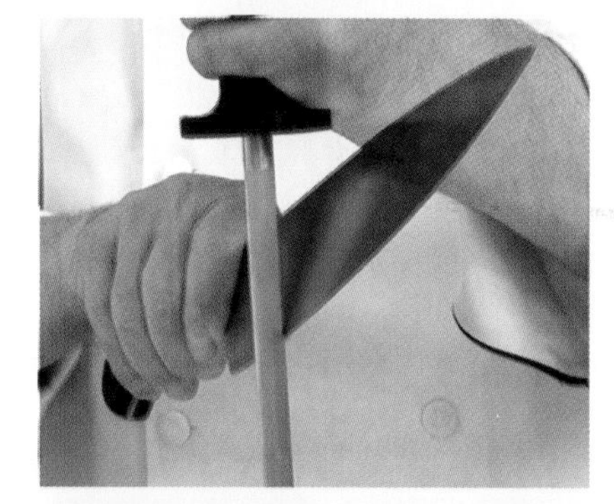

4. Repeat the honing procedure 3–5 times, using the same number of strokes on each side of the blade.

5. With the blade facing away from the body, use a clean, folded towel to wipe residue from the knife blade.

   Note: Clean and sanitize the knife prior to using.

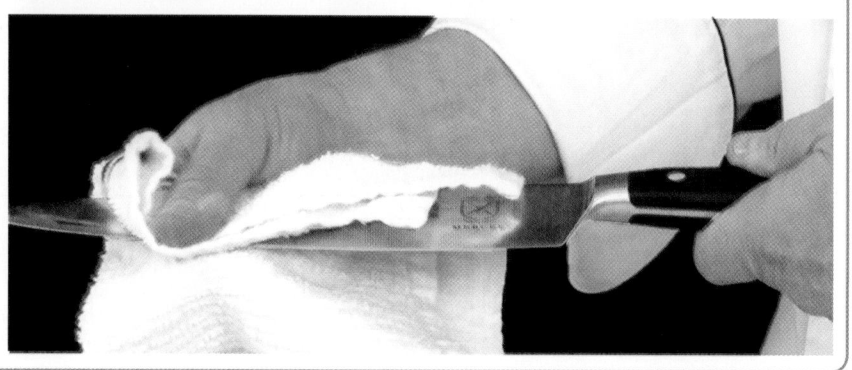

CLEANING KNIVES

Knives should be clean, sanitized, and dry between each use and before storage. Knives should never be washed in a dish machine because the heat and chemicals can ruin the handles. Instead, knives should be washed by hand with hot, soapy water and thoroughly rinsed with hot, clean water. Then the knives should be wiped down with an approved sanitizing solution and dried.

STORING KNIVES

Knives are stored in a manner that helps them maintain a sharp edge and reduces the risk of personal injury. There are a number of ways to store knives safely, including the use of knife guards, knife kits, magnetized knife holders, and slotted knife holders.

Knife Guards. Knife guards, also known as blade guards or sheaths, are individual covers that are placed over knife blades. Covering the knife blade protects the edge and allows several knives to be safely stored in one compartment.

Mercer Cutlery

Knife Kits. Knife kits come in a variety of styles, including bags, rolls, and tool-box-style cases. Knife kits are designed with individual compartments or sleeves to store knives safely and securely. Since knife kits are portable, they provide a safe and convenient way to transport knives.

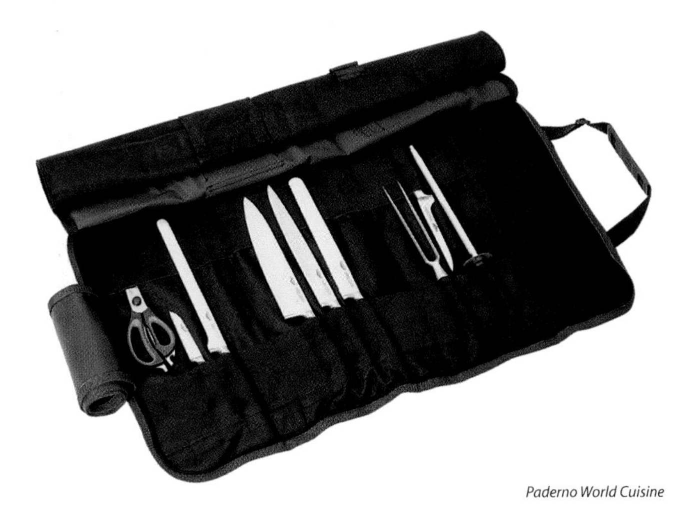

Paderno World Cuisine

Magnetized Knife Holders. Magnetized knife holders consist of magnetic strips that are mounted to a wall. The magnetic bond between steel knife blades and the magnetized strips is strong enough to hold the knives against the wall. Magnetized knife holders do not hold ceramic knives unless the manufacturer has added metal to the blade.

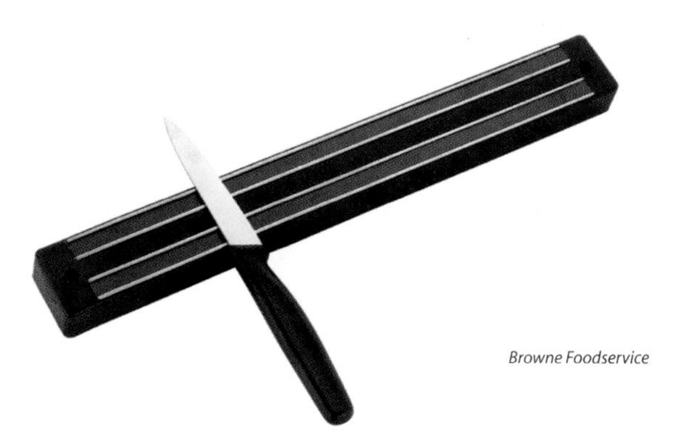

Browne Foodservice

Slotted Knife Holders. Slotted knife holders are typically containers with various sized slots that are designed to hold different sized knives. Slotted knife holders are often made of steel or wood. Steel knife holders are nonporous, which limits the risk of harboring dangerous microorganisms. Wood knife holders are typically treated to make them less porous and water resistant. Some slotted knife holders are fitted with a removable rubber insert that can be cleaned and sanitized.

Edlund Co.

QUICK TIP

An in-drawer knife holder allows knives to be stored safely and neatly in a drawer. Many in-drawer knife holders are available with specifically shaped slots to accommodate a wide range of knife styles and sizes.

HANDLING KNIVES SAFELY

Because knife usage occurs daily in most kitchens, it is imperative to handle knives safely in order to limit the risk of injury. To handle knives safely, always adhere to the following safety precautions:

- Select the appropriate knife for each task and use the knife only for its intended purpose. Never use a knife as a can, bottle, or jar opener.
- Grip the knife properly in the dominant hand to ensure safety and control.
- Always use a sharp knife and apply light pressure when cutting. Excessive pressure increases the risk of the knife slipping and causing harm.
- Clean, sanitize, and dry knives after each use. Never leave a knife in a sink, as someone might reach in and cut himself.
- Wipe a knife blade with the edge of the blade facing away from the hand and body.
- Pass a knife by laying it on a table with the knife blade edge facing away from the intended recipient and cautiously slide the knife forward.
- When walking with a knife, always keep the knife point facing down and hold the knife along the side of the body with the cutting edge facing back.
- Never attempt to catch a falling knife. Let the knife fall and come to a complete rest before picking it up by the handle.
- Store knives properly to prevent injury and to keep the knives from colliding, which can damage and dull the knives.

THE KNIFE HAND

The hand that holds the knife is referred to as the knife hand. The knife hand is typically the user's dominant hand. Therefore, the knife hand for a right-handed individual is typically the right hand, while the knife hand for a left-handed individual is typically the left hand.

The knife hand is used to properly grip the knife in order to cut foods safely, accurately, and efficiently. There are different acceptable methods for gripping a knife, but the pinch grip, choke grip, and overhand grip are typically the most frequently used knife grips.

PINCH GRIP

A pinch grip is used with both large and small knives and is the most common way to hold a chef's knife. This knife grip is called a pinch grip because the knife blade is "pinched" with the index finger and thumb at the heel of the blade near the bolster. The index finger is slightly tucked away from the knife edge and the remaining three fingers are gently curled around the handle. The pinch grip provides stability and control and helps prevent fatigue by putting less stress on the hand.

CHOKE GRIP

A choke grip is used with a paring knife or tourné knife to trim produce or create decorative patterns on foods, such as beets and potatoes. The choke grip requires the knife hand to "choke up" on the knife so that the index finger and middle finger wrap around the spine of the knife, revealing just the top third of the blade. The cutting edge is directed toward the thumb with the remaining fingers gently curled around the handle. Cutting is done by pulling the blade through the food toward the thumb.

OVERHAND GRIP

When using a stiff boning knife to fabricate bone-in cuts of meat, the knife is often held in an overhand grip. The overhand grip, also known as a fist or stabbing grip, occurs when the knife is held vertically so that the knife point is facing down toward the food item being cut. The fingers and thumb wrap around the handle like they are making a fist, and the thumb is positioned over the index finger. An overhand grip allows the knife to be held tightly so that greater force can safely be applied when cutting through bones, ligaments, and cartilage.

THE GUIDING HAND

The hand not holding the knife is referred to as the guiding hand. The guiding hand is typically the user's nondominant hand. Therefore, the guiding hand for a right-handed individual is typically the left hand, while the guiding hand for a left-handed individual is typically the right hand.

The guiding hand controls the food item being cut so that the item stays secure and does not slip. The guiding hand also makes it easier to regulate the size of the cut. The guiding hand is frequently positioned in a claw grip or parallel hold, but other guiding hand positions may be required.

CLAW GRIP

A claw grip is one of the most common positions for the guiding hand. To place the fingers of the guiding hand into a claw grip, the shape of a spider should be imitated on the table. The fingertips should all be slightly tucked, yet touch the surface of the table. A claw grip is used to safely hold the food next to the knife blade. The side of the knife blade should rest against the knuckle of the middle finger or the index finger on the guiding hand. This reduces the risk of cutting fingers.

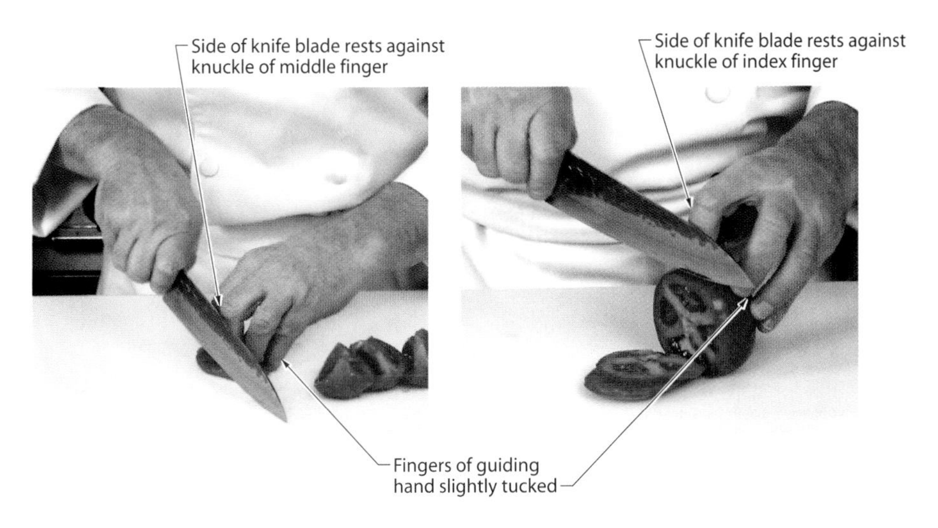

PARALLEL HOLD

A parallel hold is done by placing the guiding hand open and flat on top of the food item to be cut. A parallel hold is generally used when making a horizontal cut, such as slicing a bagel. A parallel hold is also commonly used when preparing to dice produce items, such as cucumbers or onions.

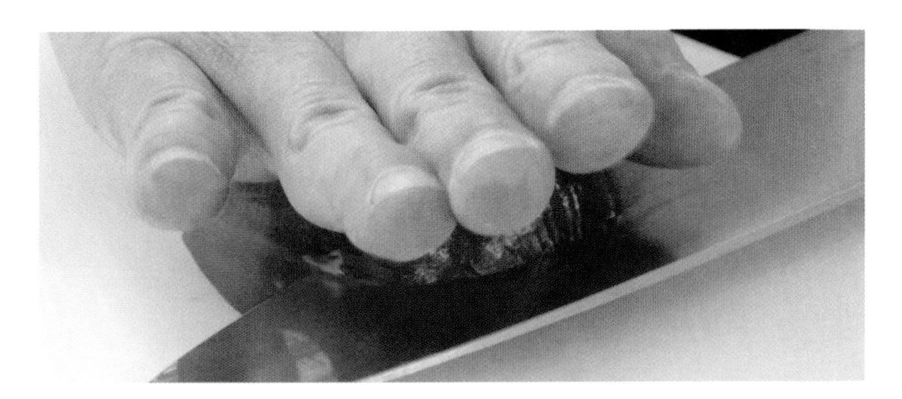

OTHER GUIDING HAND POSITIONS

The position of the guiding hand is somewhat flexible as long as both hands are safely positioned and the knife technique is safely executed. For example, the guiding hand typically holds an onion above the cutting board when the knife hand is holding a paring knife in a choke grip to peel away the skin. When this occurs, the guiding hand is responsible for holding or turning the food item. The guiding hand may also be used to push down on a chicken leg when disjointing poultry or hold a chef's fork when carving cooked meats.

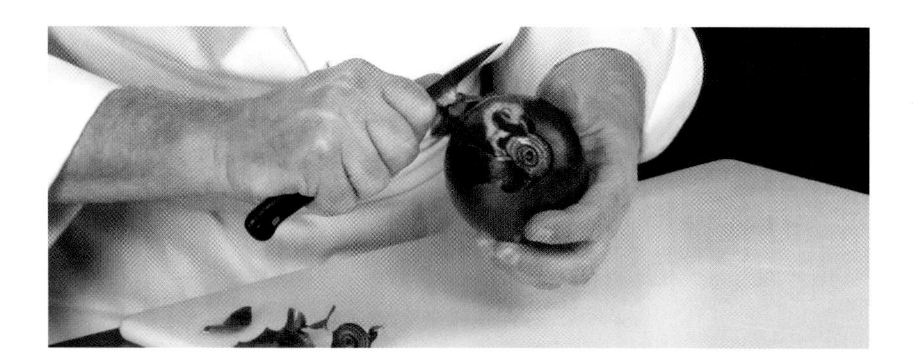

POSITIONING FOOD ITEMS ON CUTTING BOARDS

Just as important as using the knife hand and guiding hand correctly is the position of the food item being cut. Proper placement of food items on the cutting board leads to safer and more accurate knife cuts. To correctly position food items on the cutting board, the following phrases are used: knife hand side, guiding hand side, the body, and side up or side down.

KNIFE HAND SIDE

The "knife hand side" refers to the right or left side of the cutting board, depending on the user's dominant knife hand. For a right-handed individual, if the tail of a fish is facing the knife hand side, the tail would be positioned toward the right side of the cutting board. For a left-handed individual, the tail would be positioned toward the left side of the cutting board.

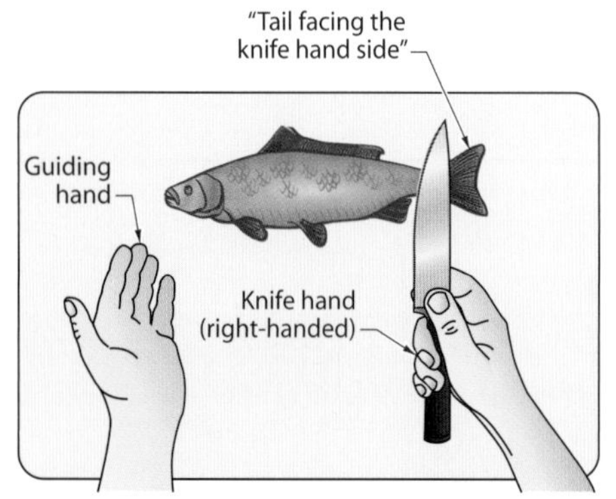

Knife Hand Side to the Right
(Right-Hand Dominant)

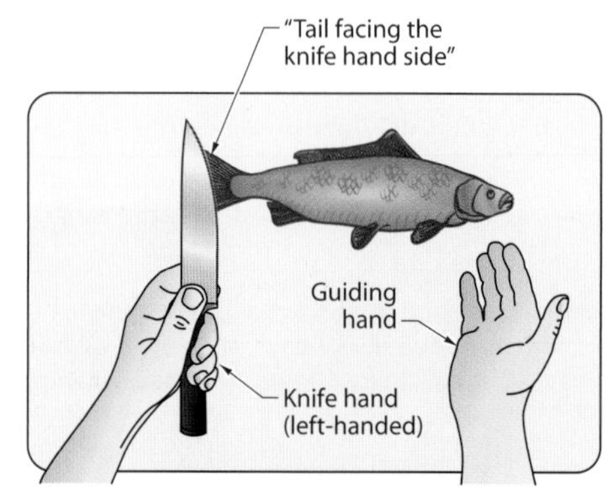

Knife Hand Side to the Left
(Left-Hand Dominant)

GUIDING HAND SIDE

The "guiding hand side" refers to the right or left side of the cutting board, depending on the user's nondominant hand. For a right-handed individual, if the head of a fish is facing the guiding hand side, the head would be positioned toward the left side of the cutting board. For a left-handed individual, the head would be positioned toward the right side of the cutting board.

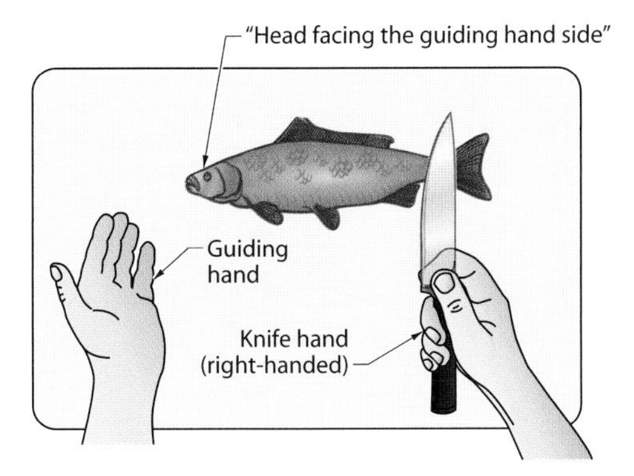

Guiding Hand Side to the Left (Right-Hand Dominant)

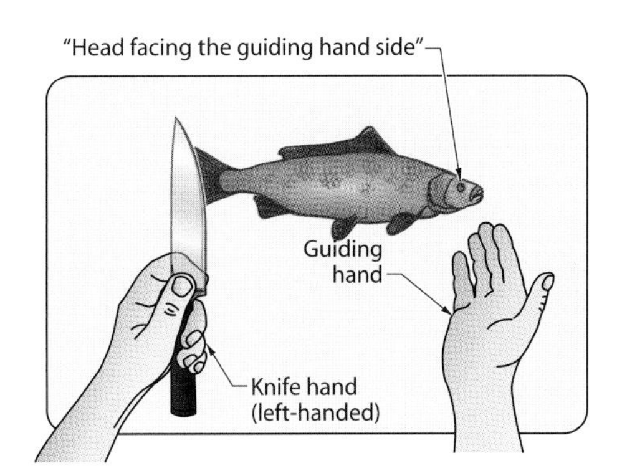

Guiding Hand Side to the Right (Left-Hand Dominant)

THE BODY

"The body" refers to the user or the individual executing the knife technique. If the tail of a fish is facing the body or closest to the body, the tail would be positioned near the side of the cutting board closest to the user. Likewise, if the tail of a fish is facing away from the body or farthest from the body, the tail is positioned near the side of the cutting board farthest from the user.

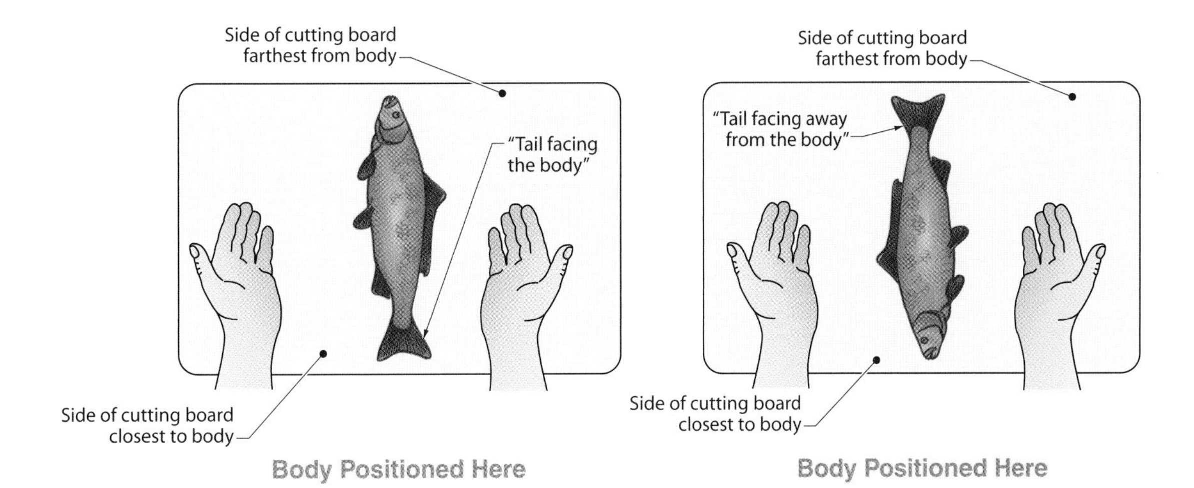

Body Positioned Here

Body Positioned Here

SIDE UP OR SIDE DOWN

A food item positioned "side up" or "side down" refers to the side of that food item the user can see. On flatfish, such as halibut or flounder, the skin on the top side is darker than the skin on the bottom side. Flatfish positioned top-side up, would allow the user to see the darker side of the fish with the lighter, bottom side against the cutting board. The same fish positioned top-side down would have the darker side against the cutting board while the user would be able to see the lighter, bottom side of the fish.

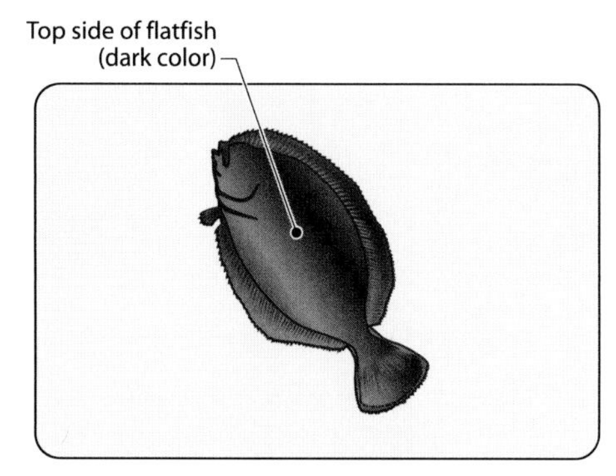

Flatfish Positioned "Top-Side Up"

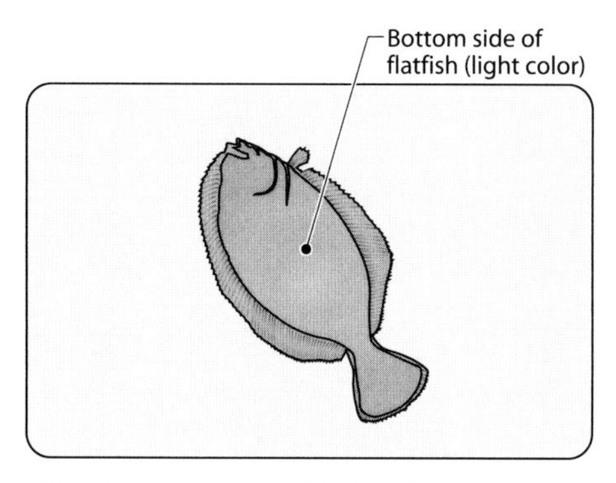

Flatfish Positioned "Top-Side Down"

POSITIONING KNIVES FOR CUTTING

Correctly positioning the knife in relation to the food item being cut or the cutting board reduces the risk of injury and promotes accuracy. Many tasks can be accomplished by positioning the knife in one of three ways: perpendicular to the food being cut, at an angle to the food being cut, or parallel to the cutting board.

PERPENDICULAR KNIFE POSITION

With a perpendicular knife position, the blade forms a 90° angle with the food being cut. The perpendicular knife position is commonly used with larger knives, such as chef's knives, santoku knives, utility knives, and slicers. A perpendicular knife position is used to prep a wide variety of foods from fruits and vegetables to meats and poultry.

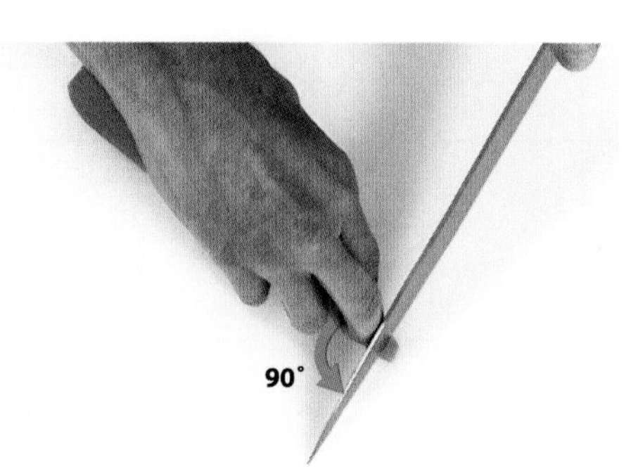

ANGLED KNIFE POSITION

With an angled knife position, the blade commonly forms a 45° angle with the food being cut. The angled knife position is often used to cut cylindrically shaped produce as well as meats, poultry, and fish on a bias (diagonal). One way to make a bias cut is to place the knife diagonally across the top of a food item so that the blade and the item form a 45° angle. A bias cut can also be made by placing the knife across the top of the food item and then tilting the knife back so that the spine of the knife makes a 45° angle with the item being cut.

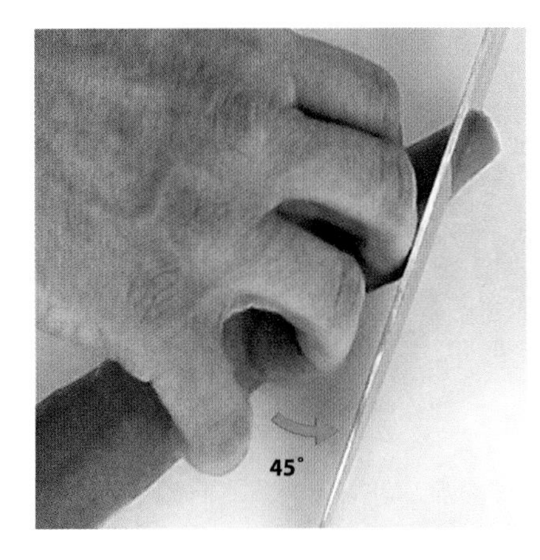

PARALLEL KNIFE POSITION

With a parallel knife position, the blade is held parallel to the cutting board. In this position, the knife does not touch the cutting board. The parallel knife position is commonly used to cut food into layers.

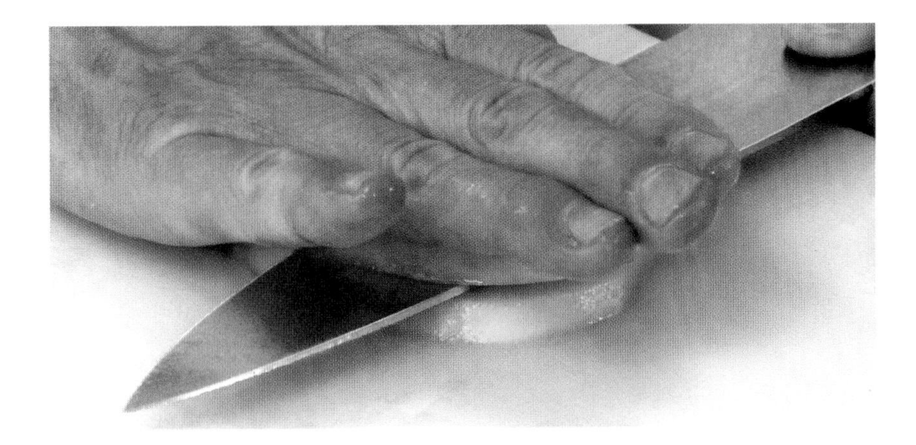

CUTTING METHODS

The tip-to-board and the tip-to-food cutting methods are the primary cutting methods used to slice through foods. With both cutting methods, the guiding hand is positioned in a claw grip and the knife (typically a chef's knife) is held in a pinch grip. The main difference between cutting methods is the placement of the knife tip. With the tip-to-board method, the knife tip remains on the cutting board while the food is cut. With the tip-to-food method, the knife tip starts on top of the food item, comes down to the cutting board, and then returns to the top of the item.

TIP-TO-BOARD CUTTING METHOD

To execute the tip-to-board cutting method, the knife tip remains on the cutting board while the blade moves in an up-and-down rocking motion to cut through the food item. The handle is brought down as the knife tip slides forward. Likewise, the handle is raised up as the knife tip slides backward. This rocking movement, coupled with the correct position of the guiding hand, creates a controlled motion that can be used to efficiently slice through food items while regulating the size of each cut.

TIP-TO-BOARD CUTTING

1. With the fingers of the guiding hand slightly tucked under, secure the food item to be cut in a claw grip.
2. With the other hand, hold the knife in a pinch grip and rest the knife blade against the knuckle of the guiding hand.
3. Keep the knife tip on the cutting board and raise the handle to approximately a 45° angle above the board.
4. Begin slicing through the item by moving the knife in a forward and downward motion until the blade has made contact with the board and cut through the item.

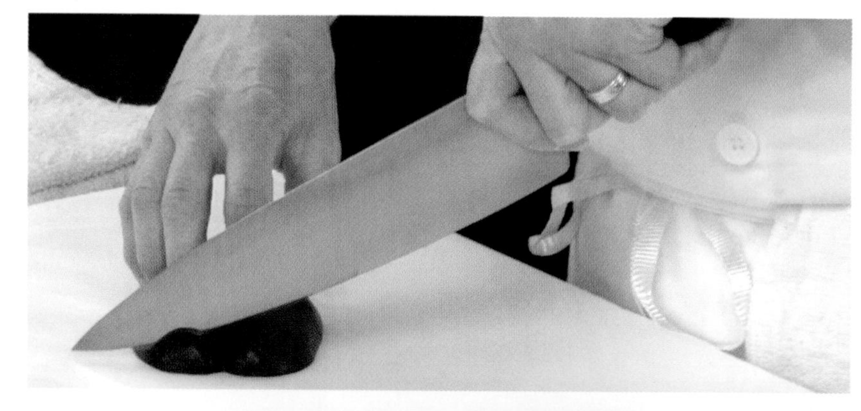

TIP-TO-BOARD CUTTING (CONTINUED)

5. Keep the knife tip against the board and raise the handle back to approximately a 45° angle while pulling the blade backward until the tip is in the starting position on the board.

6. To make another slice, keep the knife tip against the board and repeat the forward and downward rocking motion until the blade has cut through the item.

7. Repeat Steps 5–6 while moving the guiding hand back in equal increments to produce consistently sized slices.

TIP-TO-FOOD CUTTING METHOD

To execute the tip-to-food cutting method, the knife tip is placed on top of the food item to be cut. Then, a downward and forward motion is used to start cutting the item. As the knife tip touches the cutting board, the downward and forward motion continues until the rest of the blade makes contact with the board. The knife is then pulled backward, and the knife tip is again lifted and placed on top of the food item to repeat the cut as necessary.

VIDEO: KNIFE BASICS

TIP-TO-FOOD CUTTING

1. With the fingers of the guiding hand slightly tucked under, secure the food item to be cut in a claw grip.
2. With the other hand, hold the knife in a pinch grip and rest the knife blade against the knuckle of the guiding hand.
3. Keep the knife tip on top of the item to be cut and raise the handle to slightly tilt the knife tip toward the cutting board.

4. Begin slicing through the item by moving the knife in a forward and downward motion.

   Note: For some items, this step requires a slight sawing motion.

5. When the knife tip touches the board, continue to move the knife forward and downward until the blade has made contact with the board and cut through the item.

TIP-TO-FOOD CUTTING (CONTINUED)

6. Pull the knife blade backward and place the knife tip back on top of the item, raising the handle to slightly tilt the knife tip toward the cutting board.

7. Repeat Steps 4–6 while moving the guiding hand back in equal increments to produce consistently sized slices.

SECTION II
BASIC KNIFE CUTS

- RONDELLES
- DIAGONALS
- OBLIQUES
- BATONNETS AND DICE
- JULIENNES
- BRUNOISES
- PAYSANNES
- CHIFFONADE
- CHOPPING
- MINCING
- FLUTES
- TOURNÉS

LEARNER RESOURCES
ATPeResources.com/QuickLinks™
Access Code: 583241

TECHNIQUE 1

RONDELLES

VIDEO 1

A rondelle cut, also known as a round cut, is a knife technique that produces flat-sided, circular cuts. Rondelles are made by slicing cylindrically shaped food items, such as carrots, cucumbers, or zucchini, crosswise. When cutting rondelles, care should be taken to make each slice the same thickness as this promotes even cooking and enhances plated presentations. Food items cut into rondelles are often used in salads, soups, and side dishes.

Chef's Knife

— or —

Santoku Knife

CUTTING RONDELLES

* This technique was executed by a left-handed chef.

1. Place a cylindrically shaped food item, such as a carrot, perpendicular to the knife blade. With the fingers of the guiding hand tucked, hold the item in a claw grip and align the knife blade against the knuckle of the guiding hand.

   Note: See page 29 for how to position the guiding hand in a claw grip.

2. With the knife hand, use the rocking motion of the tip-to-board method to slice the item crosswise into circular cuts of equal thickness.

   Note: See pages 34–35 for the tip-to-board cutting procedure.

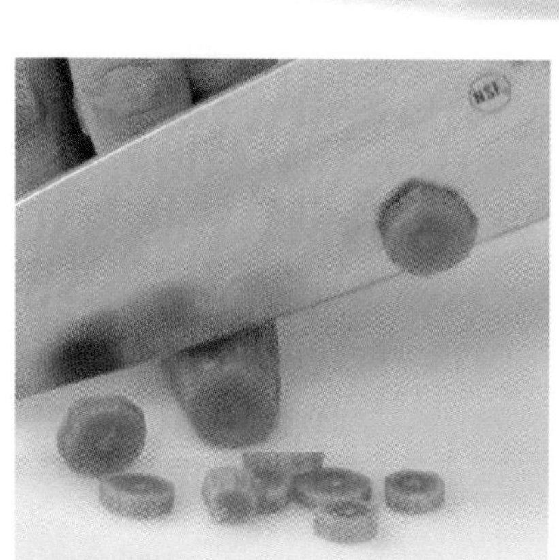

Finished rondelles are commonly ¼ inch, ⅛ inch, or 1⁄16 inch thick.

VIDEO 2

DIAGONALS

TECHNIQUE 2

A diagonal cut is a knife technique that produces flat-sided, oval, or semioval cuts. The diagonal cut is used with cylindrically shaped food items, such as carrots, scallions, and summer squash, as well as stem vegetables, such as celery and fennel. To produce a diagonal cut, the item is sliced on the bias (diagonal) at a 45° angle. Diagonal cuts are commonly used to prepare vegetables for stir-fry dishes or side dishes because the diagonal cut produces a large surface area, allowing the vegetables to cook quickly. The decorative bias edge of the diagonal cut is also used to enhance plated presentations.

Chef's Knife

— or —

Santoku Knife

CUTTING DIAGONALS

* This technique was executed by a left-handed chef.

1. Use a claw grip to hold a cylindrically shaped food item, such as a carrot. Position the knife so that the blade forms a 45° angle with the item being cut and align the blade against the knuckle of the guiding hand.

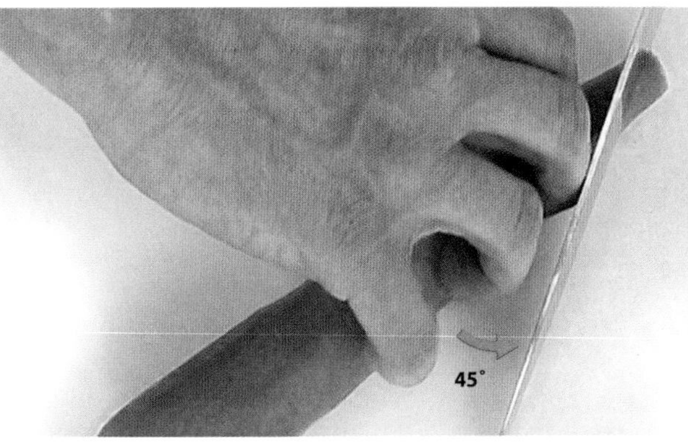

2. Keep the fingers of the guiding hand tucked and with the knife hand, use the rocking motion of the tip-to-board method to slice oval-shaped cuts of equal thickness.

Finished diagonals are typically, ½ inch, ¼ inch, ⅛ inch, or 1⁄16 inch thick.

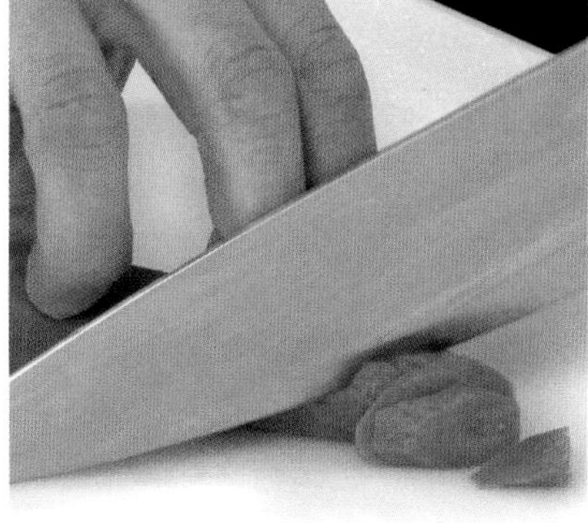

TECHNIQUE 3

OBLIQUES

VIDEO 3

An oblique cut, also known as the rolled cut, is a knife technique that produces a wedge-shaped cut with two angled sides. The oblique cut is similar to the diagonal cut because cylindrically shaped food items, such as carrots and parsnips, are cut on the bias at a 45° angle. However, the item is rolled 180° after each slice to produce the two angled sides. The oblique cut is useful when a larger surface area is desired for reducing the cooking time or adding visual appeal to plated presentations.

Chef's Knife
— or —
Santoku Knife

CUTTING OBLIQUES

* This technique was executed by a left-handed chef.

1. Use a claw grip to hold a cylindrically shaped food item, such as a carrot. Position the knife so that the blade forms a 45° angle with the item being cut and align the blade against the knuckle of the guiding hand.

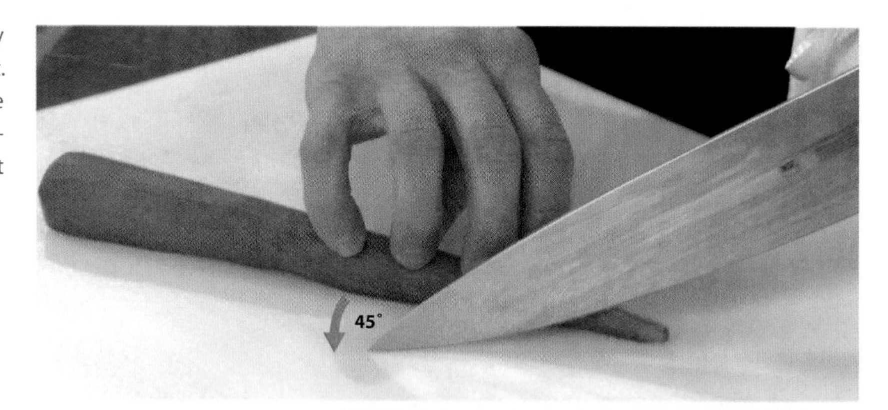

2. Keep the fingers of the guiding hand tucked and with the knife hand, use the tip-to-board method to make one slice.

   Note: The first slice does not produce an oblique cut.

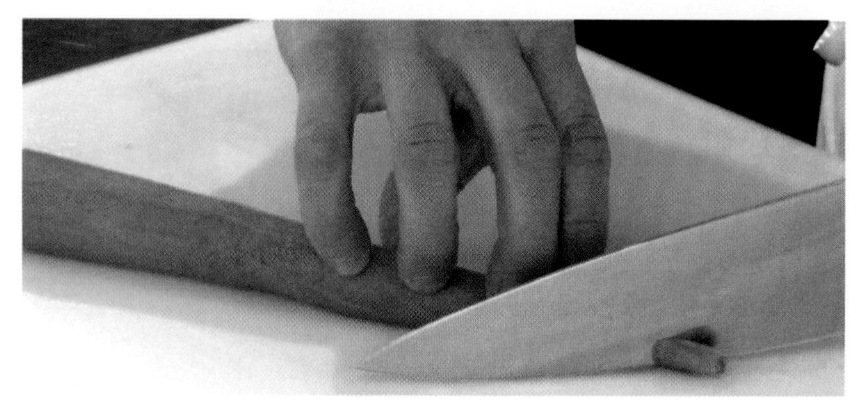

CUTTING OBLIQUES (CONTINUED)

* This technique was executed by a left-handed chef.

3. Roll the item 180°, so the portion that was facing upward is now facing the cutting board. With the guiding hand still in a claw grip and the knife blade at a 45° angle to the item, slice the item again using the tip-to-board method to produce the first oblique cut.

4. Roll the item 180° back to its original position. Still keeping the guiding hand in a claw grip and the knife blade at a 45°angle to the item, use the tip-to-board method to make another slice. Continue rolling the item 180° between slices to produce oblique cuts.

Each finished oblique cut has two angled sides.

QUICK TIP

Due to the natural shape of a carrot or parsnip, finished cuts may vary in size. For even cooking, sort the cuts into their respective sizes and cook the largest ones first.

TECHNIQUE

BATONNETS AND DICE

VIDEO 4

A batonnet cut is a knife technique that produces a stick-shaped cut ¼ inch wide × ¼ inch high × 2 inches long. *Note:* The length of stick cuts may vary depending on the desired result. A variety of vegetables and fruits can be cut into batonnets including carrots, jicama, pears, and melons. Potatoes are also frequently cut into batonnets for French fries. The uniform appearance of a batonnet cut gives a refined look to salads, side dishes, and crudité platters.

A dice cut is a knife technique that produces precise cubes cut from uniform stick-shaped cuts. Common dice cuts include large dice, medium dice, small dice, brunoise, and fine brunoise. To produce a dice cut of a desired size, a stick cut with the appropriate dimension is sliced into cubes with six equal sides.

Note: Disposable gloves are typically worn in the foodservice industry when preparing ready-to-eat food items.

Chef's Knife

— or —

Santoku Knife

CUTTING BATONNETS

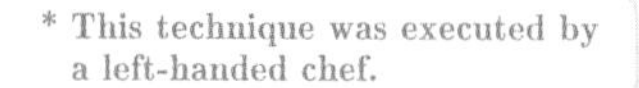

* This technique was executed by a left-handed chef.

1. Slice a washed and peeled food item, such as a carrot, into pieces 2 inches in length.

2. Secure the item lengthwise with a claw grip. Keep the fingers of the guiding hand tucked and use the tip-to-food method to slice the item lengthwise, squaring off three sides. Reserve sides for later use.

   Note: See pages 36–37 for the tip-to-food cutting procedure.

* This technique was executed by a left-handed chef.

3. Position the rounded side of the item so that it faces the guiding hand. Secure the item lengthwise with a claw grip and use the tip-to-food method to slice ¼ inch planks.

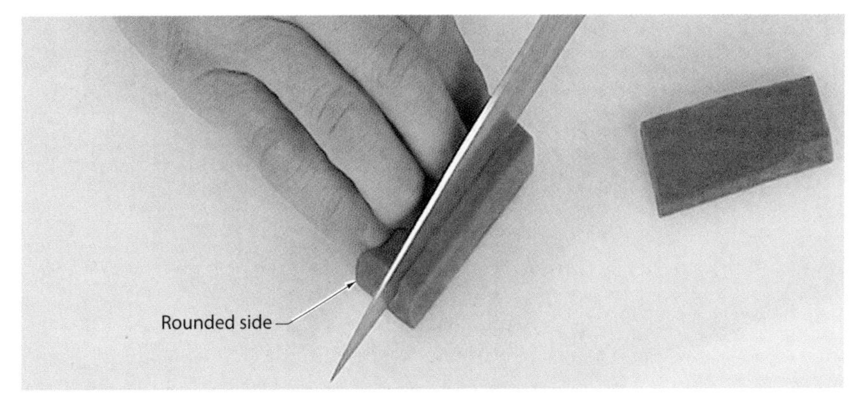

4. Stack several ¼ inch planks.

5. Continue using a claw grip and the tip-to-food method to slice lengthwise through the stack of planks at ¼ inch intervals to produce consistently sized sticks.

Finished batonnets measure ¼ inch wide × ¼ inch high × 2 inches long.

CUTTING MEDIUM DICE

* This technique was executed by a left-handed chef.

1. Slice a food item, such as beets, into stick cuts measuring ½ inch wide × ½ inch high × 2 inches long.

   Note: Lengths of stick cuts may vary.

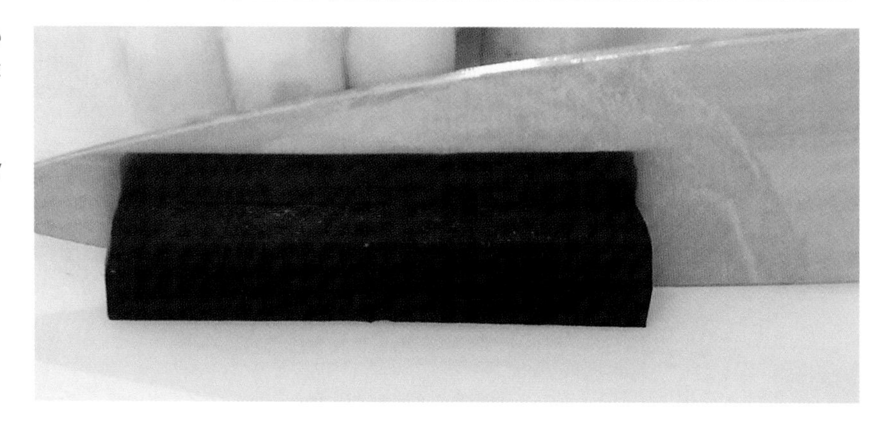

QUICK TIP

Small dice cuts are made from ¼ wide × ¼ high × 2 inch long stick cuts. Large dice cuts are made from ¾ wide × ¾ high × 2 inch long stick cuts. In both examples, the length of the stick cuts may vary.

2. Place a small bundle of sticks perpendicular to the knife blade. Use the side of the knife blade to align the ends of the stick bundle.

   Note: When cutting beets, gloves are commonly worn to prevent the hands from getting stained.

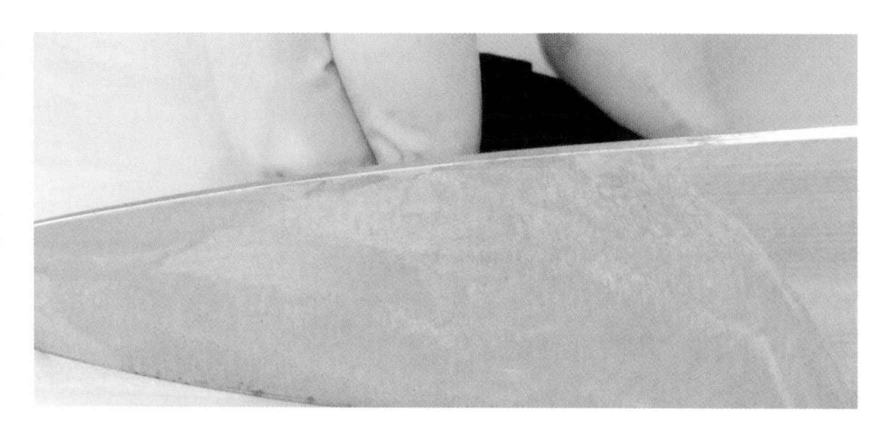

QUICK TIP

If sticks that will be diced do not align evenly against the knife blade, they can be trimmed to produce a flat, even surface before being diced.

3. Secure the bundle with a claw grip. Keep the fingers of the guiding hand tucked and use the tip-to-food method to slice crosswise through the bundle at ½ inch intervals to produce consistently sized cubes.

   Finished medium dice cubes have six equal sides measuring ½ inch each.

VIDEO 5

TECHNIQUE 5

JULIENNES

A julienne cut, also known as a matchstick cut, is a knife technique that produces a stick-shaped cut ⅛ inch wide × ⅛ inch high × 2 inches long. A fine julienne cut is a knife technique that produces a stick-shaped cut 1⁄16 inch wide × 1⁄16 inch high × 2 inches long. There are many vegetables and fruits suitable for julienne and fine julienne cuts, including carrots, daikon, and apples. The uniform look of julienne and fine julienne cuts elevates the appearance of soups, salads, and sandwiches while adding a layer of texture and flavor.

Chef's Knife

— or —

Santoku Knife

Note: The length of stick cuts may vary depending on the desired result.

CUTTING JULIENNES

* This technique was executed by a left-handed chef.

1. Slice a washed and peeled food item, such as a carrot, into pieces 2 inches in length.

2. Secure the item lengthwise with a claw grip. Keep the fingers of the guiding hand tucked and use the tip-to-food method to slice the item lengthwise, squaring off three sides. Reserve sides for later use.

* This technique was executed by a left-handed chef.

3. Position the rounded side of the item so that it faces the guiding hand. Secure the item lengthwise with a claw grip and use the tip-to-food method to slice ⅛ inch planks.

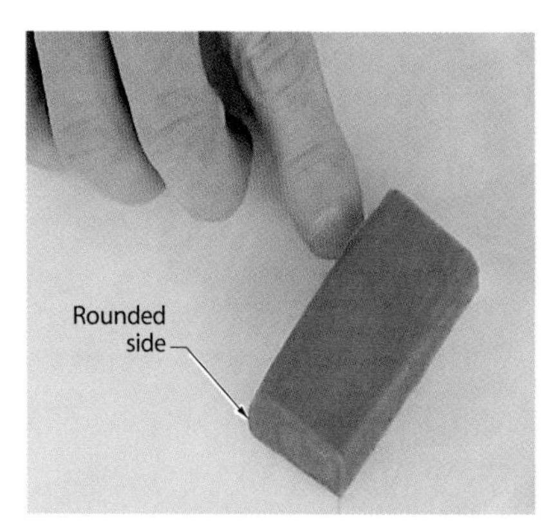

4. Stack several ⅛ inch planks.

5. Continue using the claw grip and the tip-to-food method to slice lengthwise through the stack of planks at ⅛ inch intervals to produce consistently sized sticks.

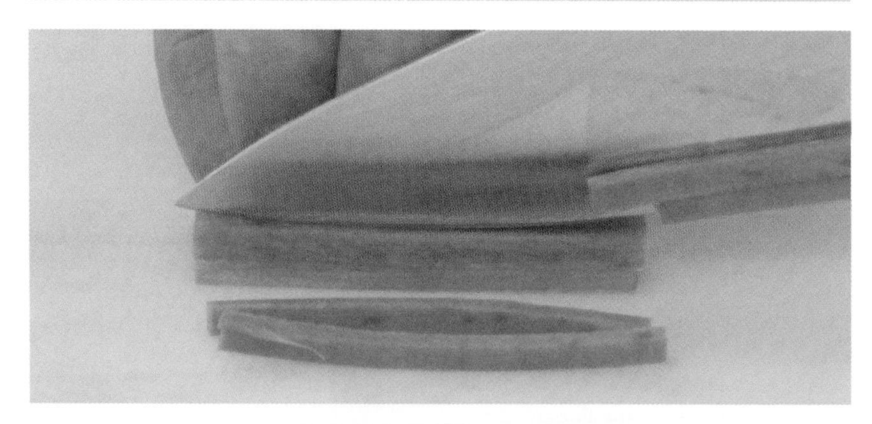

Finished julienne cuts are ⅛ inch wide × ⅛ inch high × 2 inches long.

CUTTING FINE JULIENNES

* This technique was executed by a left-handed chef.

1. Start with a 2 inch long food item that has been squared off on three sides. With the rounded side facing the guiding hand, secure the item lengthwise in a claw grip and use the tip-to-food method to slice the item into 1⁄16 inch planks.

2. Stack several 1⁄16 inch planks and slice lengthwise through the stack at 1⁄16 inch intervals.

Finished fine julienne cuts are 1⁄16 inch wide × 1⁄16 inch high × 2 inches long.

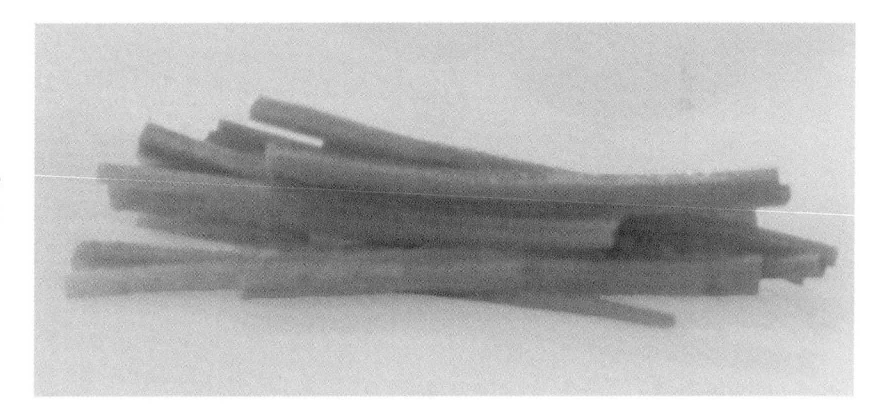

TECHNIQUE 6

BRUNOISES

VIDEO 6

A brunoise cut is a knife technique that produces a dice-shaped cube with six equal sides measuring ⅛ inch each. A fine brunoise cut is a knife technique that produces a dice-shaped cube with six equal sides measuring 1⁄16 inch each. Julienne (stick-shaped) cuts are used to produce brunoise cuts, and fine julienne cuts are used to produce fine brunoise cuts. Items with a brunoise or fine brunoise cut are generally used as garnishes to elevate the presentation of dishes, including hors d'oeuvres and consommés.

Chef's Knife

— or —

Santoku Knife

CUTTING BRUNOISES

* This technique was executed by a left-handed chef.

1. Place a small bundle of julienne sticks perpendicular to the knife blade. Use the side of the knife blade to align the ends of the stick bundle.

   Note: Refer to Technique 5: Juliennes for the procedure on cutting juliennes.

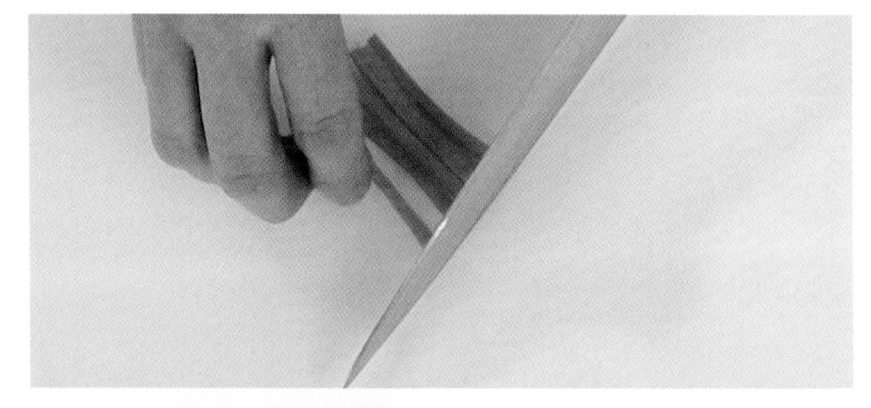

2. Secure the bundle with a claw grip. Keep the fingers of the guiding hand tucked and with the knife hand, use the rocking motion of the tip-to-board method to slice crosswise through the bundle at ⅛ inch intervals to produce consistently sized cubes.

   Finished brunoise cuts have six equal sides measuring ⅛ inch each.

CUTTING FINE BRUNOISES

* This technique was executed by a left-handed chef.

1. Start with an aligned bundle of fine julienne sticks.

   Note: Refer to Technique 5: Juliennes for the procedure on cutting juliennes.

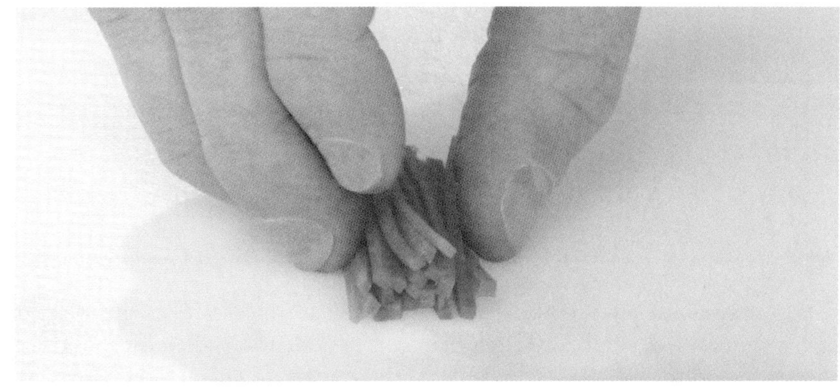

2. Place the bundle of fine julienne sticks perpendicular to the knife blade and secure with a claw grip. Use the tip-to-board method to slice crosswise through the bundle at 1⁄16 inch intervals to produce consistently sized cubes.

QUICK TIP

To secure a bundle of stick cuts in a claw grip, it is helpful to use the pinky finger and thumb to hold the sticks together. The ring and index finger stabilize the sticks from the top, and the middle finger becomes the guide finger.

Finished fine brunoise cuts have six equal sides measuring 1⁄16 inch each.

TECHNIQUE 7

PAYSANNES

VIDEO 7

A paysanne cut is a knife technique that produces a flat, tile-shaped square, circular, or triangular cut ½ inch wide × ½ inch high × ⅛ inch thick. For square tiles, the paysanne cut is similar to the medium dice except each cut is sliced at ⅛ inch intervals instead of ½ inch intervals. For circular tiles, the paysanne cut is similar to the rondelle cut except the paysanne cut has more specific dimensions. Paysanne cuts are typically made with root vegetables and are used to produce decorative garnishes.

Chef's Knife

— or —

Santoku Knife

CUTTING PAYSANNES

1. For square tiles, use a stick-cut food item, such as a carrot, that measures ½ inch wide × ½ inch high × 2 inches long. For circular tiles, use a cylindrical item with a diameter (distance across a circle) of approximately ½ inch.

   Note: The length of the item may vary depending on the desired result.

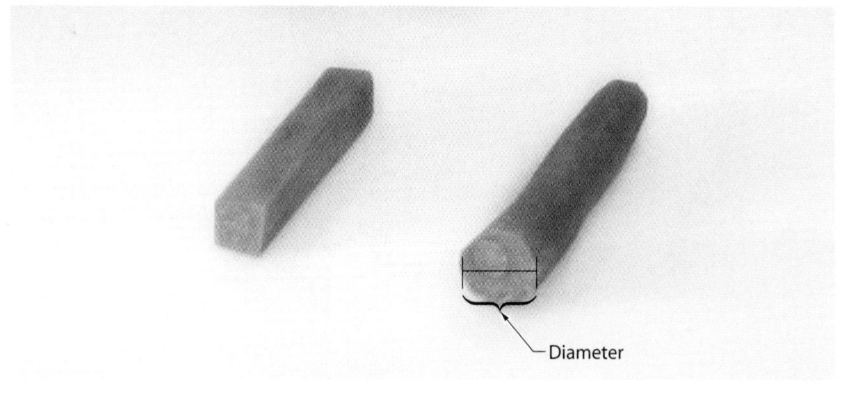

2. Place the item perpendicular to the knife blade. Hold the item with a claw grip and align the knife blade against the knuckle of the guiding hand.

CUTTING PAYSANNES (CONTINUED)

3. Use the tip-to-food method to slice crosswise through the item at ⅛ inch intervals to produce flat tiles of the desired shape.

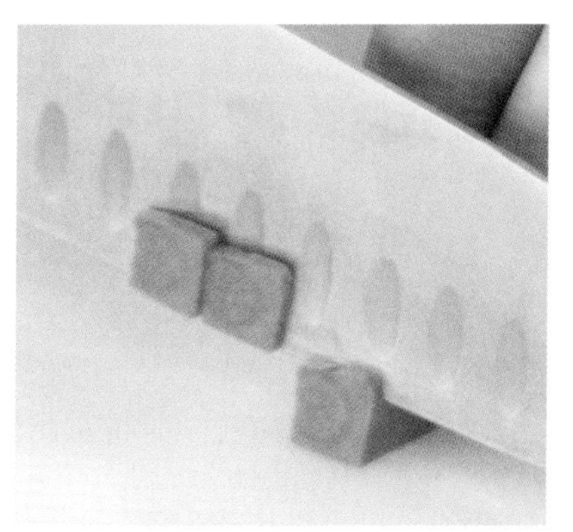

Finished paysannes are ½ inch wide × ½ inch high × ⅛ inch thick.

QUICK TIP

To make triangular tiles, first cut square-shaped tiles from a stick-cut item measuring 1 inch wide x 1 inch high x 2 inches long. Then, slice the square-shaped tiles diagonally to yield triangular tiles.

TECHNIQUE 8

CHIFFONADE

VIDEO 8

A chiffonade cut is a knife technique that produces long strips of herbs or leafy greens. Chiffonade-cut herbs are typically sliced thin and used as flavorful garnishes. When preparing a chiffonade of herbs, the tip-to-board method is used to slice the leaves. Using the tip-to-board method helps prevent the leaves from bruising and turning brown.

Leafy greens, such as kale, collards, and Swiss chard, have rigid stems that are typically removed before cutting. Once the stems are removed, greens are commonly cut into strips. The size of the strips depends on the application. For example, thinner strips are often served raw in salads and sandwiches or added to soups and pasta dishes during the final stages of cooking. Thicker strips are used in applications such as braising, which requires a longer cooking time.

Note: Disposable gloves are typically worn in the foodservice industry when preparing ready-to-eat food items.

CHIFFONADE HERBS

* This technique was executed by a left-handed chef.

1. Place washed, dry leaves of herbs in a neat stack. If the leaves have stems, place the stems so that they are all facing the same direction or remove the stems prior to cutting.

2. Roll the stack into a tight cylinder with the dull side of the leaves exposed. Place the cylinder perpendicular to the blade of a chef's knife.

   Note: Rolling herbs into a cylinder with their dull side exposed makes cutting easier by creating extra friction against the knife. It also does less damage to the herb.

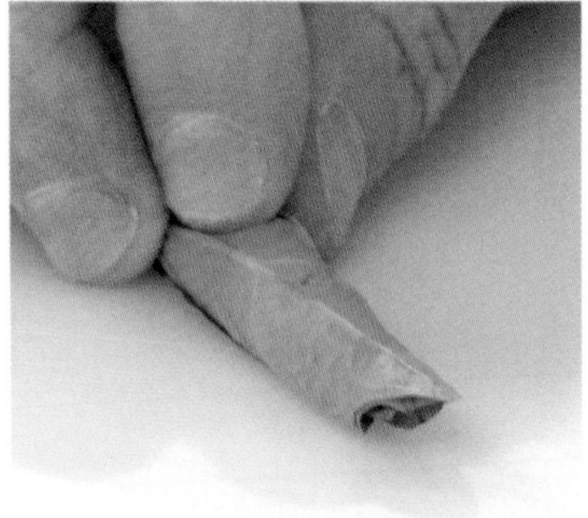

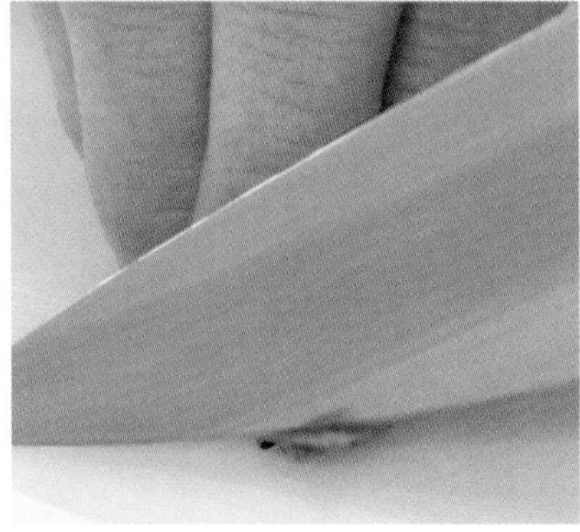

CHIFFONADE HERBS (CONTINUED)

* This technique was executed by a left-handed chef.

3. With the fingers of the guiding hand tucked, secure the herbs in a claw grip. With the knife hand, use the rocking motion of the tip-to-board method to thinly slice the leaves crosswise. If stems are present, stop cutting when the stems become thicker than desired.

QUICK TIP

Leftover stems of herbs can be composted or used in sachets to flavor stocks, sauces, or soups.

A finished chiffonade produces finely sliced herbs that resist browning.

CHIFFONADE LEAFY GREENS

1. Use Method A or Method B to remove the thick, rigid part of the stem from each leaf. The thin, tender section of the stem can remain with the leaf.

Method A:

- Fold the leaves in half along the stem so that the dull side of the leaf is exposed and the stem is visible.
- Hold the leafy green in the guiding hand and with the other hand, pinch along the sides of the stem while pulling it away from the leaf.

CHIFFONADE LEAFY GREENS (CONTINUED)

Method B:

- Place the leafy green flat on a cutting board with the dull side of the leaf facing upward.
- Use the tip of a paring knife to cut along each side of the thick, rigid part of the stem. Pull the stem away from the leaf to remove it completely.

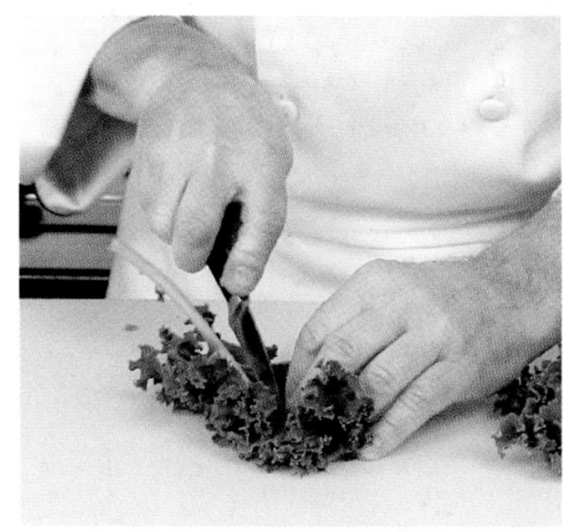

QUICK TIP

The stems of many leafy greens are edible and can be cooked a few minutes before the leaves so that both are tender and done cooking at the same time. The stems can also be composted or discarded.

2. Stack several leaves and roll the stack into a tight cylinder so that the dull side of the leaves are exposed.

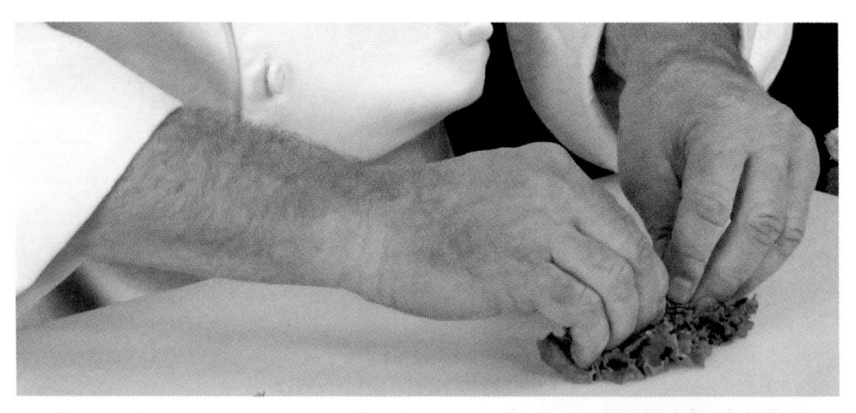

3. Place the cylinder of leafy greens perpendicular to the blade of a chef's knife. With the fingers of the guiding hand tucked, secure the greens in a claw grip. With the knife hand, use the rocking motion of the tip-to-board method to slice the greens crosswise.

   A finished chiffonade of leafy greens can be made as thin or as thick as desired.

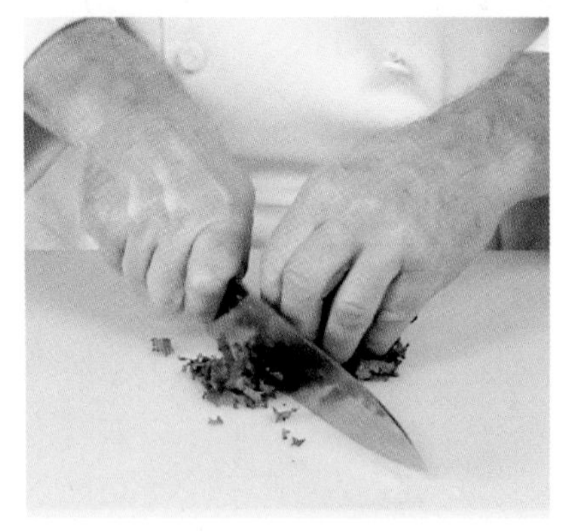

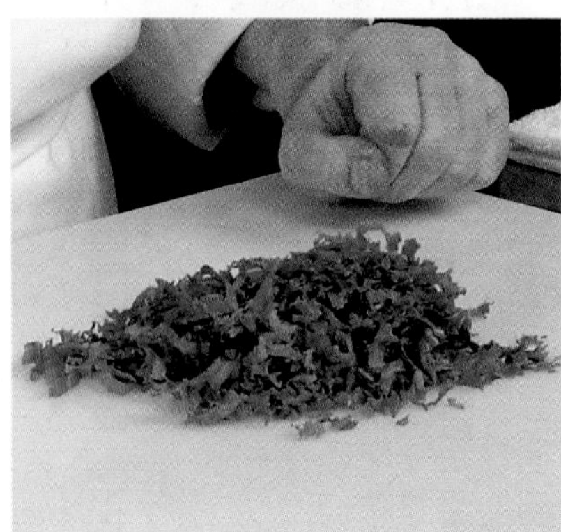

CHOPPING

TECHNIQUE 9

Chopping is a knife technique that involves rough-cutting an item into small pieces that lack uniformity in shape and size. For applications in which uniform dimensions are not required, items such as nuts, hard-cooked eggs, and herbs are commonly chopped.

Commonly chopped herbs include parsley, cilantro, mint, and rosemary. The leaves of small leaf herbs, such as parsley, are commonly pulled from the stem before chopping. The leaves of woody herbs, such as rosemary, are typically removed by gently running the thumb and index finger down the stems. Chopped herbs are often used as a garnish or to add a layer of aroma and flavor to a dish.

Note: Disposable gloves are typically worn in the foodservice industry when preparing ready-to-eat food items.

CHOPPING HERBS

* This technique was executed by a left-handed chef.

1. Gather washed, dry leaves of herbs, such as parsley, into a tight bundle with the guiding hand.

2. With the fingers of the guiding hand tucked, secure the leaves in a claw grip and use the rocking motion of the tip-to-board method to slice the leaves into thin strips.

* This technique was executed by a left-handed chef.

3. Point the edge of the knife blade away from the body and use the guiding hand to carefully remove any leaves remaining on the blade. Gather the leaves into a pile.

4. Place the guiding hand, opened and flat, on the spine of the knife near the tip. Keep the edge of the tip on the cutting board and rock the blade up and down while simultaneously pivoting the blade back and forth.

5. Gather the leaves into a pile again and repeat the chopping process until the shavings are fine.

Finished chopped herbs lack uniformity in shape and size.

VIDEO 10

MINCING

Mincing is a knife technique that involves finely chopping an item to yield very small pieces that are not entirely uniform in shape. Although mincing is done frequently in the kitchen, it is only performed on a narrow range of items, such as shallots, garlic, ginger, hot peppers, and fresh herbs. Minced items are commonly added to sauces, marinades, and dips in which evenly distributed flavor and minimal texture are desired. In addition to flavor, minced herbs often serve as a garnish, adding color and interest to plated presentations.

Paring Knife

Chef's Knife

— or —

Santoku Knife

MINCING SHALLOTS

1. Hold a shallot in the guiding hand with the stem end angled upward toward the knife hand side. Grasp a paring knife in a choke grip and slightly trim the stem end.

   Note: See page 28 for how to position the hand and knife in a choke grip.

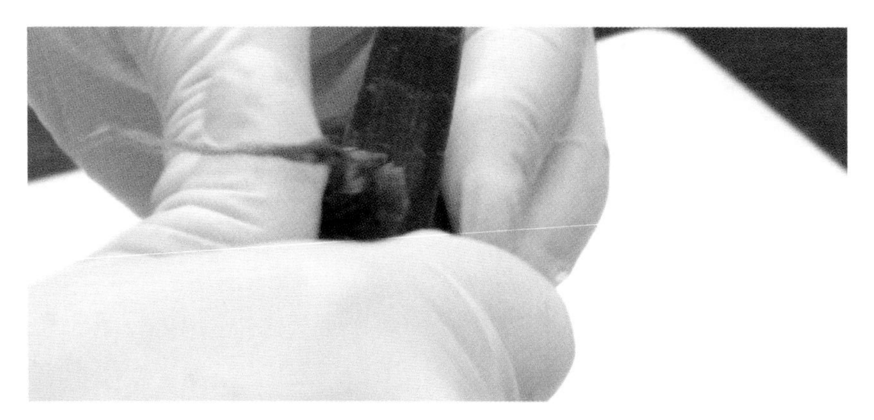

2. Turn the shallot so that the root end is angled upward toward the knife hand side. Grasp the paring knife in a choke grip and slightly trim the root end, keeping some of the root intact.

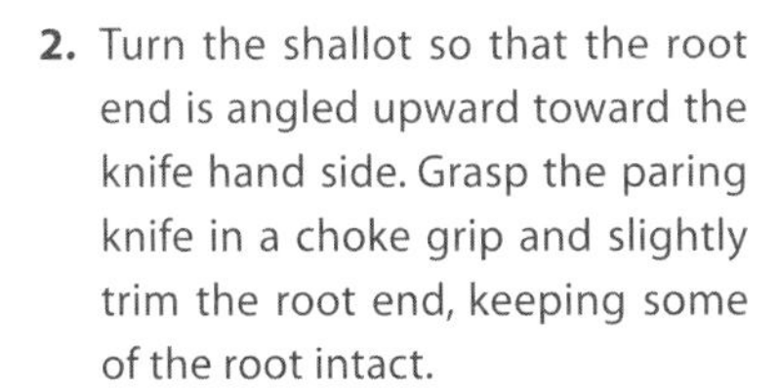

   Note: The root is not sliced off completely because it holds the layers of the shallot in place and prevents it from falling apart.

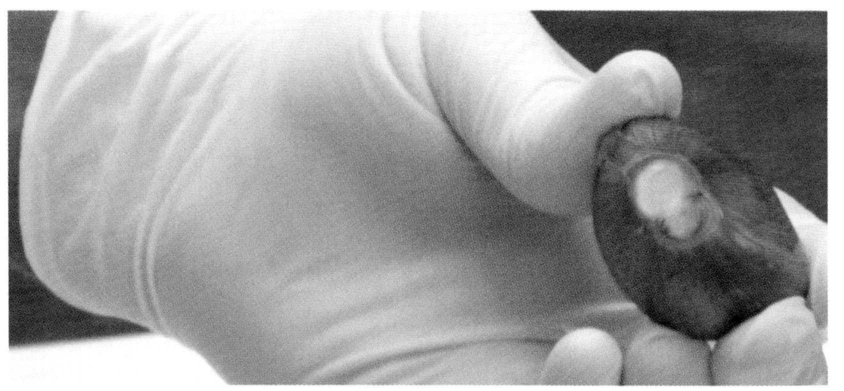

MINCING SHALLOTS (CONTINUED)

3. Starting at the stem end, hold the paring knife in a choke grip and insert the heel of the blade under the first layer of skin beneath the papery layer. Use the knife hand thumb to keep the layers pressed against the side of the knife blade and pull down to remove a section of skin. Continue this process until the shallot is entirely peeled.

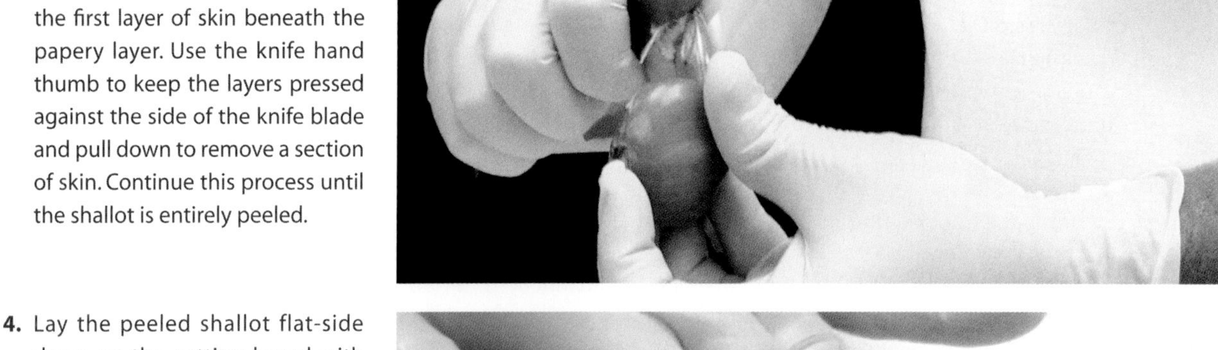

4. Lay the peeled shallot flat-side down on the cutting board with the stem end facing the knife hand side. Hold the shallot in place with the guiding hand and make three or four equally spaced horizontal cuts most of the way through the shallot. Leave the root end intact.

   Note: If the shallot is very rounded, cut it in half from end to end to create a flat surface. Then follow Step 4, making only two or three horizontal cuts.

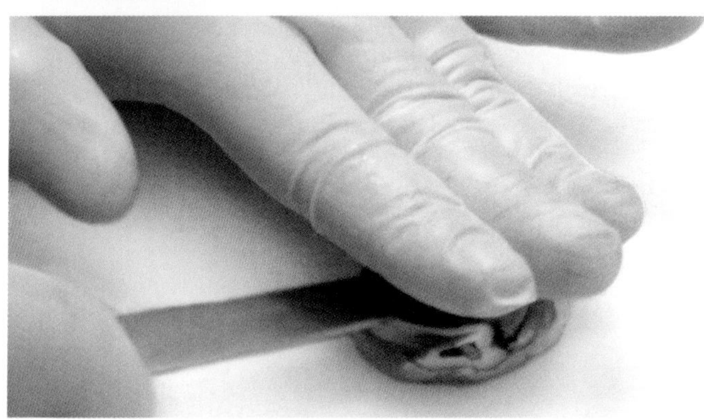

QUICK TIP

Along with onions and garlic, shallots are part of the allium family. Shallots have a sweeter, milder flavor than onions with a hint of garlic.

5. Rotate the shallot 90° so that the stem end faces the body and secure it with the guiding hand. Use the tip-to-food method to make a series of vertical slices ⅛ – ¼ inch apart, again leaving the root end intact.

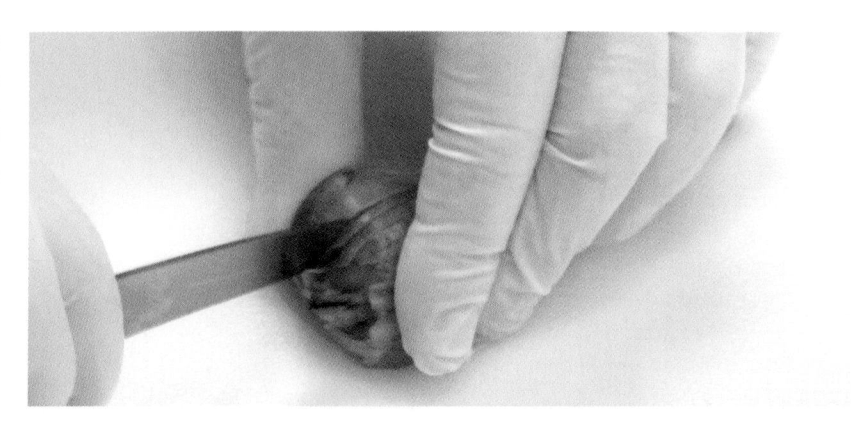

MINCING SHALLOTS (CONTINUED)

6. Rotate the shallot 90° back to its previous position and place the blade of a chef's knife perpendicular to the stem end. Secure the shallot with a claw grip and use the tip-to-board method to slice the shallot crosswise at ⅛ inch intervals until only the root remains.

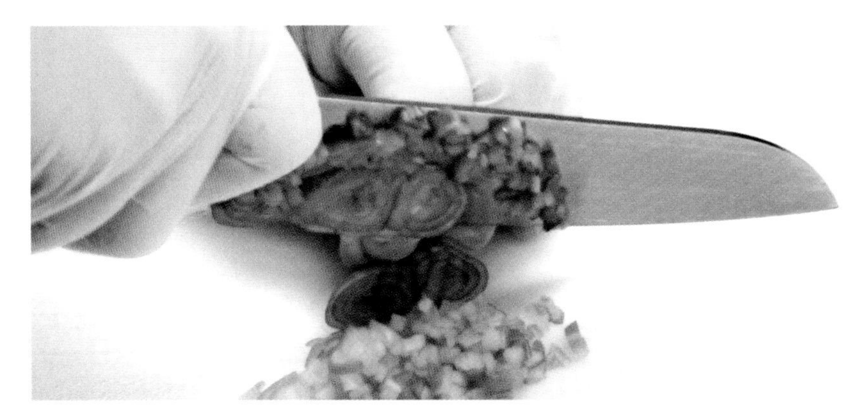

7. Point the edge of the knife blade away from the body and use the guiding hand to carefully remove any minced shallots remaining on the blade.

8. For finely minced shallots, place the guiding hand, opened and flat, on the spine of the knife near the tip. Keep the edge of the tip on the cutting board and rock the blade up and down while simultaneously pivoting the blade back and forth until the shallots are finely minced.

Finished minced shallots are small pieces that vary slightly in shape.

FLUTES

VIDEO 11

A fluted cut is a knife technique that produces a decorative spiral pattern on the surface of an item by removing only a sliver of the item with each cut. Button mushrooms are often fluted and used as a garnish in presentations that call for a more refined look. Fluted cuts require strong hand-eye coordination because as each sliver is cut away, the item is rotated slightly toward the knife hand before making the next cut from the same central point on top of the item.

Paring Knife

Note: When cutting flutes, it is helpful to identify the 12 o'clock and 7 o'clock knife positions.

Knife Position at 12 O'clock

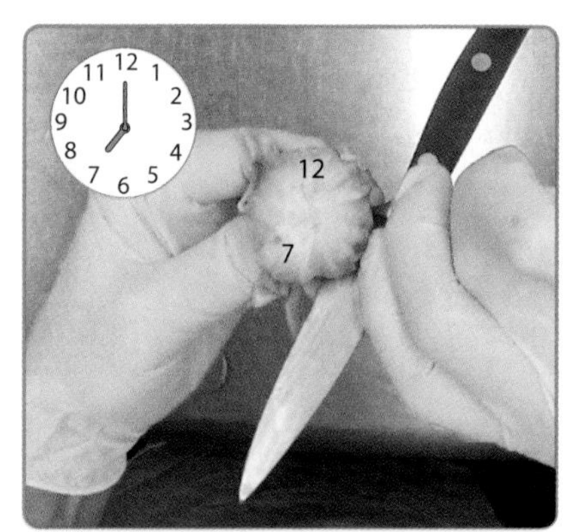

Knife Position at 7 O'clock

CUTTING FLUTES

1. Pinch the heel of a paring knife near the bolster with the thumb and index finger of the dominant hand. Slightly tuck the remaining fingers.

CUTTING FLUTES (CONTINUED)

2. Hold a mushroom in between the thumb and index finger of the guiding hand with the cap-side up. Place the side of the knife blade on top of the mushroom with the knife tip facing 12 o'clock and the edge of the heel centered on the mushroom.

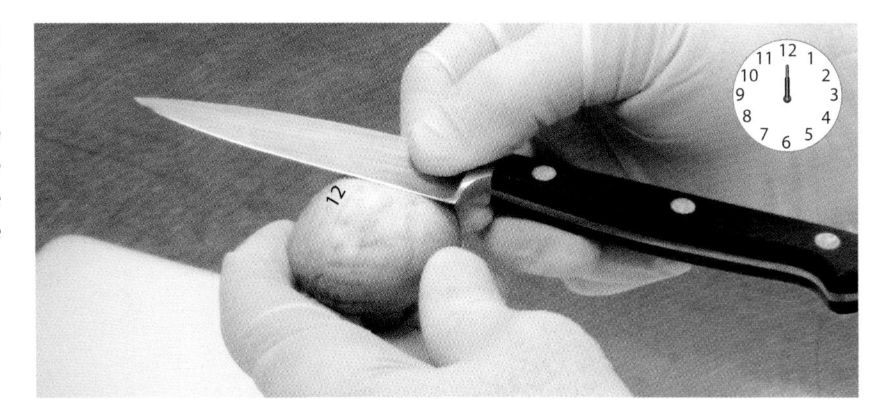

3. If the knife is in the right hand, carve a sliver from the mushroom by turning the knife tip to 7 o'clock and slightly rotating the mushroom clockwise (toward the knife hand).

   If the knife is in the left hand, carve a sliver from the mushroom by turning the knife tip to 5 o'clock and slightly rotating the mushroom counter-clockwise (toward the knife hand).

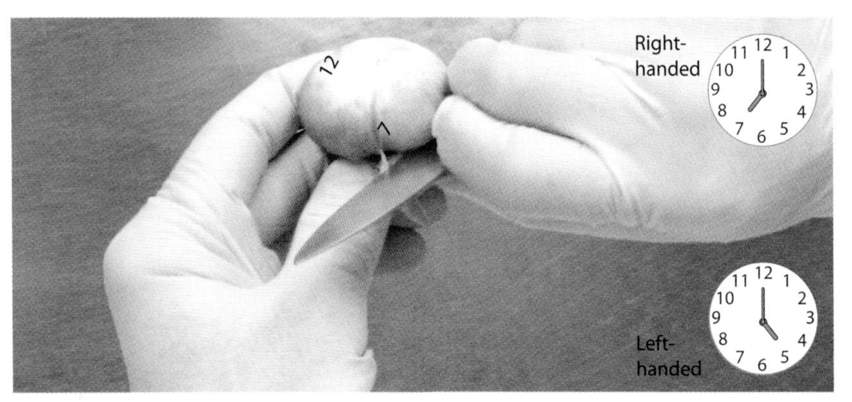

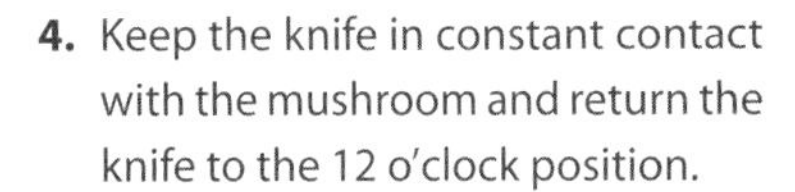

4. Keep the knife in constant contact with the mushroom and return the knife to the 12 o'clock position.

   Note: A thin strip of mushroom will be removed with each cut, leaving a fluted edge.

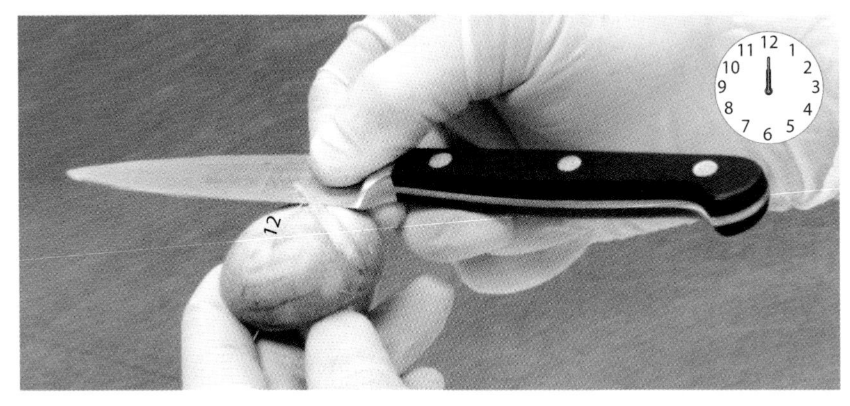

QUICK TIP

A cold button mushroom that has been wiped clean with a damp towel keeps the mushroom firm and, therefore, easier to cut.

5. Flute the entire mushroom cap by continuing to slightly rotate the mushroom toward the knife hand while moving the knife tip from 12 o'clock to 7 o'clock (right-handed) or 12 o'clock to 5 o'clock (left-handed) and back to 12 o'clock. Pull off mushroom slivers that remain attached to the cap.

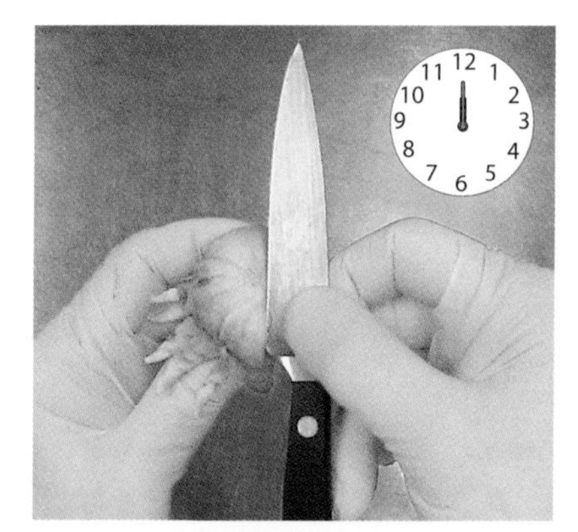

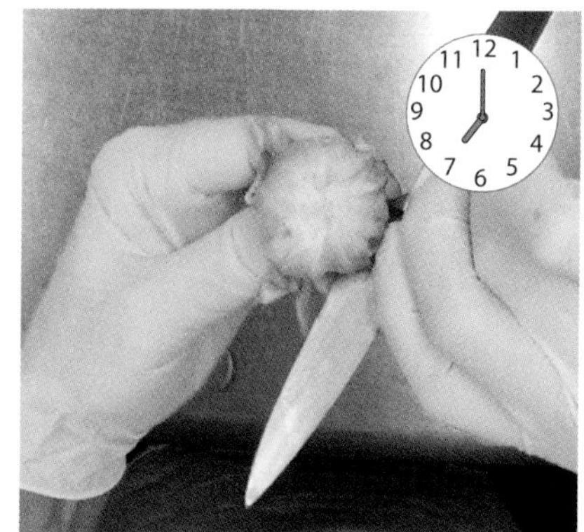

Fluting is complete when the mushroom has made one full rotation (360°) and a spiral pattern covers the mushroom cap.

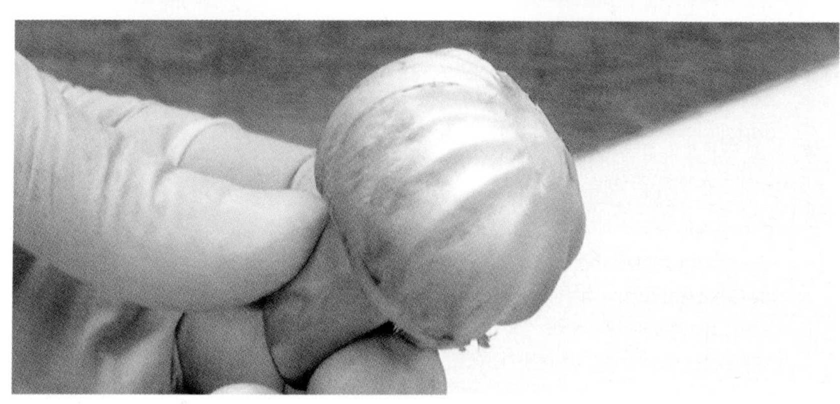

VIDEO 12

TOURNÉS

TECHNIQUE 12

A tourné cut is a knife technique that produces a seven-sided, football-shaped item with small flat ends. The word tourné means "turned" in French, and subsequently a food that is tournéed is turned after each cut. Tourné cuts work well with tubers and root vegetables such as potatoes, turnips, carrots, parsnips, and beets. Tournéed items are generally used to add a design element to plated presentations.

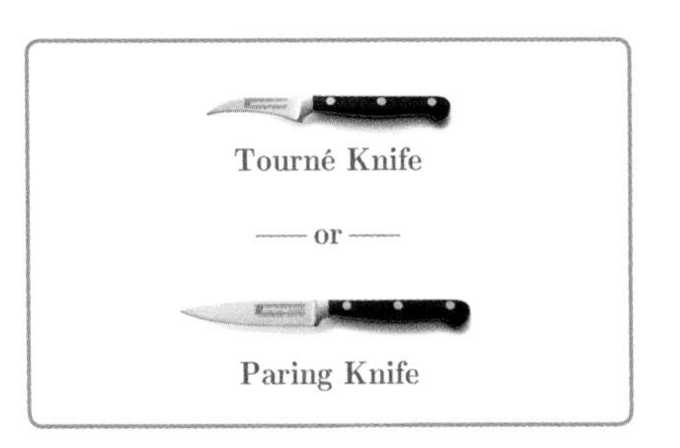

Tourné Knife

—— or ——

Paring Knife

CUTTING TOURNÉS

* This technique was executed by a left-handed chef.

1. Start with a stick-cut tuber or root vegetable such as a potato or beet that measures 1 inch × 1 inch × 2 inches.

2. Position the stick-cut item so that an edge side is facing upward. Hold the item lengthwise between the index finger and thumb of the guiding hand. Rest the remaining fingers underneath the item.

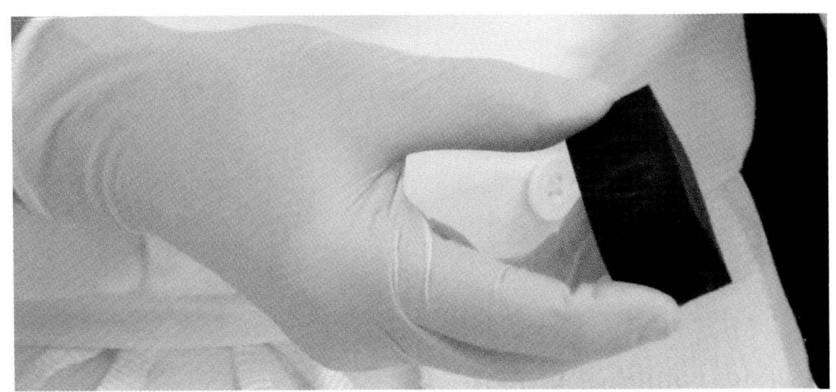

QUICK TIP

When cutting beets, wear gloves to prevent the hands from getting stained.

* This technique was executed by a left-handed chef.

3. Hold a tourné knife in a choke grip and position the thumb of the knife hand on the top corner of the item that is closest to the body. Position the knife blade approximately ¼ inch below the top corner of the item that is farthest from the body.

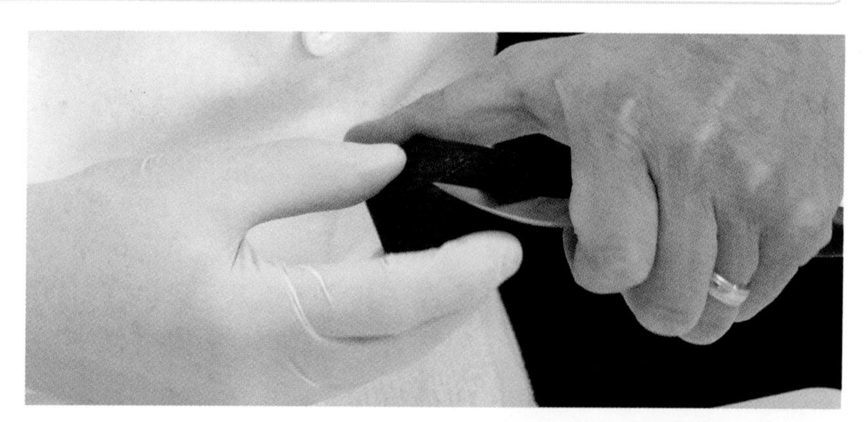

4. Pull the knife blade toward the thumb in a smooth, continuous arc to create a slightly rounded surface.

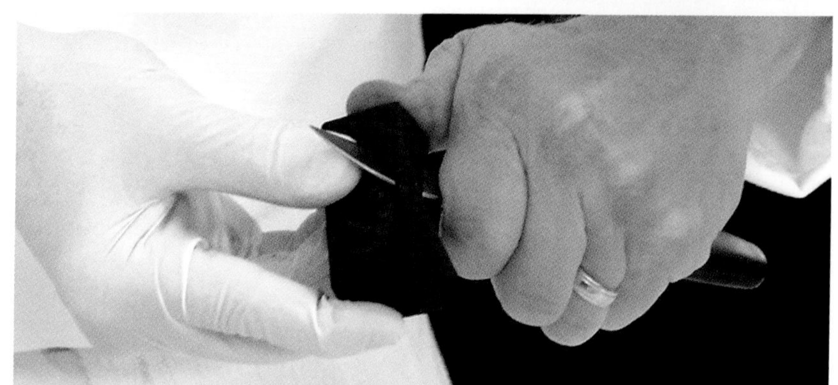

5. Rotate the item slightly toward the guiding hand and carve the item again, starting at a point farthest from the body and pulling the knife blade toward the thumb in a smooth, continuous arc to create another rounded surface.

Note: As each slice is carved, the ends will become narrower than the middle.

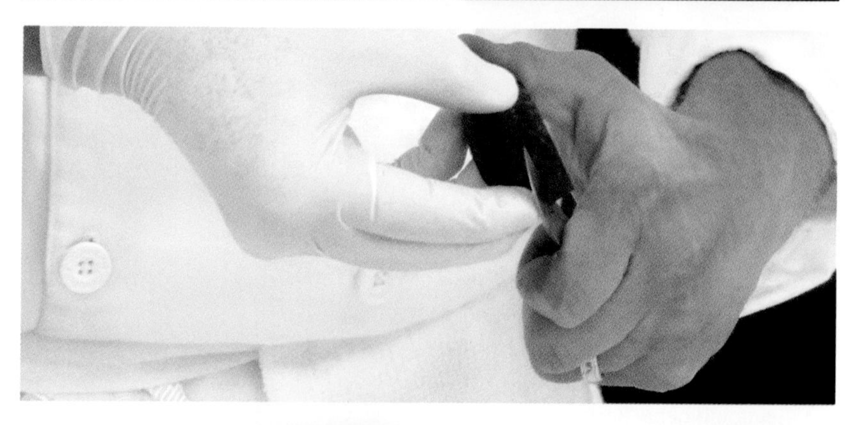

6. Continue to slightly rotate the item, making smooth, curved strokes with each slight rotation until a seven-sided football shape is achieved.

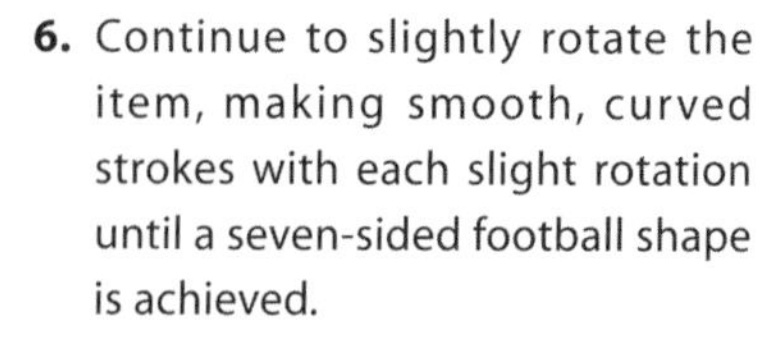

A finished tourné has seven sides with small, flat ends and measures approximately ½ inch wide in the middle and ⅛ – ¼ inch wide at the ends.

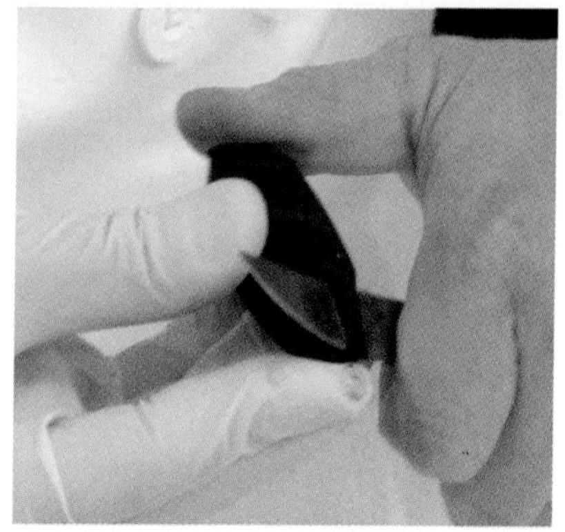

SECTION III
CUTTING PRODUCE

- SLICING MUSHROOMS
- CUTTING TOMATOES
- CUTTING SPEARS AND FLORETS
- SHREDDING HEAD VEGETABLES
- SLICING AND DICING ONIONS
- CUTTING AND CREAMING GARLIC
- SLICING AND DICING CYLINDRICAL FRUIT-VEGETABLES
- CUTTING SWEET PEPPERS
- PEELING AND CORING POMES
- CUTTING AVOCADOS
- CUTTING MANGOES
- CUTTING MELONS
- CUTTING PINEAPPLES
- CUTTING CITRUS SUPREMES
- CUTTING ARTICHOKES

LEARNER RESOURCES
ATPeResources.com/QuickLinks™
Access Code: 583241

TECHNIQUE 13

SLICING MUSHROOMS

VIDEO 13

Although they are classified as fungi, mushrooms are used in the same manner as vegetables in dishes ranging from appetizers and soups to salads and entrées. Popular cultivated mushrooms, including button mushrooms, creminis, portobellos, and shiitakes have a "meaty" texture and a savory taste known as umami.

The way a mushroom is cut depends on the type of mushroom and the intended dish. For example, button mushrooms are commonly quartered and sautéed and served as a side dish or added to soup. Quartering mushrooms helps them retain a firm texture. When a softer texture is desired, mushrooms are often thinly sliced. Thinly sliced mushrooms are frequently served on top of pizzas, meats, and poultry, or used in salads, sandwiches, and pasta dishes.

Note: Disposable gloves are typically worn in the foodservice industry when preparing ready-to-eat food items.

QUICK TIP

Wipe mushrooms with a damp towel to clean them before use. Mushrooms should never be cleaned by soaking them in water because they will absorb the water and become mushy.

QUARTERING BUTTON MUSHROOMS

1. Hold a mushroom cap in the guiding hand with the stem angled upward toward the knife hand. Grasp a paring knife in a choke grip and cut approximately ¼ inch off the stem to remove the dry, tough end.

   Note: For Step 1, it is also acceptable to place the mushroom on the cutting board and remove the stem with the tip-to-food method and a chef's knife.

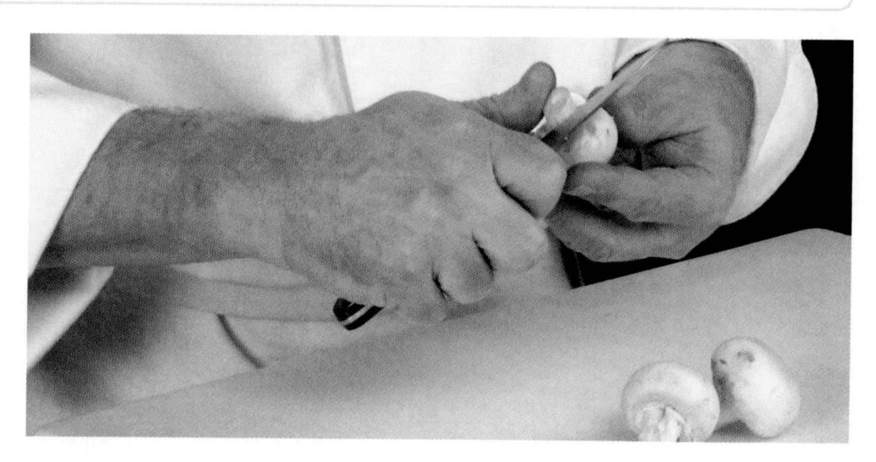

QUARTERING BUTTON MUSHROOMS (CONTINUED)

2. Place the mushroom on the cutting board and secure the mushroom cap with a claw grip. Use a chef's knife and the tip-to-food method to slice the mushroom in half from the cap end to the stem end.

3. Position the mushroom halves flat-side down and secure a mushroom cap with a claw grip. Use the tip-to-food method to slice the mushroom in half again, cutting from the cap end to the stem end. Repeat this cut with the other mushroom half to produce a quartered mushroom.

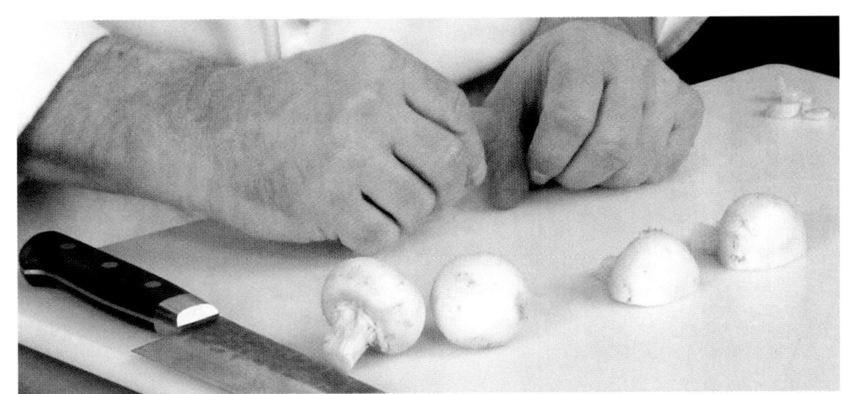

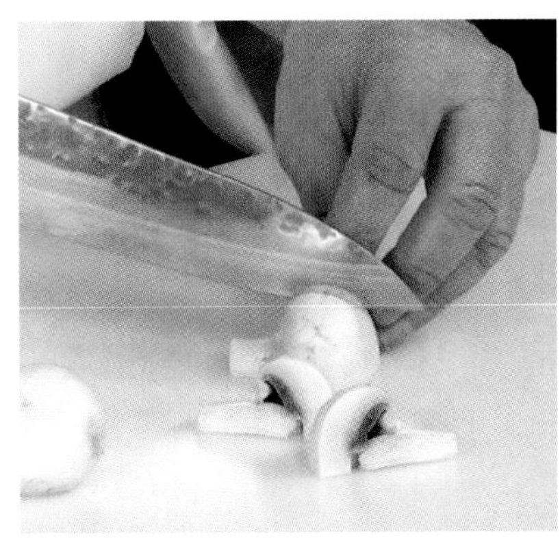

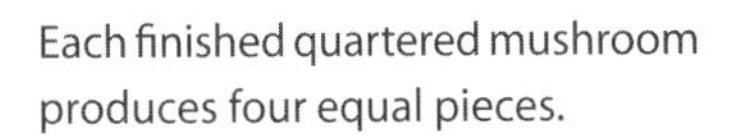

Each finished quartered mushroom produces four equal pieces.

SLICING CREMINI MUSHROOMS

1. Hold a mushroom cap in the guiding hand with the stem angled upward toward the knife hand. Grasp a paring knife in a choke grip and cut approximately ¼ inch off the stem to remove the dry, tough end.

   Note: For Step 1, it is also acceptable to place the mushroom on the cutting board and remove the stem with the tip-to-food method and a chef's knife.

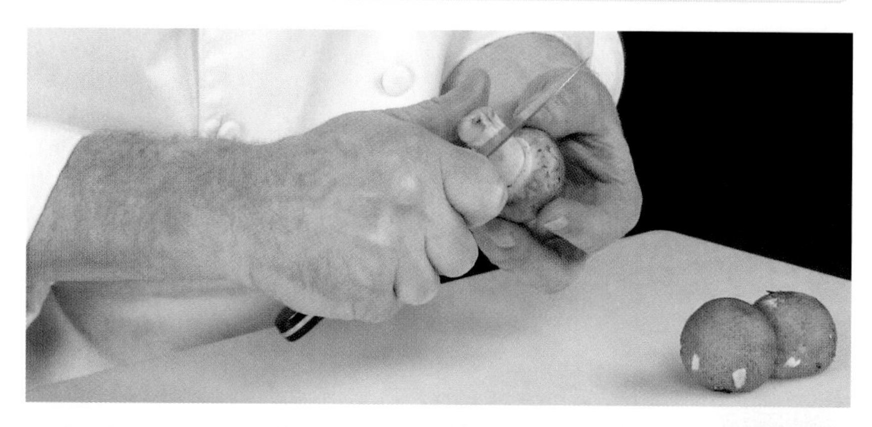

2. Place the mushroom on the cutting board and secure the mushroom cap with a claw grip. Starting at the side of the mushroom closest to the knife hand, use a chef's knife and the tip-to-food method to make slices at equally spaced intervals until reaching the other side of the mushroom.

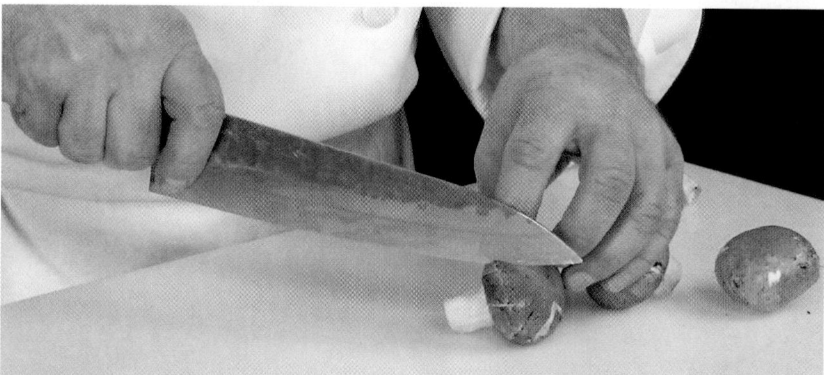

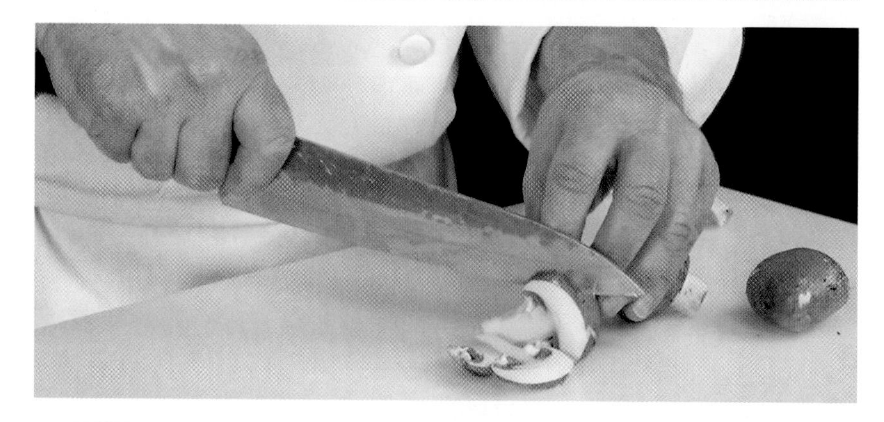

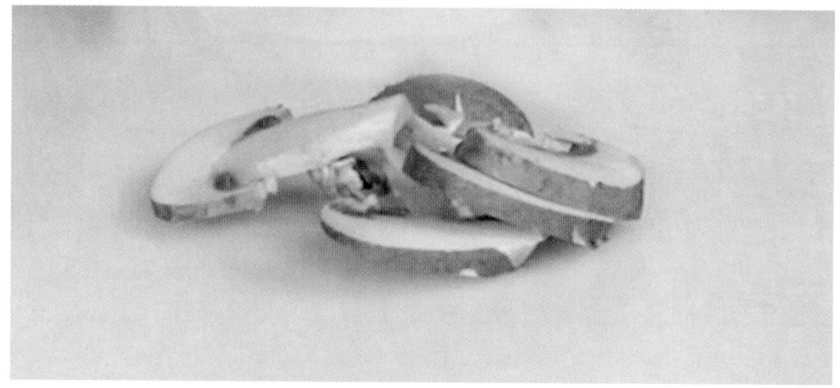

Finished sliced mushrooms have a uniform thickness.

CUTTING TOMATOES

VIDEO 14

TECHNIQUE 14

Tomatoes are often diced, cut into wedges, or sliced. A properly sliced tomato has a circular or oval shape that adds both color and flavor to dishes. To slice a tomato correctly, it is essential to use a sharp knife and the right amount of pressure. Using a dull knife or applying too much pressure compresses the tomato, changing its shape and causing its juices and seeds to escape.

While tomatoes are frequently served raw, they are also cooked in items such as soups and sauces. During the cooking process, the tomato skins and seeds typically do not break down, causing an unpleasant texture. Therefore, tomatoes are often prepared concassé. Concassé is a preparation method in which a tomato is peeled, seeded, and then diced or chopped. To easily peel a tomato, a paring knife is used to make a shallow "X" (approximately ½ inch wide by ½ inch high) into the bottom or blossom end of the tomato. The tomato is then blanched by placing it in a pot of boiling water for 15–30 seconds or until the skin around the "X" starts to pull away from the tomato. Next, the tomato is immediately removed from the boiling water and placed in a bowl of ice water to stop the cooking process. At this point, the tomato is ready to be peeled.

Note: Disposable gloves are typically worn in the foodservice industry when preparing ready-to-eat food items.

CORING AND SLICING TOMATOES

1. Hold a tomato in the guiding hand with the stem end angled upward toward the knife hand side. Grasp a paring knife in a choke grip and insert the knife tip at a slight angle 1 inch into the side of the stem.

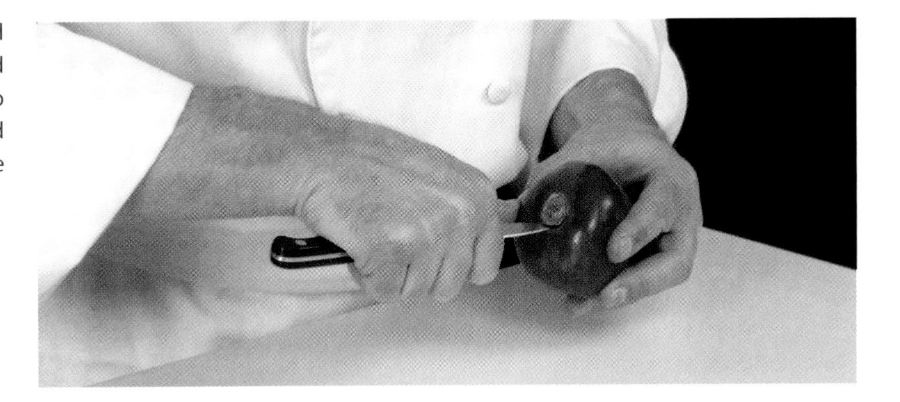

QUICK TIP

Paderno World Cuisine

A tomato corer is a tool with sharp, serrated teeth that can quickly and effectively remove the core from tomatoes.

2. Cut with a slight sawing motion while using the guiding hand to rotate the tomato around the knife until the stem and a small portion of the core are removed.

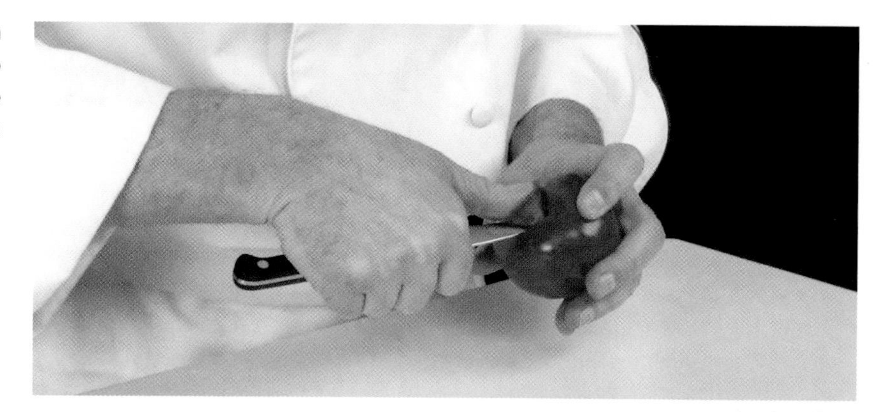

3. Place the tomato on the cutting board with the stem end facing the knife hand side. Secure the tomato with a claw grip and position the tip of a chef's knife on top of the tomato near the stem end.

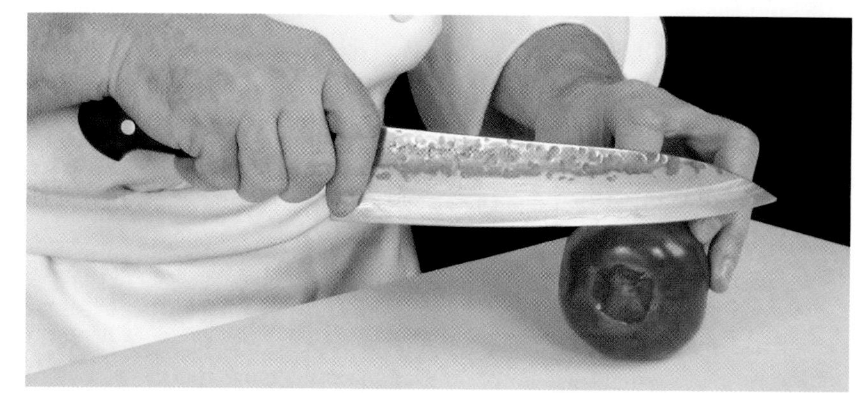

4. Use a sawing motion with the tip-to-food method to slice the tomato, making the first slice thick enough to remove the portion that contained the stem.

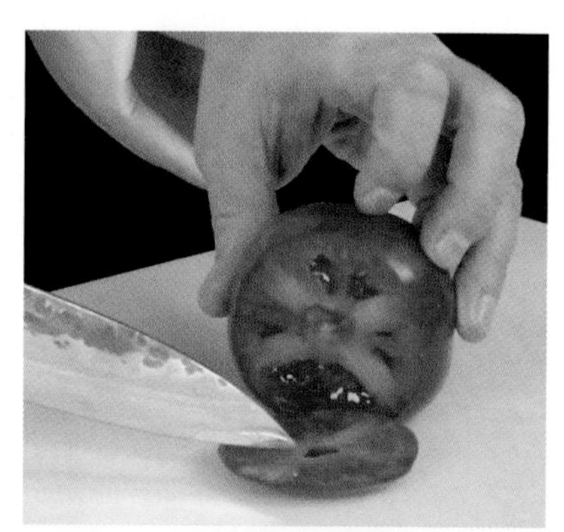

5. Continue using a sawing motion with the tip-to-food method to slice the tomato from end to end, keeping the slices a consistent thickness.

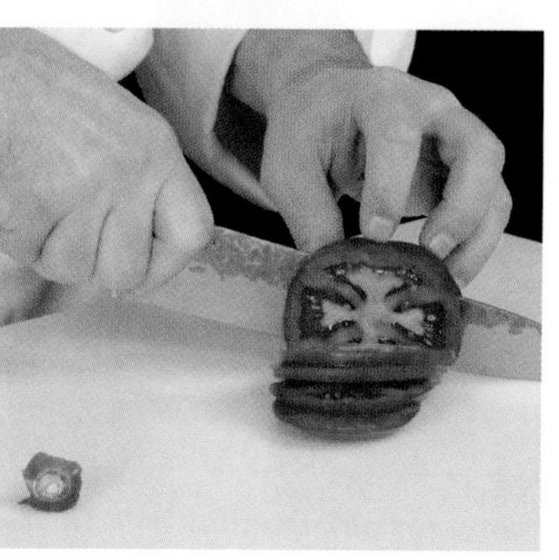

Finished tomato slices have a circular or oval shape and the juice and seeds remain intact.

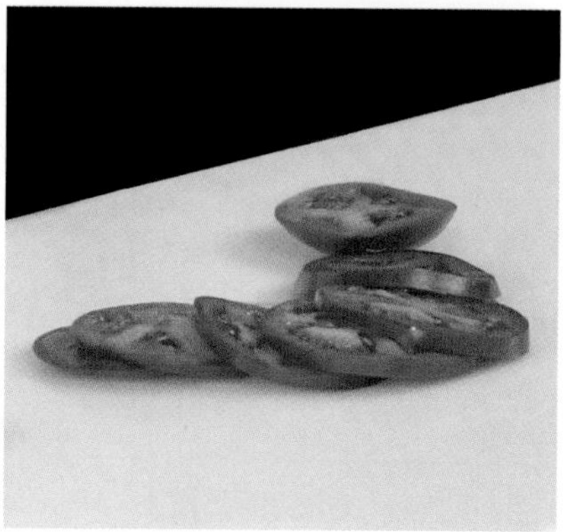

CUTTING TOMATOES CONCASSÉ

1. Start with a blanched tomato that has the stem removed and a shallow "X" cut on the bottom side. Hold the tomato in the guiding hand so that the "X" cut is angled upward toward the knife hand side.

   Note: Refer to the introduction of this technique for how to blanch a tomato.

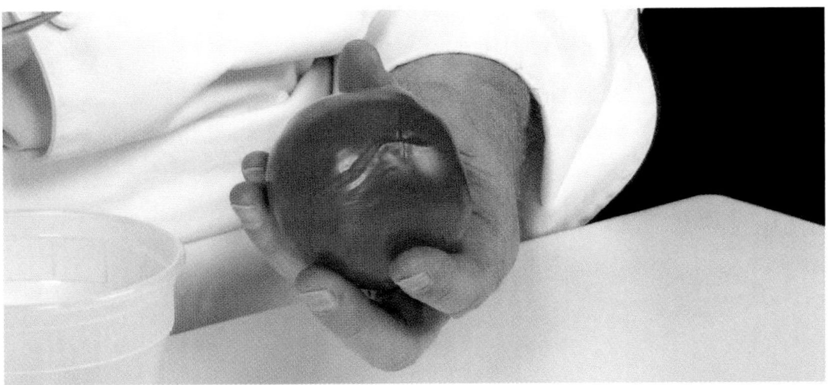

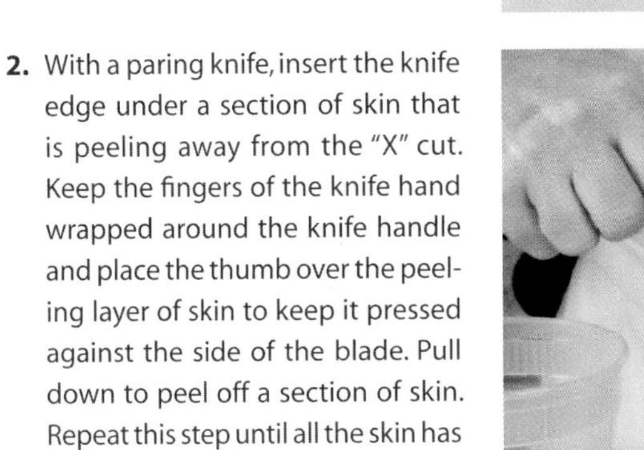

2. With a paring knife, insert the knife edge under a section of skin that is peeling away from the "X" cut. Keep the fingers of the knife hand wrapped around the knife handle and place the thumb over the peeling layer of skin to keep it pressed against the side of the blade. Pull down to peel off a section of skin. Repeat this step until all the skin has been removed.

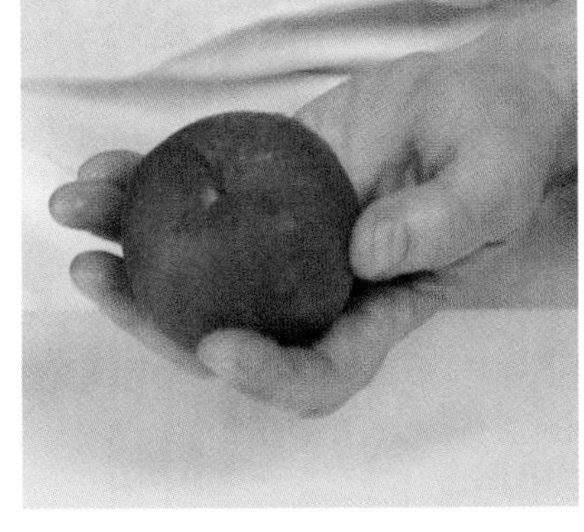

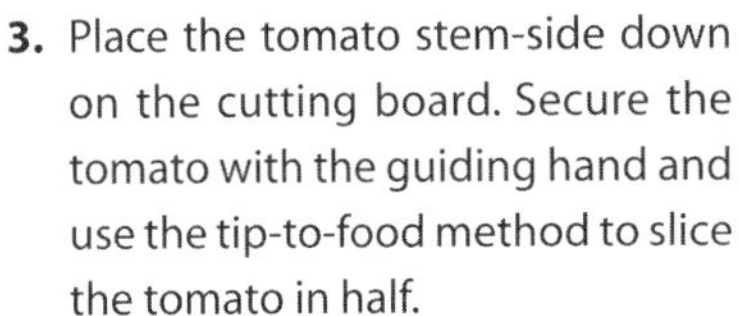

3. Place the tomato stem-side down on the cutting board. Secure the tomato with the guiding hand and use the tip-to-food method to slice the tomato in half.

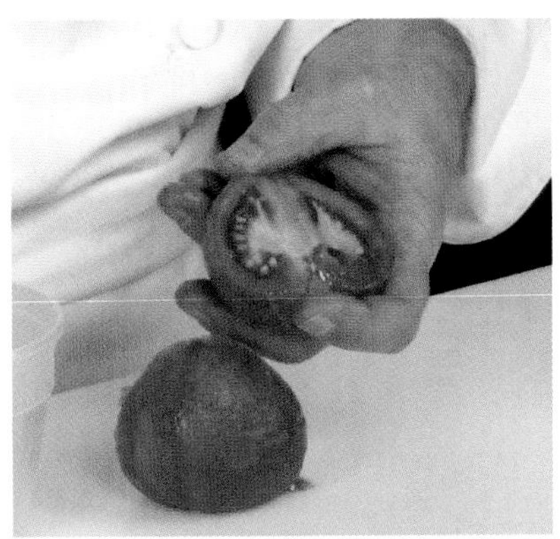

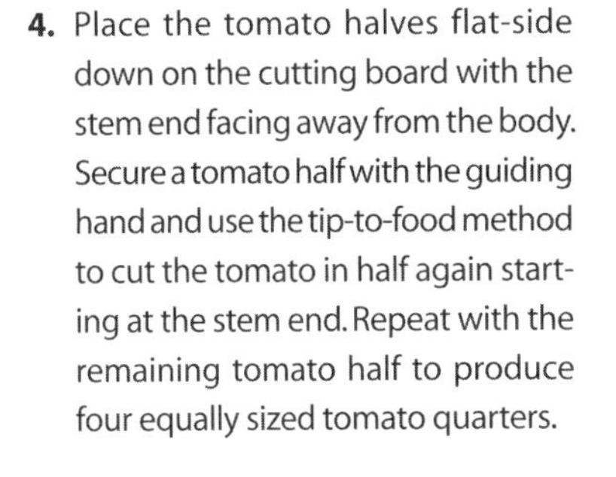

4. Place the tomato halves flat-side down on the cutting board with the stem end facing away from the body. Secure a tomato half with the guiding hand and use the tip-to-food method to cut the tomato in half again starting at the stem end. Repeat with the remaining tomato half to produce four equally sized tomato quarters.

   Note: For Steps 3–4, it is also acceptable to use a chef's knife instead of a paring knife.

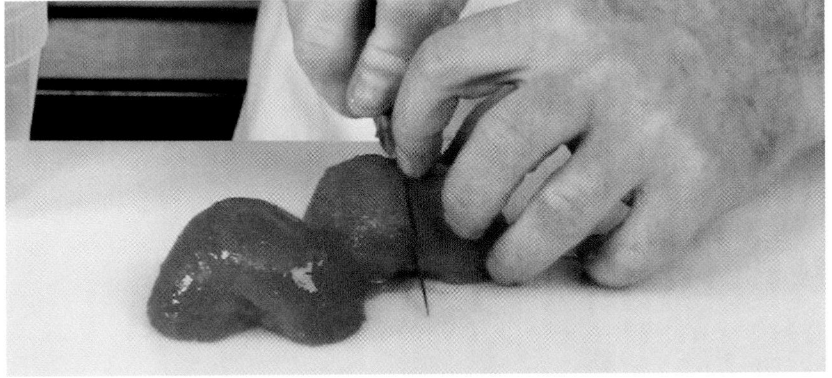

5. Use the hands to remove the seeds and membranes from each quartered tomato.

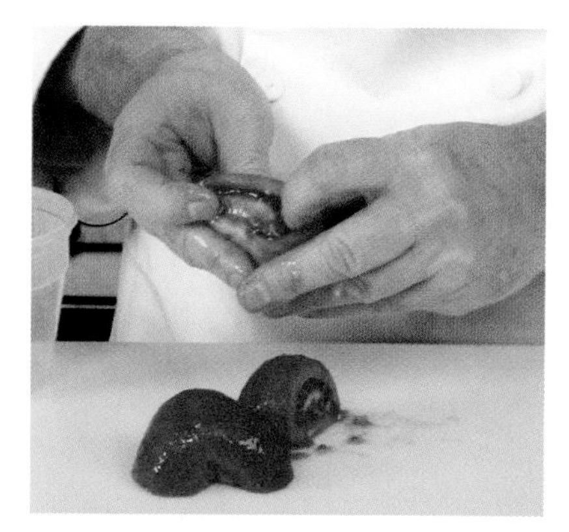

6. To make consistently sized tomato strips, place a quartered tomato lengthwise on the cutting board with the peeled-side up and hold the tomato in a claw grip. With a chef's knife, start at the side closest to the knife hand and slice the tomato at approximately ⅛–¼ inch intervals.

7. Rotate the tomato strips 90° and place them perpendicular to the knife blade. Use the side of the knife blade to align the ends of the tomato strips.

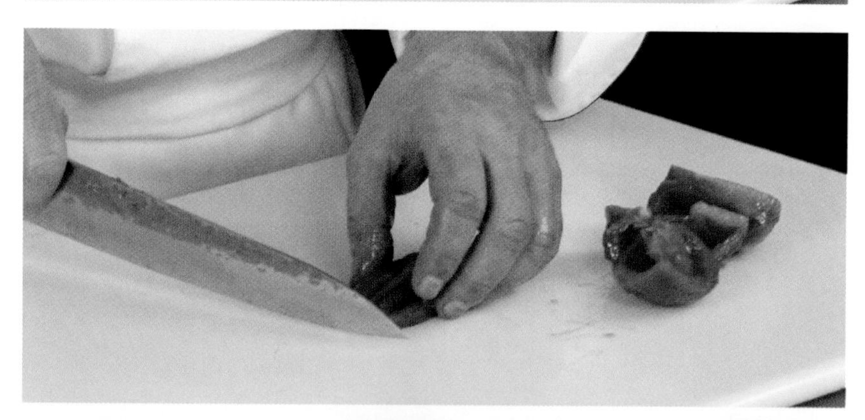

8. Secure the tomato strips with a claw grip and use the tip-to-board method to slice the bundle crosswise at approximately ⅛–¼ inch intervals to produce rough dice cuts.

CUTTING TOMATOES CONCASSÉ (CONTINUED)

9. If chopped tomatoes are desired, place the palm of the guiding hand on the spine of the knife near the tip. Keep the knife tip on the cutting board and rock the blade up and down over the rough dice cuts while pivoting the blade back and forth until the tomatoes are chopped.

Cutting tomatoes concassé commonly produces rough dice cuts or chopped pieces of peeled and deseeded tomato.

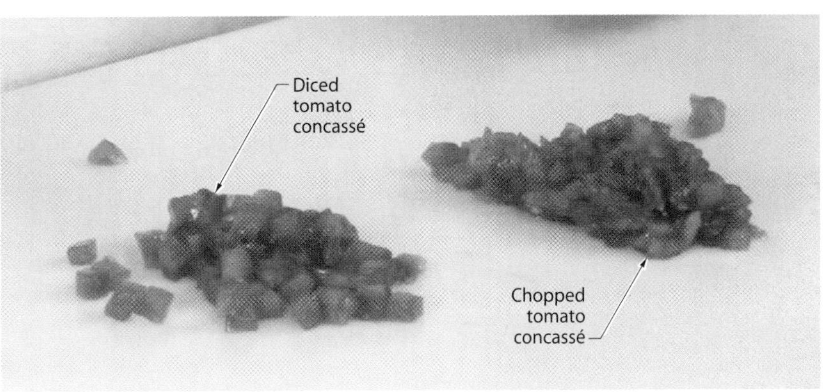

TECHNIQUE 15

CUTTING SPEARS AND FLORETS

VIDEO 15

Broccoli, cauliflower, and Broccoflower® (broccoli crossed with cauliflower) have a central stalk attached to a head of florets. Because the stalk is mostly tough and fibrous, recipes generally call for the stalk to be removed from the florets. In the case of broccoli, as the stalk approaches the florets, it branches outward, becoming thinner and less fibrous. For this reason, 2–3 inches of the stalk are left attached to the individual florets when cutting broccoli spears. Broccoli spears offer a contrast of texture between the firm stalk and the tender floret and are often served as a side dish.

Note: Disposable gloves are typically worn in the foodservice industry when preparing ready-to-eat food items.

Paring Knife

— or —

Utility Knife

Chef's Knife

— or —

Santoku Knife

CUTTING BROCCOLI SPEARS

1. Place a head of broccoli on the cutting board with the stalk facing the knife hand side. Hold the floret end in the guiding hand and tilt the broccoli until the stalk angles slightly upward. Grasp a paring knife in a choke grip and trim the stalk so that 2–3 inches of the stalk remain attached to the florets.

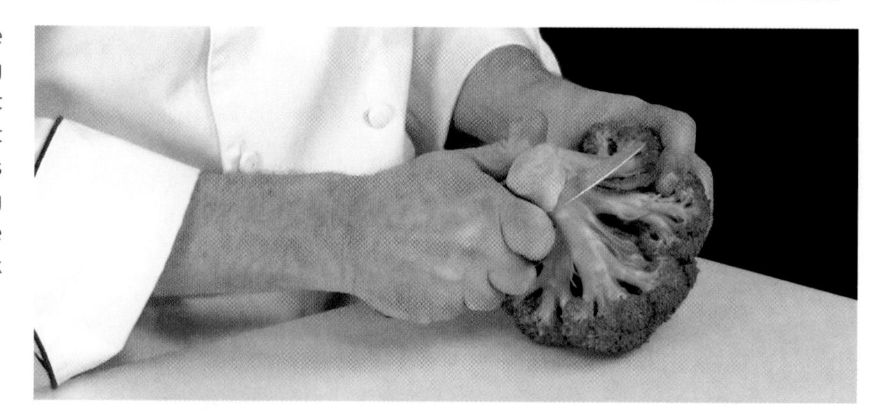

QUICK TIP

Vegetables high in the green pigment chlorophyll, such as broccoli, turn a drab green color when prepared with an acidic ingredient, such as lemon juice. To help preserve color, blanch vegetables containing chlorophyll in salted water.

CUTTING BROCCOLI SPEARS (CONTINUED)

2. Turn the broccoli so that the stalk angles slightly upward toward the body. Hold the paring knife in a choke grip and place the knife tip where the stalk begins to branch toward a floret. To cut a broccoli spear, place the thumb at the end of the stalk and pull the knife toward the thumb. Repeat this step on the remaining broccoli head.

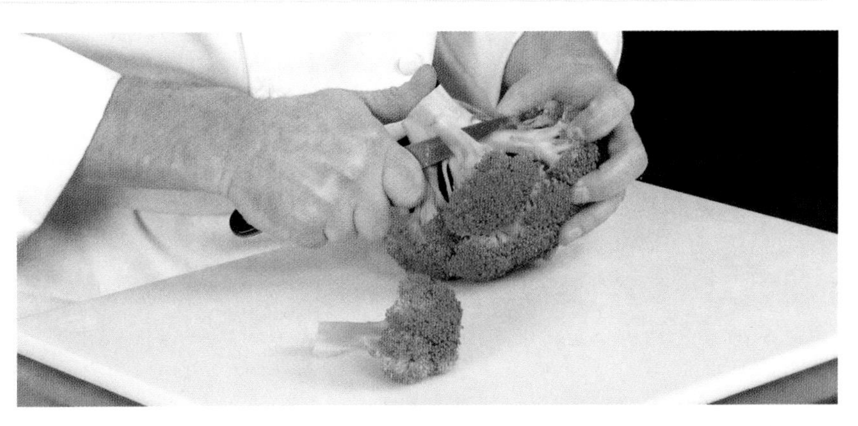

3. If a spear is wider than desired, place the spear on the cutting board and secure the floret end with a claw grip. Slice the spear lengthwise from the floret end to the stalk end. Repeat with the remaining broccoli spears to produce consistently sized pieces.

   Note: For Steps 1–3, it is also acceptable to place the broccoli on the cutting board and use the tip-to-food method with a chef's knife to remove the stem and cut the spears.

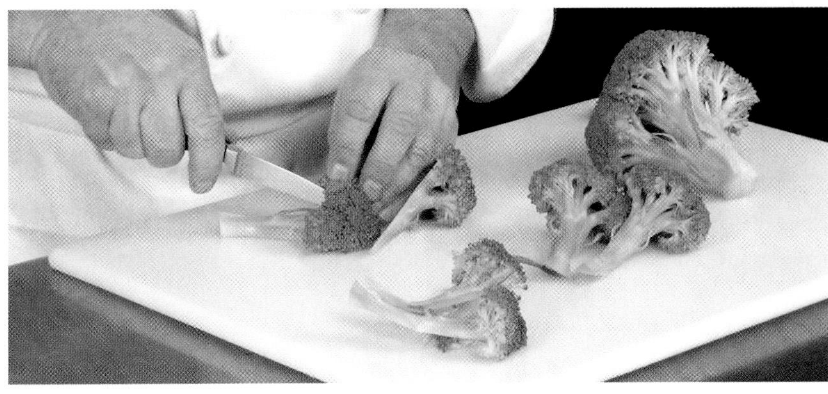

QUICK TIP

When preparing broccoli spears, if an extra-tender stalk is desired, use a peeler to trim the top layer of broccoli from the stalk.

Finished broccoli spears include a 2–3 inch portion of the stalk with each floret.

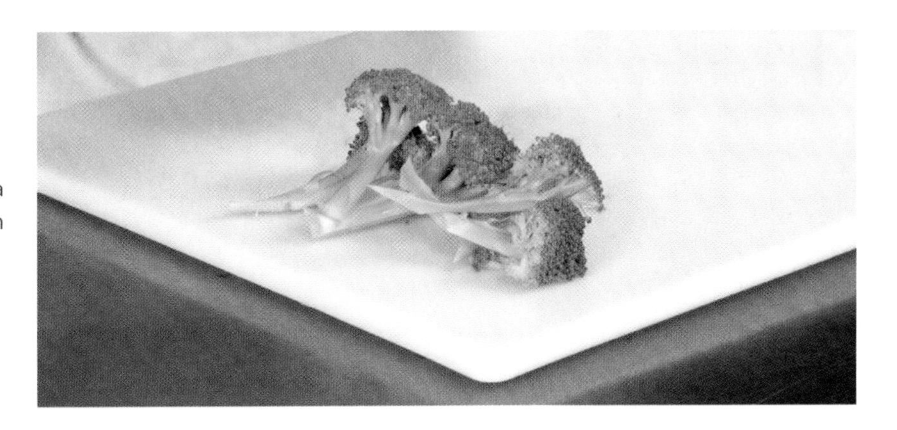

CUTTING BROCCOLI FLORETS

1. Place a head of broccoli on the cutting board with the stalk facing the knife hand side and secure the florets with the guiding hand. Use a paring knife or chef's knife to cut through the branched area of the stalk to release the florets.

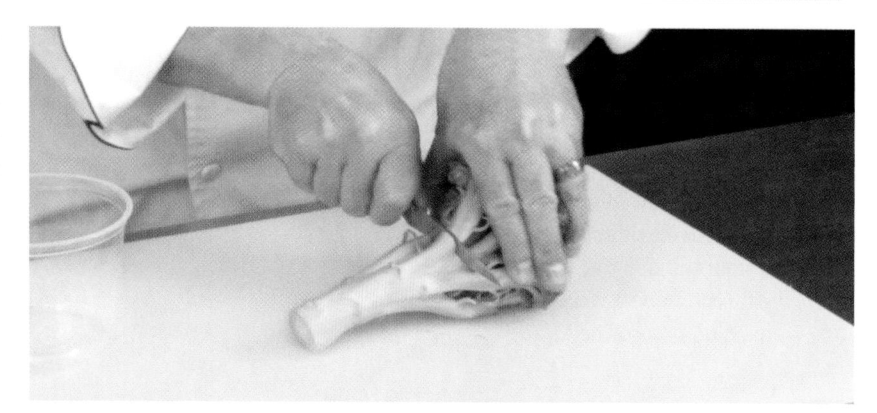

2. If a floret is larger than desired, place the floret on the cutting board and secure the floret end with a claw grip. Slice lengthwise through the floret. Repeat with the remaining florets to produce consistently sized pieces.

Finished broccoli florets have a uniform size based on the desired application.

CUTTING CAULIFLOWER FLORETS

1. Place a head of cauliflower on the cutting board with the stalk facing the knife hand side and hold the floret end in a claw grip. With a chef's knife, use a sawing motion to cut off the stalk just before it reaches the florets.

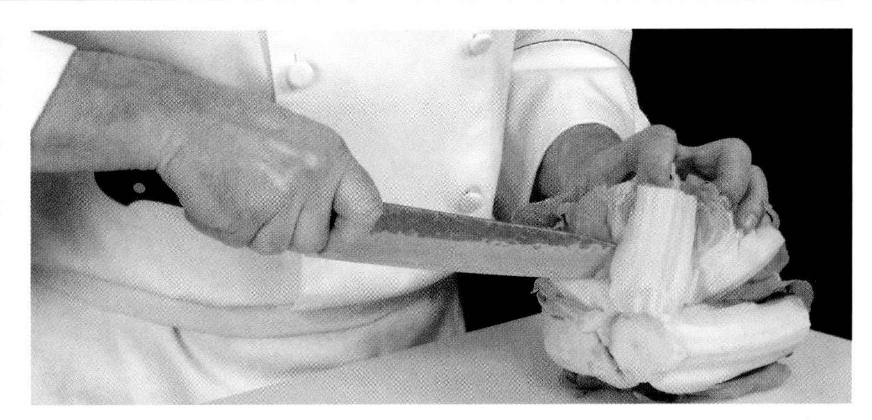

2. Place the cauliflower stalk-side up and use a choke grip with a paring knife to cut around the core. Pull the knife blade toward the thumb until 1–2 inches of the core is removed.

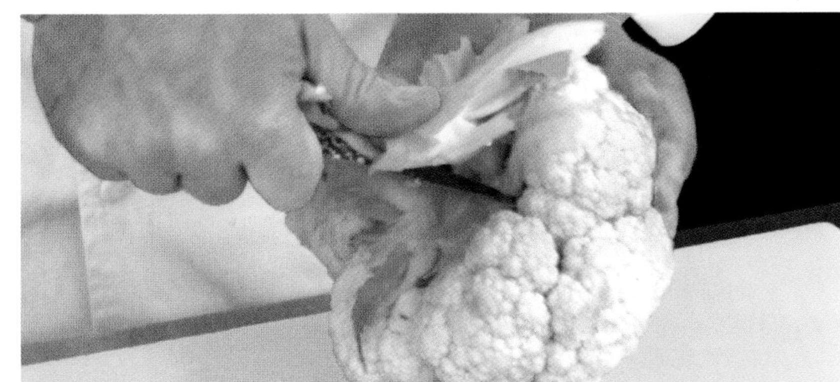

3. Place the cauliflower stalk-side down and secure with a claw grip. Using the chef's knife and the tip-to-food method, cut the cauliflower in half.

4. Position a cauliflower half flat-side down on the cutting board and secure with a claw grip. Use the tip-to-food method to cut the cauliflower in half again. Repeat this cut with the other cauliflower half to produce a quartered cauliflower.

CUTTING CAULIFLOWER FLORETS (CONTINUED)

5. Hold the florets of a quartered cauliflower in the guiding hand and stand the end of the stalk on the cutting board so that the exposed stalk faces the knife hand side. With the cauliflower tilted back at an angle, position a utility or chef's knife at the top of the stalk and cut in a downward motion to release the florets from the stalk. Repeat this cut with the remaining cauliflower quarters.

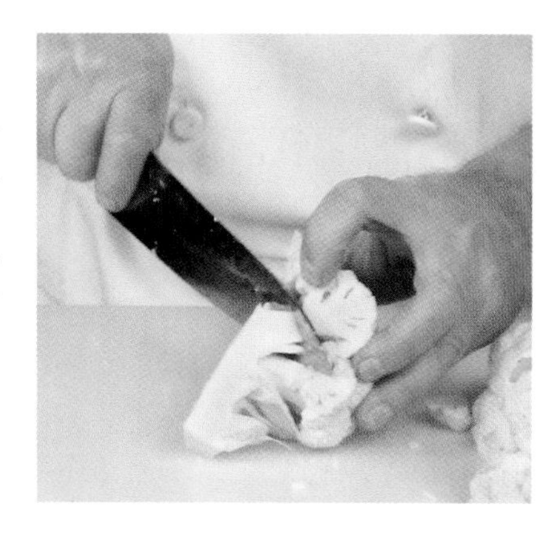

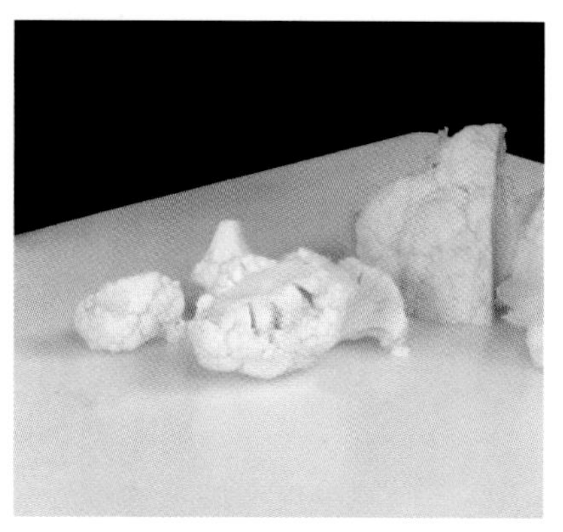

QUICK TIP

Florets that are cut to a uniform size cook more evenly, therefore producing a consistent texture.

6. If a floret is larger than desired, place the floret on the cutting board and use a utility or paring knife to cut the floret to the desired size. Repeat with the remaining florets to produce consistently sized pieces.

Finished cauliflower florets have a uniform size with most of the stalk removed from the florets.

QUICK TIP

Broccoflower® is cut in the same manner as cauliflower to produce florets.

VIDEO 16

SHREDDING HEAD VEGETABLES

TECHNIQUE 16

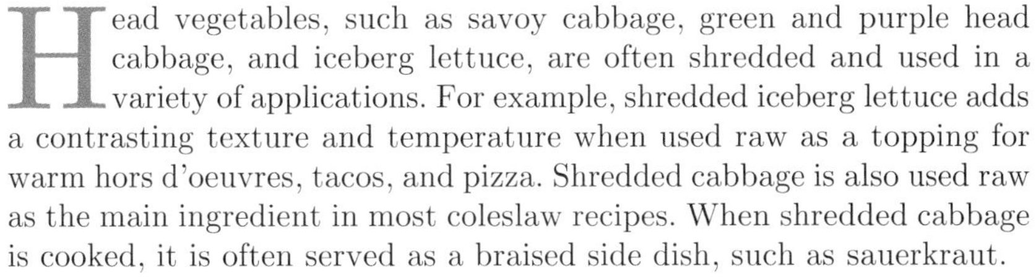

Head vegetables, such as savoy cabbage, green and purple head cabbage, and iceberg lettuce, are often shredded and used in a variety of applications. For example, shredded iceberg lettuce adds a contrasting texture and temperature when used raw as a topping for warm hors d'oeuvres, tacos, and pizza. Shredded cabbage is also used raw as the main ingredient in most coleslaw recipes. When shredded cabbage is cooked, it is often served as a braised side dish, such as sauerkraut.

Chef's Knife

Paring Knife

Note: Disposable gloves are typically worn in the foodservice industry when preparing ready-to-eat food items.

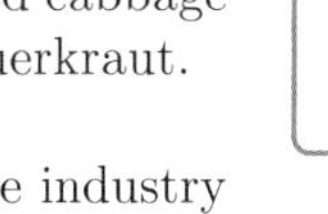

SHREDDING CABBAGE

1. Place a head of cabbage on the cutting board with the root end facing away from the body and secure the cabbage in a claw grip. Use a chef's knife and the tip-to-food method to cut the cabbage in half starting at the root end.

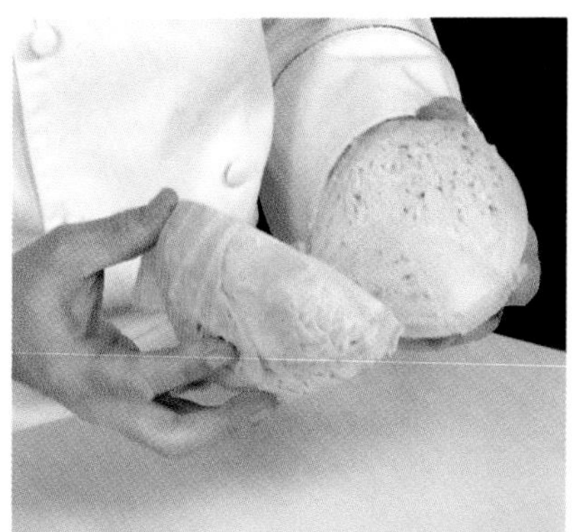

2. Position a cabbage half flat-side down on the cutting board and secure with a claw grip. Use the tip-to-food method to cut the cabbage in half again. Repeat this cut with the other cabbage half to produce a quartered cabbage.

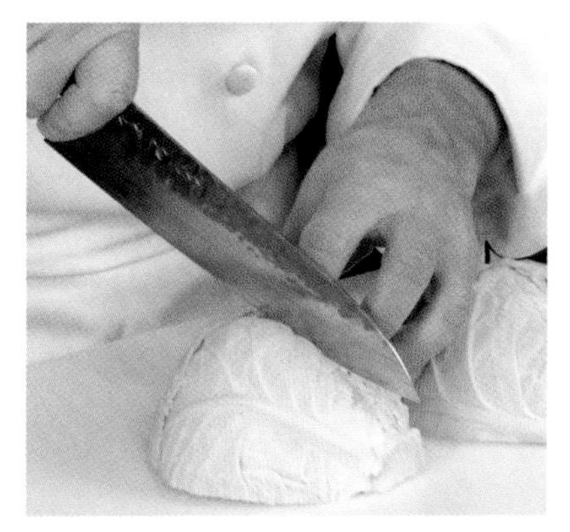

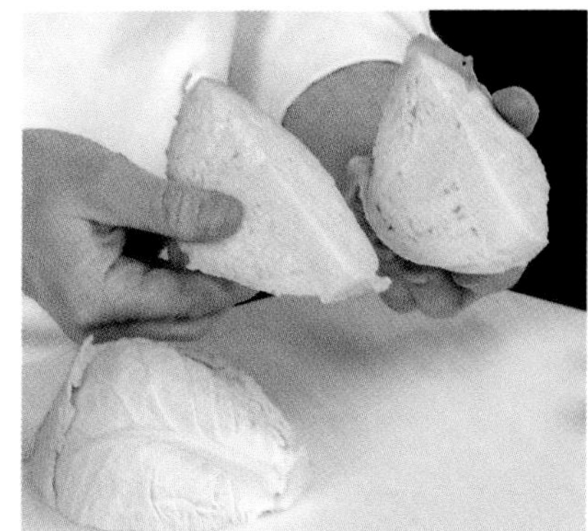

SHREDDING CABBAGE (CONTINUED)

3. Hold the top portion of a quartered cabbage in the guiding hand and stand the root end on the cutting board with the core facing the knife hand side. With the cabbage tilted back at a slight angle, position a paring knife or chef's knife on the top portion of the core and cut in a downward motion to remove the core. Repeat this cut with the remaining cabbage quarters.

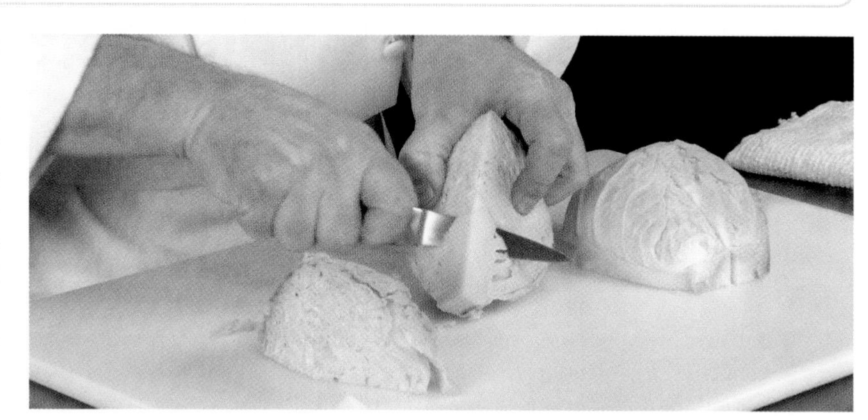

4. Place a quartered cabbage flat-side down on the cutting board with the tapered end facing the knife hand and secure the cabbage in a claw grip. With a chef's knife, use the tip-to-food method to thinly slice the cabbage from end to end, keeping each slice a consistent thickness. Repeat this cut with the remaining cored cabbage.

Finished shredded cabbage consists of narrow strips that vary in length but have a uniform width.

QUICK TIP

If cabbage is shredded prior to use, store the shreds in ice water to keep the cabbage crisp and prevent it from wilting.

TECHNIQUE 17

VIDEO 17

SLICING AND DICING ONIONS

Onions are a versatile kitchen staple recognized for both their aromatic qualities and flavor. Onions are often sliced into rings and used on salads and sandwiches or deep fried and served as a crunchy appetizer or side. Likewise, onion strips enhance many dishes including fajitas and stir-fries. In addition to slices, onions are often diced into rough (imprecise) cubes. Proper dice cuts result in precise cubes. However, rough dice cuts are often acceptable based on the desired application. Diced onions are commonly used raw in salsas and relishes or sautéed and used as a flavorful base in items such as sauces, soups, and grain dishes.

Note: Disposable gloves are typically worn in the foodservice industry when preparing ready-to-eat food items.

Paring Knife

Chef's Knife

— or —

Santoku Knife

PEELING AND TRIMMING ONIONS

1. Hold an onion in the guiding hand with the stem end angled upward toward the knife hand side. Grasp a paring knife in a choke grip and slightly trim the stem end.

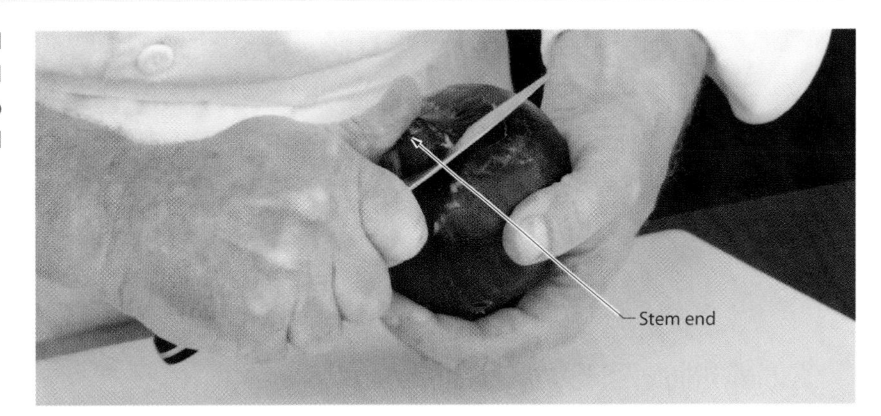

2. Turn the onion so that the root end is angled upward toward the knife hand side. Grasp the paring knife in a choke grip and slightly trim the root end, keeping some of the root intact.

   Note: The root is not sliced off completely because it holds the layers of the onion in place and prevents it from falling apart.

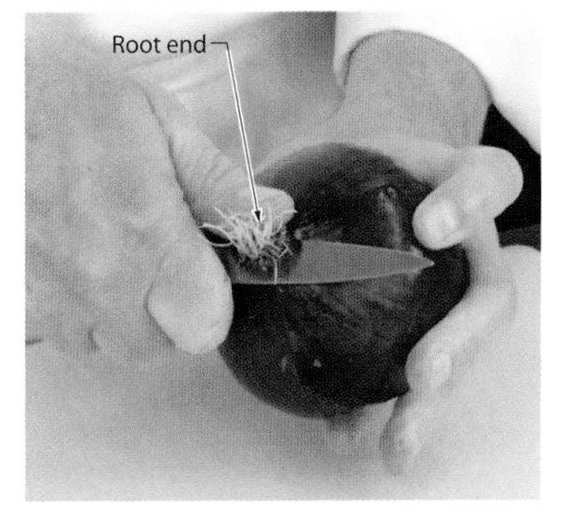

PEELING AND TRIMMING ONIONS (CONTINUED)

> **QUICK TIP**
>
> *Reserve the trimmed stem end from an onion for making stocks, but discard or compost the trimmed root end as it can harbor dirt and grit that is difficult to remove completely.*

3. Use Method A or Method B to peel the skin from the onion.

Method A:

- Starting at the stem end, hold the paring knife in a choke grip and insert the heel of the blade under the first layer of skin beneath the papery layer. Use the knife hand thumb to keep the onion layers pressed against the side of the knife blade and pull down to remove a section of skin. Continue this process until the papery layer and next layer of skin are completely removed from the onion.

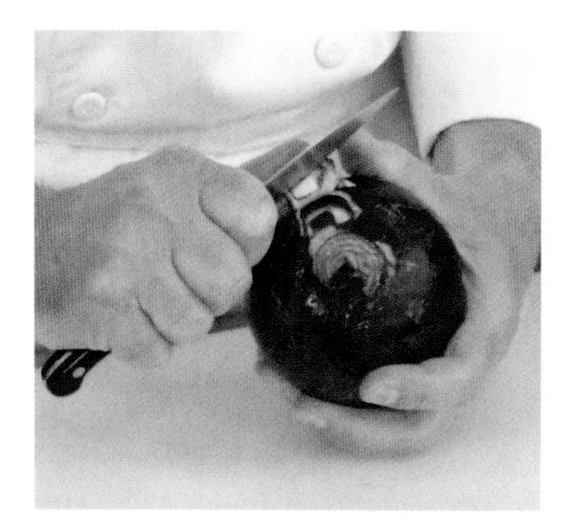

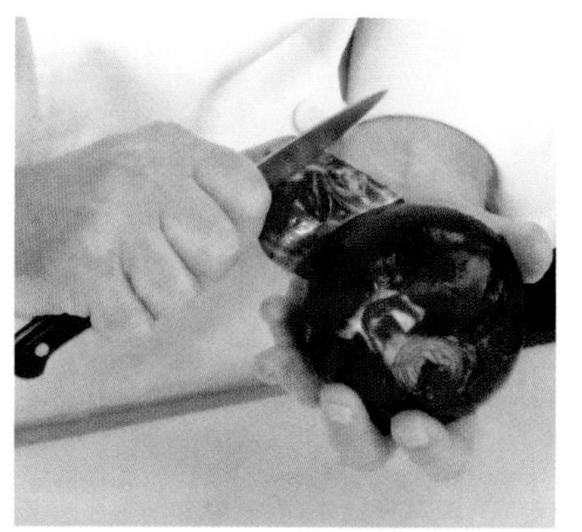

Method B:

- Use the hands to remove both the papery layer and the next layer of onion skin until the onion is completely peeled.

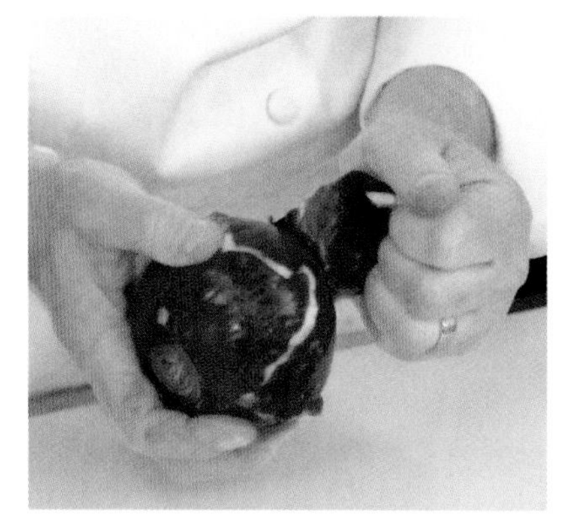

SLICING ONION RINGS

1. Place a peeled and trimmed onion on the cutting board so that the stem end faces the knife hand side and secure the onion in a claw grip. Position the tip of a chef's knife on top of the onion near the stem end.

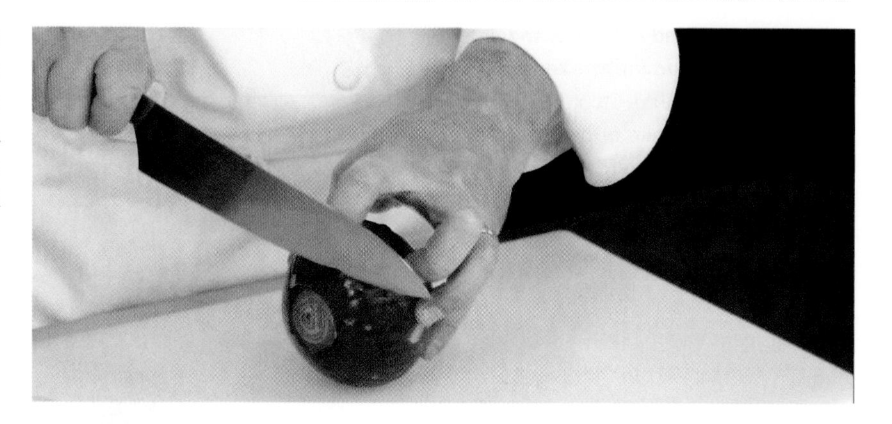

SLICING ONION RINGS (CONTINUED)

2. Use a sawing motion with the tip-to-food method to slice the onion from the stem end to the root end at equally spaced intervals until only the root remains.

   Note: If the first slice of the onion is dry, reserve that slice for composting or making stocks.

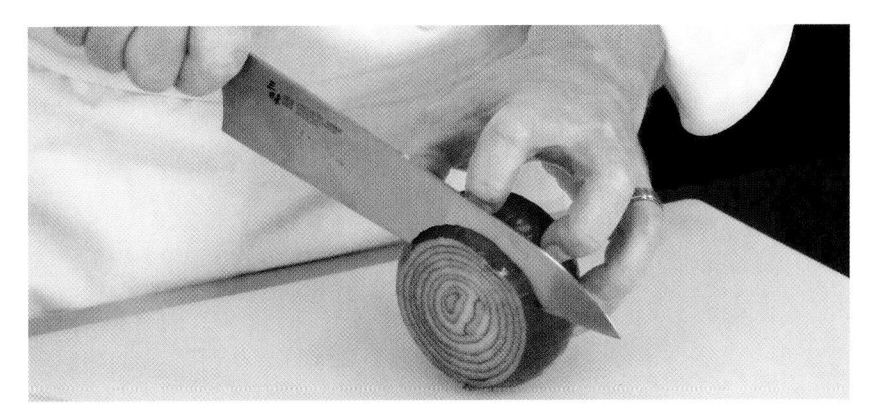

Finished sliced onion rings contain multiple rings in each slice that can be left intact or separated before use.

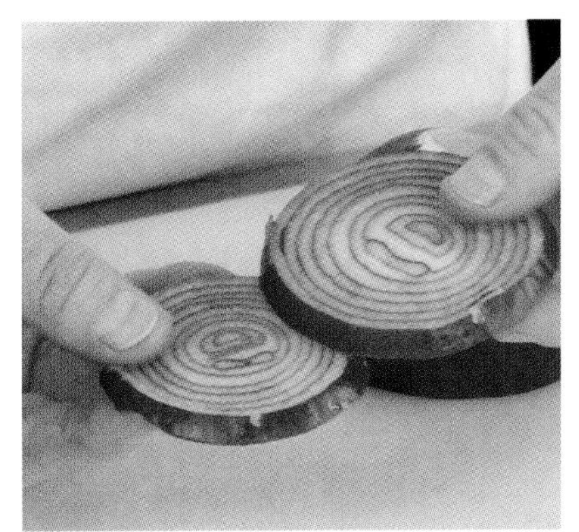

QUICK TIP

Cut onions can be placed in an air-tight container and stored in a refrigerator set to 40° or below for up to seven days.

SLICING ONION STRIPS

1. Place a peeled and trimmed onion stem-side down on the cutting board. Secure the onion with a claw grip and center the edge of a chef's knife on the root side of the onion. Use the tip-to-food method to cut the onion in half.

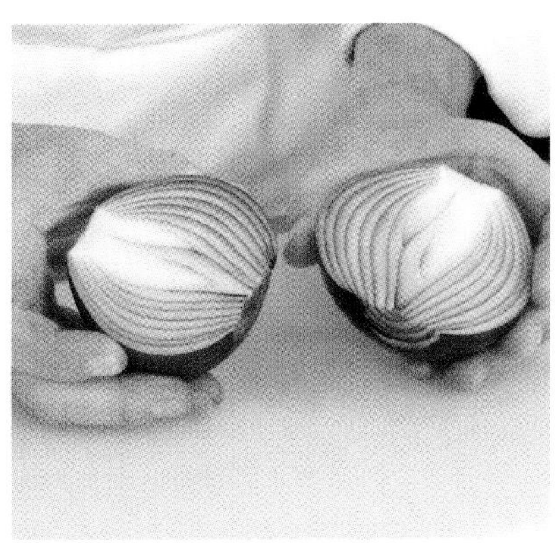

SLICING ONION STRIPS (CONTINUED)

2. Place a halved onion flat-side down on the cutting board with the root end facing the knife hand side. Secure the onion with a claw grip and position the knife tip slightly above the root at a 45° angle. Cut downward at an angle to remove the root.

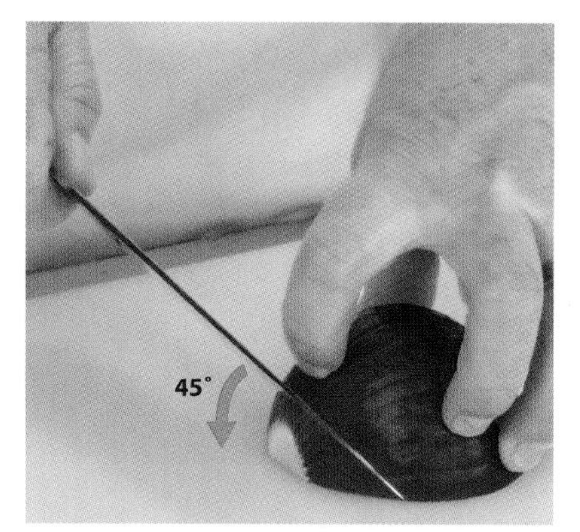

3. Rotate the onion 90° and secure it with a claw grip. Use the tip-to-food method to trim off the side of the onion to the desired thickness. Rotate the onion 180° and repeat this cut on the other side of the onion. Reserve both pieces.

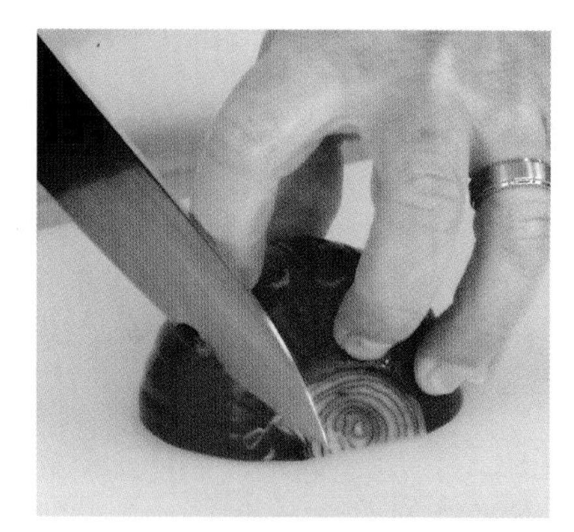

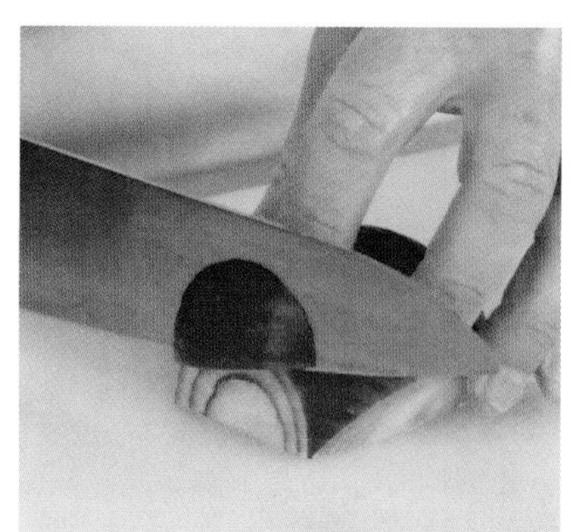

4. Secure the onion with a claw grip. Start at the side closest to the knife hand and use the tip-to-food method to slice the onion into consistently sized strips. Repeat this cut on the reserved side pieces of onion from Step 3.

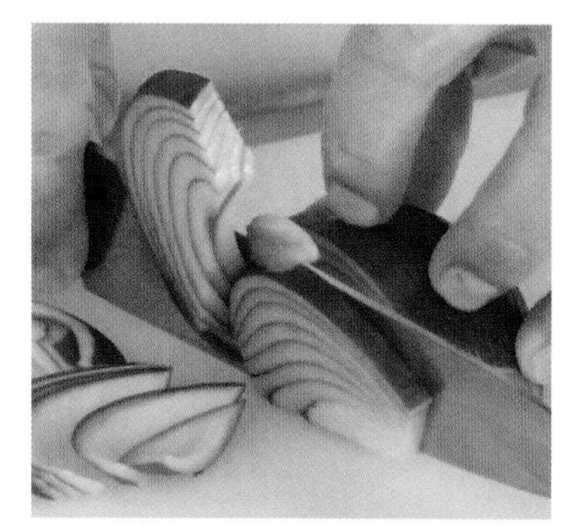

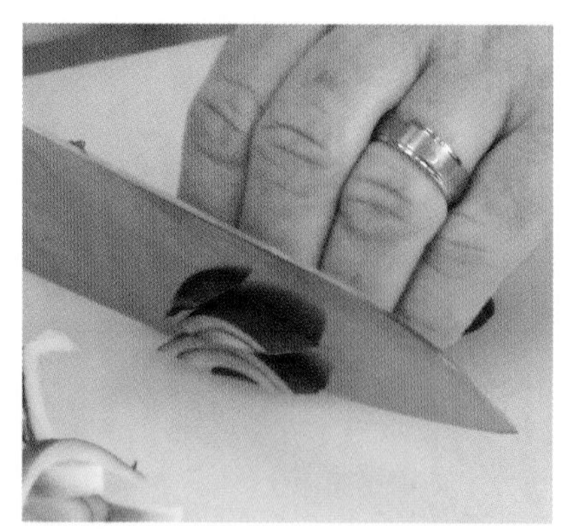

Finished onion strips have a slight arch and are typically separated before use.

DICING ONIONS

* This technique was executed by a left-handed chef.

1. Place a peeled and trimmed onion that has been cut in half from the root end to the stem end on the cutting board with the flat-side down and the stem end facing the knife hand side.

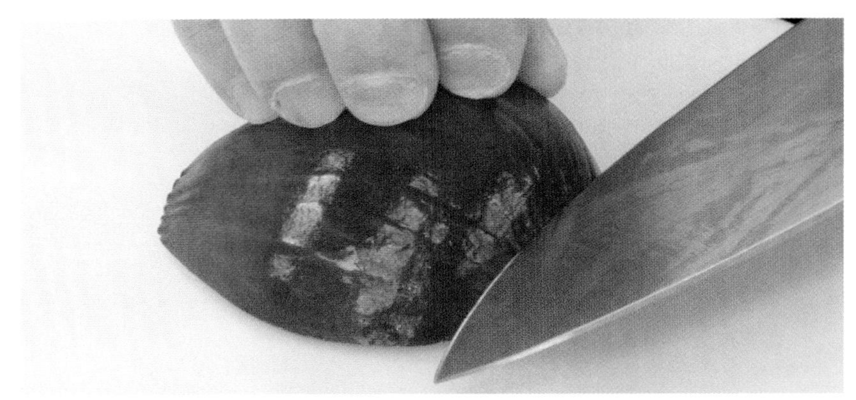

2. Set the palm of the guiding hand on top of the onion and position a chef's knife so that the blade is parallel to the cutting board at the stem end of the onion. Make two or three equally spaced horizontal cuts most of the way through the onion, leaving the root intact.

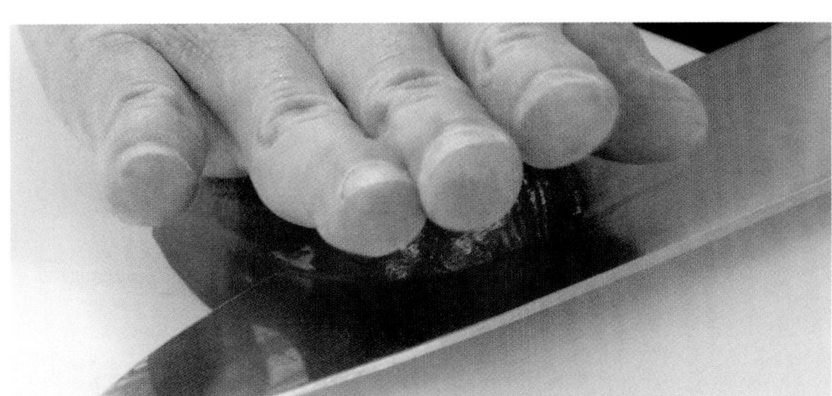

3. Rotate the onion approximately 90° so that the root end is positioned farthest away from the body and secure the onion with a claw grip. Starting at the side of the onion closest to the knife hand, use the tip-to-food method to make vertical (lengthwise) slices at equally spaced intervals until reaching the other side of the onion, again leaving the root intact.

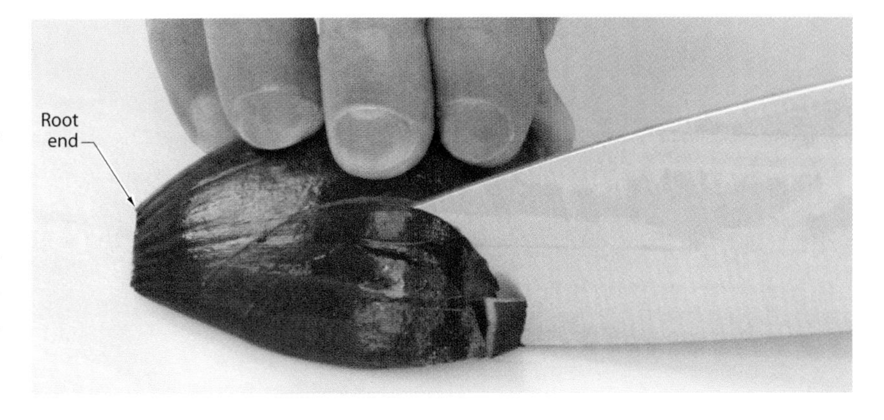

* This technique was executed by a left-handed chef.

4. Rotate the onion back to its previous position with the stem end facing the knife hand side and secure the onion with a claw grip. Use the tip-to-food method to slice the onion crosswise from stem end to root end at equally spaced intervals until only the root remains.

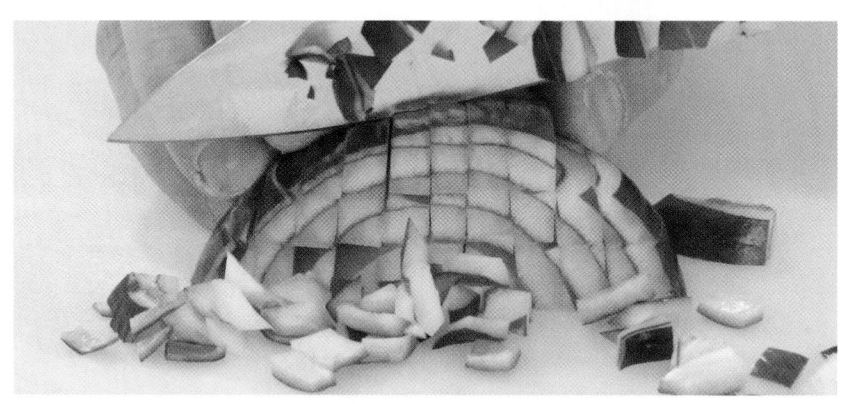

Finished diced onions are imprecise cubes due to the natural contour of the onion. The size of the finished dice is controlled by the number of horizontal, vertical, and crosswise slices. The closer together the slices, the smaller the finished dice.

QUICK TIP

Rub a stainless steel spoon over your hands while rinsing them under cold water to remove the smell of onion.

VIDEO 18

CUTTING AND CREAMING GARLIC

TECHNIQUE 18

When garlic is cut, its cells rupture to release flavorful compounds. Therefore, the more garlic is broken down from its original state, the stronger its flavor. Subsequently, a chopped garlic clove will infuse a sauce with more flavor than a whole garlic clove. Whether garlic is to be used whole, sliced, diced, chopped, minced, or creamed depends on the item being prepared. For example, biting down on a piece of raw, diced garlic in a vinaigrette can leave an unappealing flavor. However, the smooth paste of creamed garlic will evenly disperse throughout a vinaigrette to infuse a robust balanced flavor.

Chef's Knife

— or —

Santoku Knife

DICING, CHOPPING, AND MINCING GARLIC

1. Hold a head of garlic in one hand. With the other hand, start at the stem end and pry the garlic cloves away from the head.

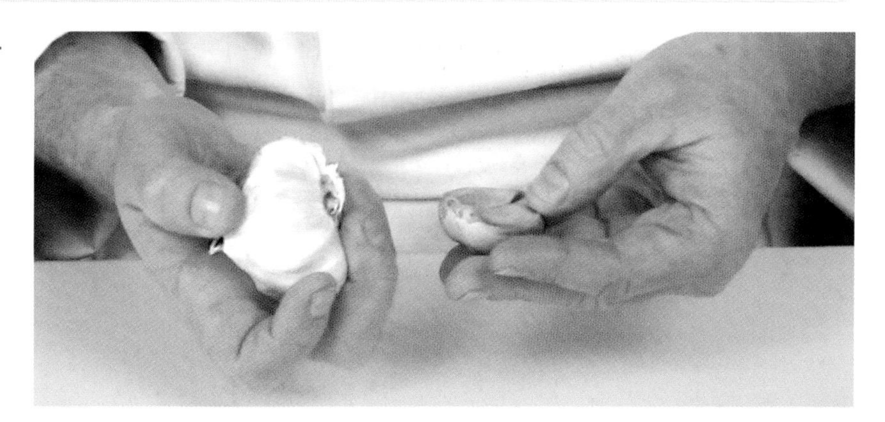

2. To remove the dry peel from a garlic clove, place the clove flat-side down on the cutting board and position the side of the knife blade on top of the clove. Place the palm of the guiding hand on top of the blade and apply slight pressure to the blade until the peel loosens enough to be removed with the fingers.

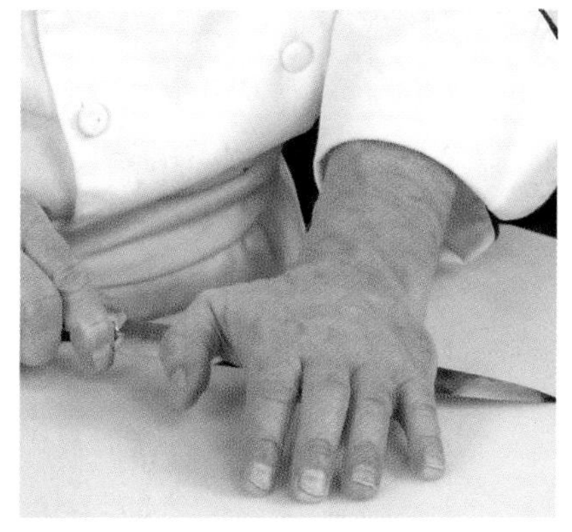

DICING, CHOPPING, AND MINCING GARLIC (CONTINUED)

3. To dice the garlic, place the peeled clove flat-side down on the cutting board with the stem end facing the knife hand side. Secure the clove with the guiding hand and make two to three equally spaced horizontal cuts most of the way through the garlic clove, leaving the root end intact.

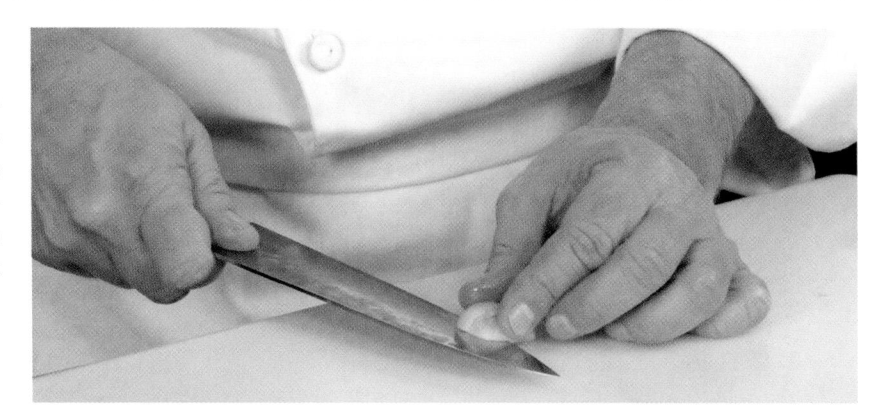

4. Rotate the clove 90° with the root end farthest from the body and secure with a claw grip. Starting at the side of garlic closest to the knife hand, make slices at equally spaced intervals until reaching the other side of the garlic, leaving the root end intact.

   Note: For Steps 3–4, it is also acceptable to use a paring knife instead of a chef's knife.

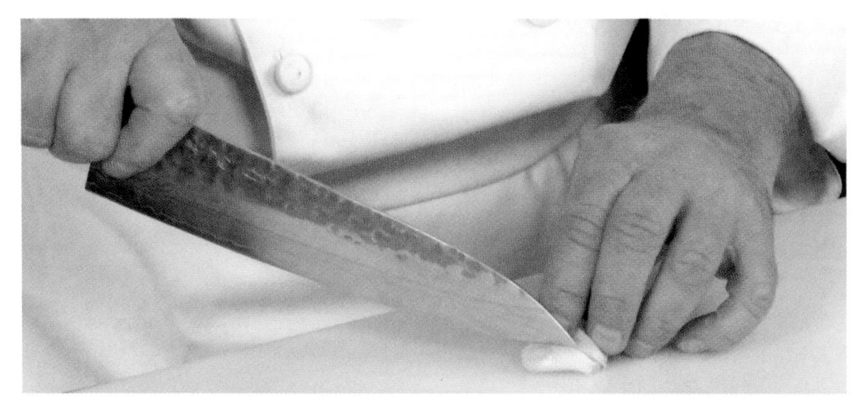

5. Rotate the clove back to its previous position with the stem end facing the knife hand side and secure the clove with a claw grip. Use the tip-to-board method to slice the garlic from stem end to root end at equally spaced intervals until only the root remains. Discard the root.

6. If a smaller cut is desired, place the palm of the guiding hand on the spine of the knife near the tip. Keep the knife tip on the cutting board and rock the blade up and down while pivoting the blade back and forth over the garlic until it is chopped or minced as desired.

   Finished diced garlic has a cubic shape, while chopped or minced garlic lacks uniformity.

CREAMING GARLIC

1. Sprinkle a generous pinch of kosher salt over chopped or minced garlic.

   Note: The salt acts as an abrasive substance that helps break down the garlic into a paste.

2. Place a chef's knife flat on the cutting board with the knife blade edge facing the garlic mound. Rotate the knife hand so that the spine of the knife comes in contact with the board and the knife blade edge tilts upward to form a slight angle with the board. Maintain this position and use the guiding hand to slide the knife closer to the garlic so that the tip of the blade hovers over the garlic.

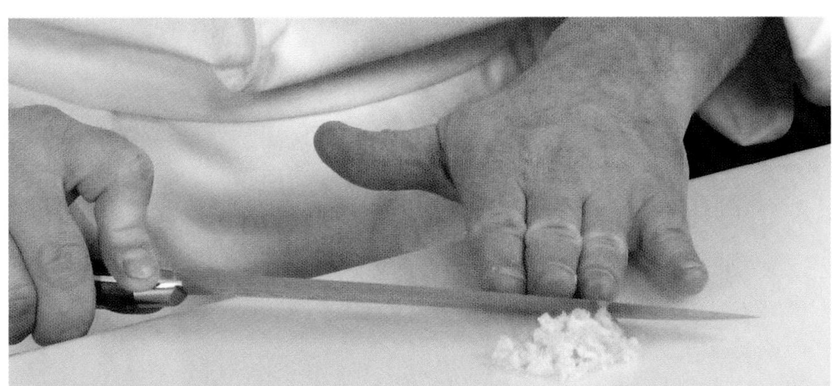

3. Use the guiding hand to push down on the knife blade so that the side and edge of the blade begin to crush the garlic.

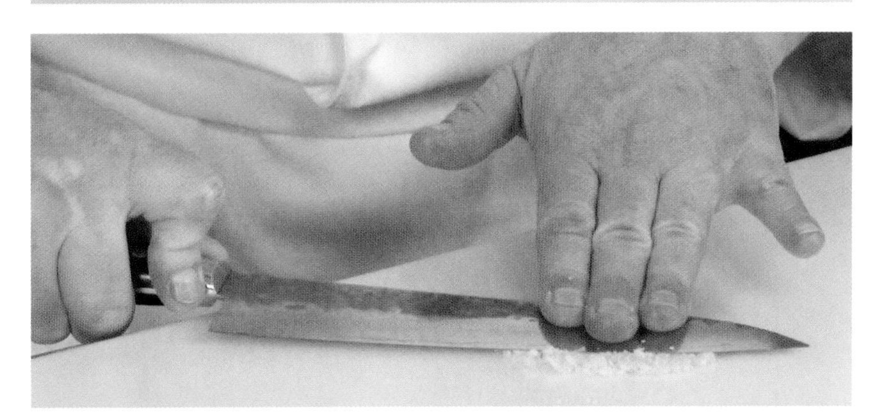

4. Pull the knife blade back toward the body while rotating the knife so that the spine of the knife forms a slight angle with the cutting board and the knife blade edge spreads the garlic on the cutting board.

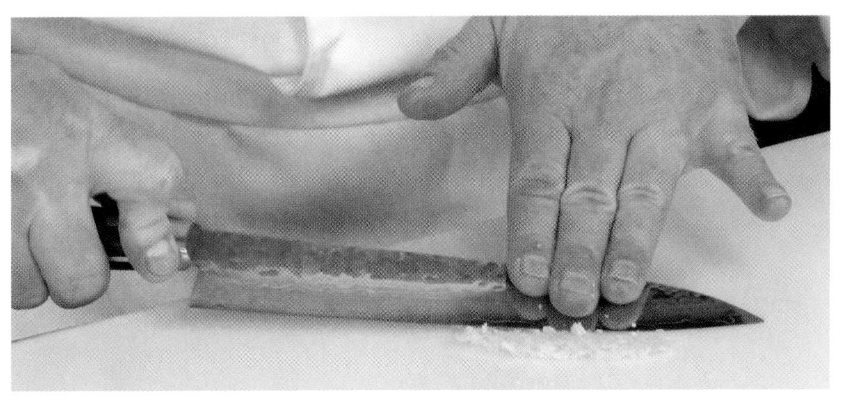

CREAMING GARLIC (CONTINUED)

5. Rotate the knife so that the knife blade edge again forms a slight angle with the board. Use the guiding hand to push the spine of the knife blade forward while simultaneously pushing down on the blade so that the side and edge of the blade continue to crush the garlic.

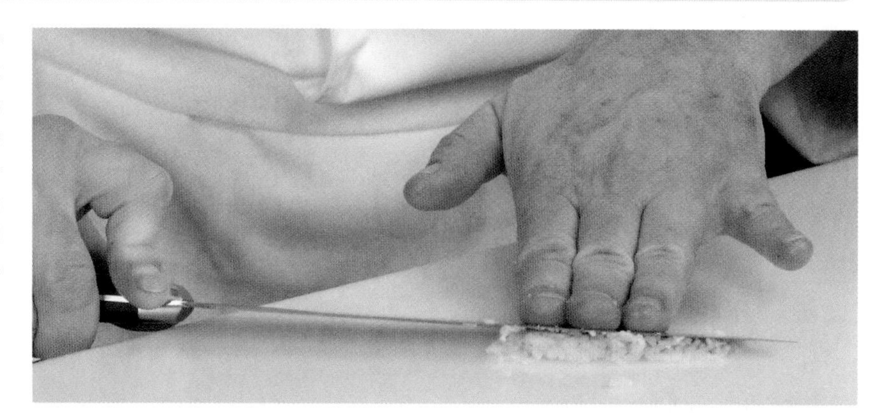

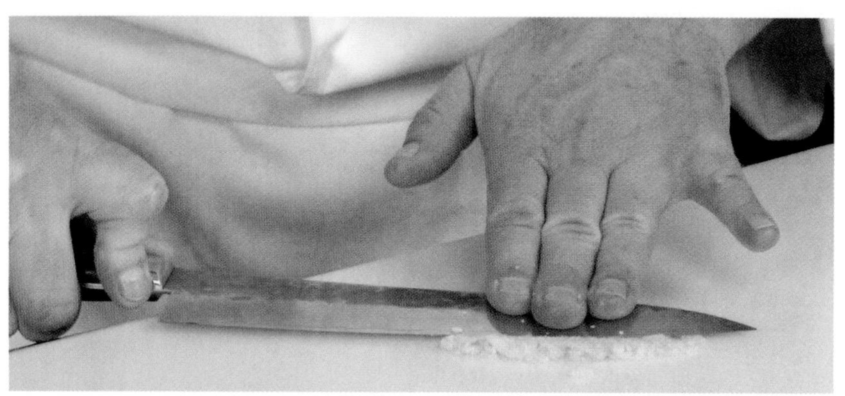

6. Repeat Steps 4–5, using the pushing and pulling motion with the knife blade to crush and spread the garlic onto the cutting board until it forms a creamy paste.

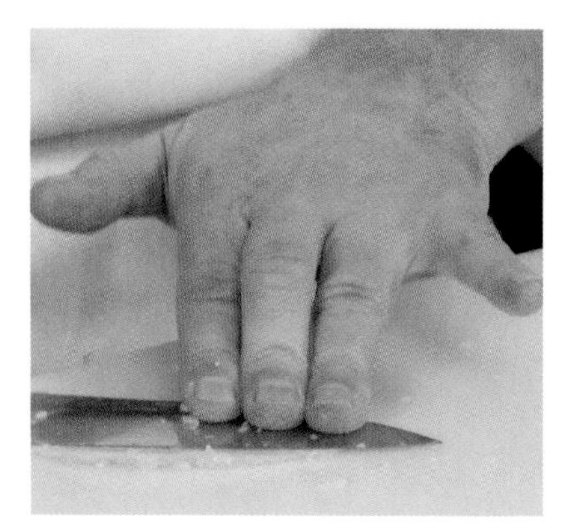

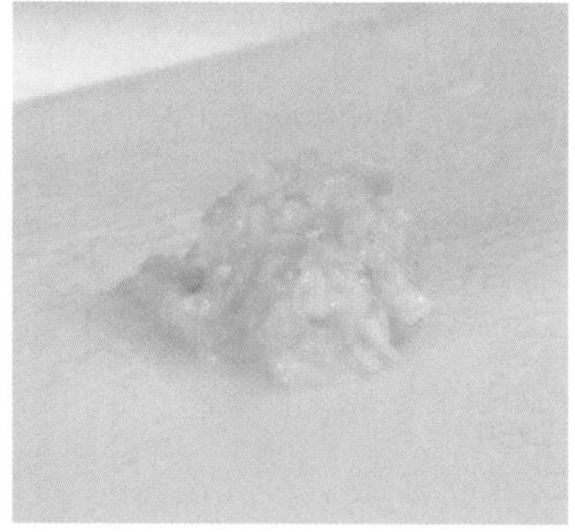

QUICK TIP

Creamed garlic is an effective flavor enhancer in items such as dips, spreads, soups, sauces, and salad dressings.

VIDEO 19

TECHNIQUE 19

SLICING AND DICING CYLINDRICAL FRUIT-VEGETABLES

A fruit-vegetable is a botanical fruit that is sold, prepared, and served as a vegetable. Cylindrically shaped fruit-vegetables include cucumbers, zucchini, and yellow squash. Fruit-vegetables may be peeled and deseeded before serving. This is often the case with cucumbers because cucumber skin is slightly bitter and may be coated in a food-grade wax that is difficult to remove. Cucumber seeds are also commonly removed to limit moisture from seeping into prepared dishes.

When preparing fruit-vegetables, they are often cut in half lengthwise and then sliced crosswise to produce visually appealing semicircular shapes, commonly used in soups, salads, sides, and pastas. Cylindrical fruit-vegetables are also frequently diced, forming imprecise cubes that enhance the texture, flavor, and presentation of salsas, relishes, and garnishes.

Note: Disposable gloves are typically worn in the foodservice industry when preparing ready-to-eat food items.

QUICK TIP

To keep the cutting board clean and make cleaning up easy, place a sheet of plastic wrap on the board before peeling and seeding fruits and vegetables. Use the plastic wrap to catch the peelings and seeds, wrap them up, and discard or compost appropriately.

SLICING CUCUMBERS

1. In the guiding hand, hold a cucumber lengthwise above the cutting board. With the dominant hand, position a peeler at the end of the cucumber closest to the body. Slide the peeler down the entire length of the cucumber to remove a strip of skin. Repeat this motion, rotating the cucumber as necessary until all of the skin has been removed.

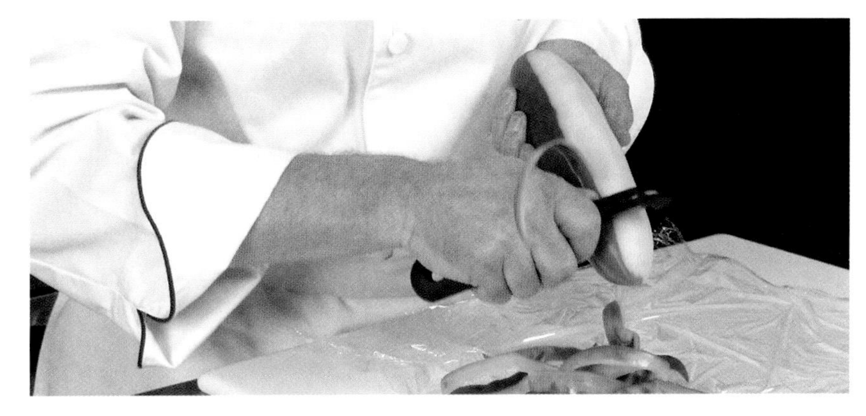

SLICING CUCUMBERS (CONTINUED)

2. Hold the peeled cucumber in the guiding hand so that an end is angled upward toward the knife hand side. Grasp a paring knife in a choke grip and slightly trim the end of the cucumber closest to the knife hand. Turn the cucumber around and repeat this cut on the other end.

3. Place the cucumber lengthwise on the cutting board. Hold the cucumber with the guiding hand and use the paring knife to cut the cucumber in half lengthwise, starting at the far end of the cucumber and pulling the knife toward the body.

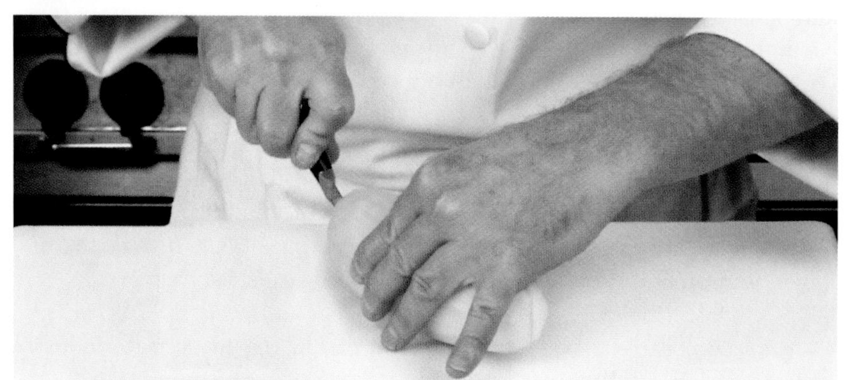

4. If seeds are to be removed, hold a cucumber half in the guiding hand with the seeds facing upward. With the dominant hand, use a spoon to scrape and scoop out the seeds from the center of the cucumber. Repeat on the remaining cucumber half.

5. Place a halved cucumber (with or without seeds) flat-side down across the cutting board and secure the cucumber with a claw grip. With a chef's knife, slice the cucumber crosswise using either Method A or Method B.

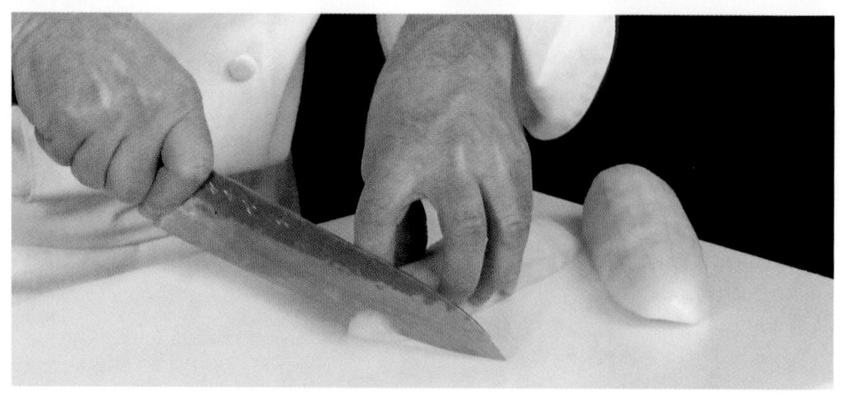

SLICING CUCUMBERS (CONTINUED)

Method A:

- Use the rocking motion of the tip-to-board method to cut the cucumber into consistently sized slices.

  Note: Point the edge of the knife blade away from the body and use the guiding hand to carefully remove any cucumber slices that may stick to the blade.

Method B:

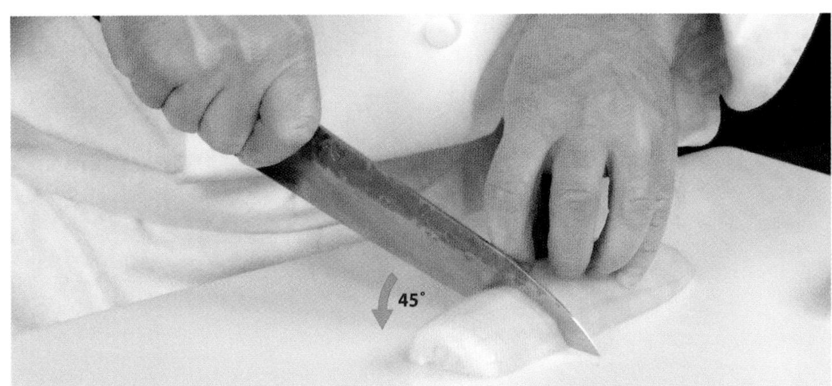

- To make slices that do not stick to the blade, keep the knife handle angled approximately 45° above the cutting board and slice through the cucumber. As the tip of the blade makes contact with the cutting board, pull the knife straight back toward the body while maintaining the handle's 45° angle with the cutting board. Repeat this procedure to make consistently sized slices that keep the shape of the cucumber intact.

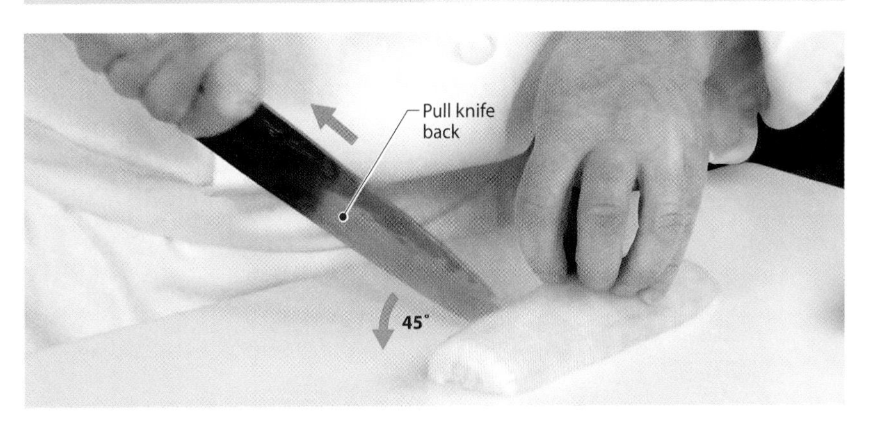

Finished sliced cucumbers without seeds are consistently sized arc-shaped pieces, while finished sliced cucumbers with seeds are consistently sized semicircular pieces.

DICING CUCUMBERS

Note: Cucumbers can be diced with the skin and seeds intact or removed.

1. Start with a cucumber that has been cut in half lengthwise and crosswise. Place the cucumber flat-side down across the cutting board with the widest end facing the knife hand side. Set the palm of the guiding hand on top of the cucumber and position a chef's knife so that the blade is parallel to the cutting board.

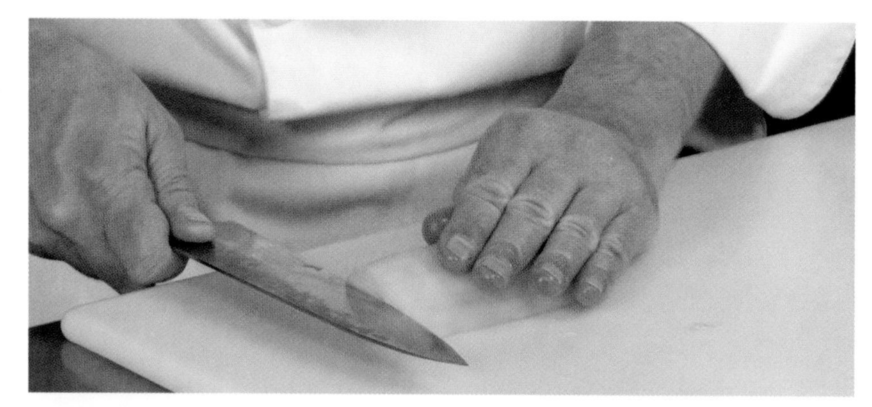

2. Make a horizontal cut through the center of the cucumber.

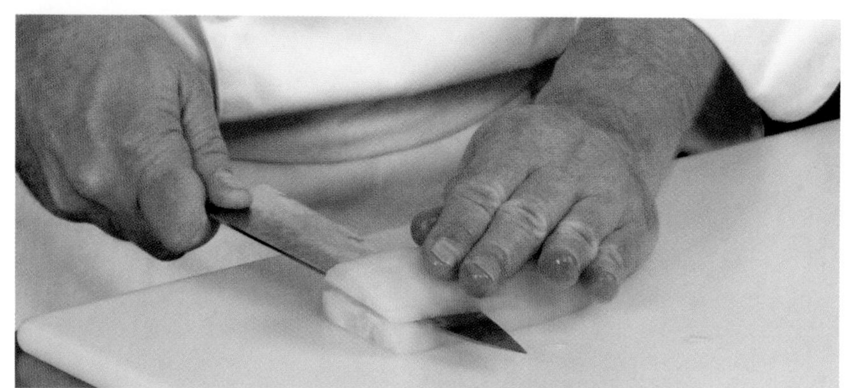

3. Rotate the cucumber approximately 90° so that the tapered end is farthest from the body and secure the cucumber with a claw grip. Starting at the side of the cucumber closest to the knife hand, use the tip-to-food method to make vertical (lengthwise) slices at equally spaced intervals until reaching the other side of the cucumber.

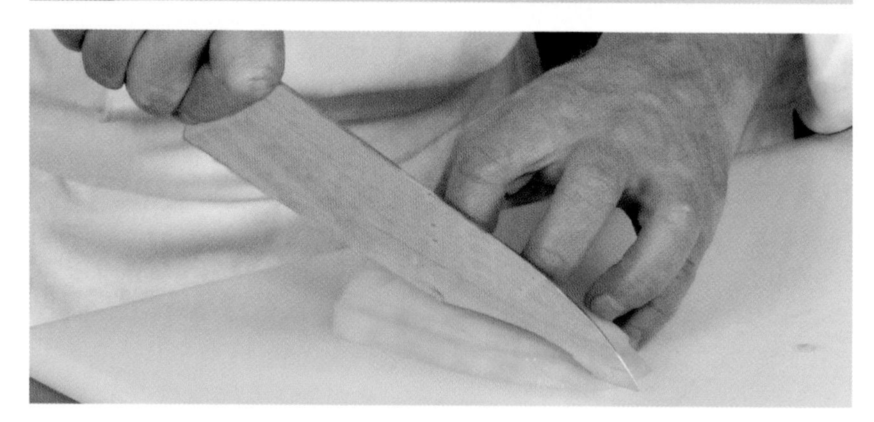

QUICK TIP

English cucumbers, also known as hot house or greenhouse cucumbers, are virtually seedless and grow approximately 1–2 feet long. English cucumbers also have a thinner skin and are milder in flavor than more common cucumber varieties.

DICING CUCUMBERS (CONTINUED)

4. Rotate the cucumber back to its previous position with the widest end of the cucumber facing the knife hand side and align the cucumber slices against the side of the knife blade. Secure the cucumber in a claw grip and use the tip-to-board method to slice the cucumber crosswise at equally spaced intervals to produce rough dice cuts.

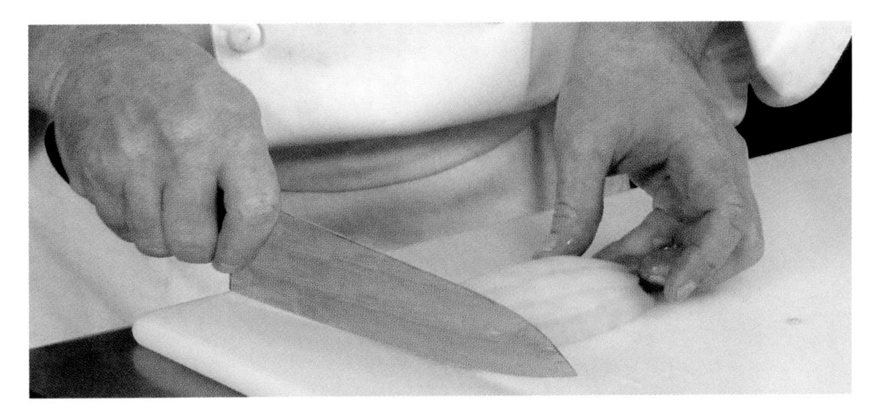

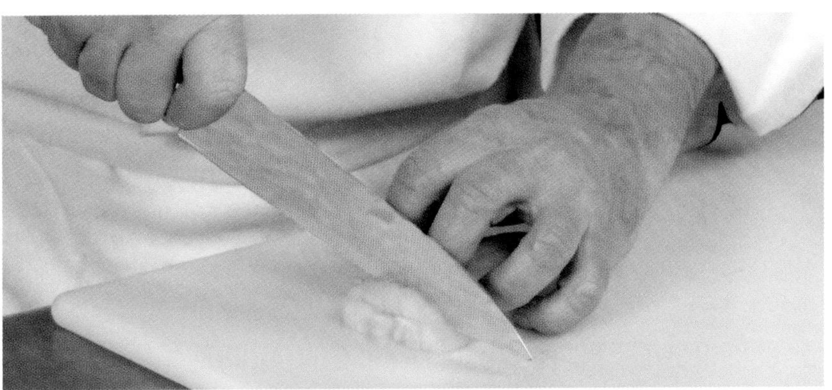

Finished diced cucumbers are regular and slightly irregular cube shapes that are close to uniform in size.

TECHNIQUE 20

CUTTING SWEET PEPPERS

VIDEO 20

Because of their unique shape and inner core, it can be challenging to cut sweet (bell) peppers into consistently shaped pieces. An effective way to remove the core is by using either a horizontal cut or a vertical cut. The horizontal cut produces a long rectangular piece of sweet pepper, while the vertical cut produces four rectangular pieces, both of which make it easy to create batonnet, julienne, or dice cuts. The resulting stick- or cube-shaped peppers are served raw or cooked in a variety of items including breakfast dishes, appetizers, salads, sandwiches, and entrées. Stick-cut peppers also add appealing color and flavor to crudité platters.

Chef's Knife

— or —

Santoku Knife

Note: Disposable gloves are typically worn in the foodservice industry when preparing ready-to-eat food items.

CORING SWEET PEPPERS USING A HORIZONTAL CUT

* This technique was executed by a left-handed chef.

1. Place a sweet pepper on the cutting board with the stem end facing the knife hand side. Secure the pepper with a claw grip and use the tip-to-food method to slice off the top of the pepper just below the stem. Discard or compost the stem and reserve the top piece.

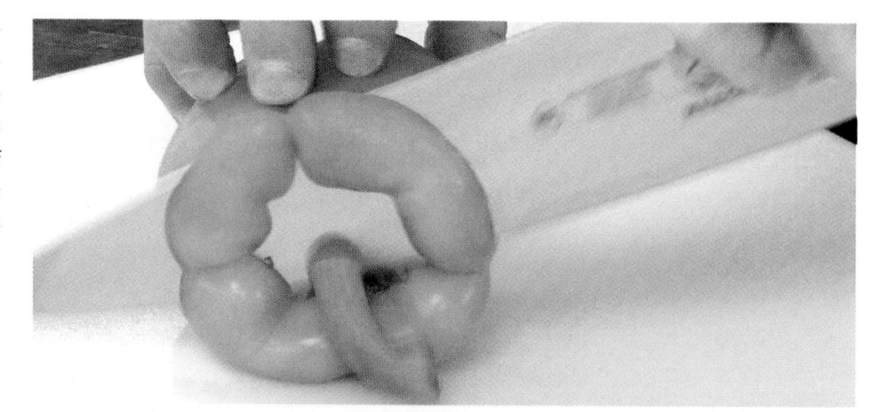

2. Rotate the pepper so that the bottom (blossom end) of the pepper faces the knife hand side. Secure the pepper with a claw grip and use the tip-to-food method to slice approximately ½ inch off the bottom of the pepper. Reserve this piece.

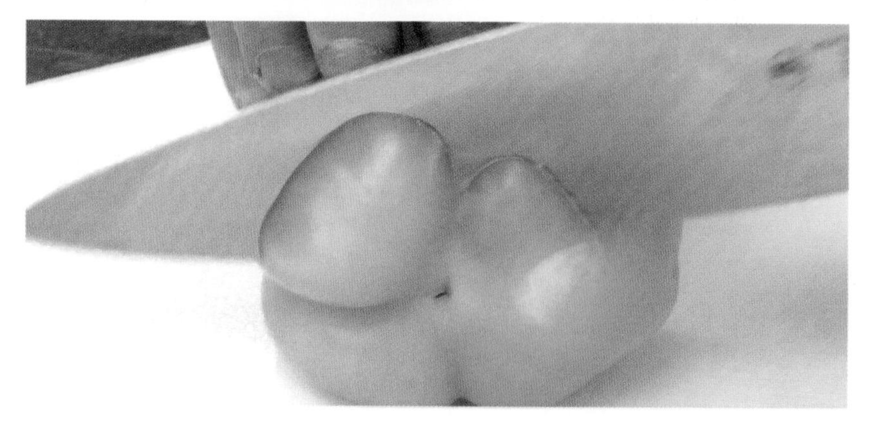

CORING SWEET PEPPERS USING A HORIZONTAL CUT (CONTINUED)

* This technique was executed by a left-handed chef.

> **QUICK TIP**
> *The pieces that are trimmed from the pepper prior to being cored can be sliced into rough stick- or cube-shaped pieces and added to items such as salsas, soups, salads, and stir-fry dishes.*

3. Stand the pepper upright with the bottom end against the cutting board and hold the top of the pepper with the guiding hand. Avoiding the ribs, use the knife tip to make a vertical cut from the top to the bottom of the pepper that is just deep enough to cut through the flesh on one side of the pepper.

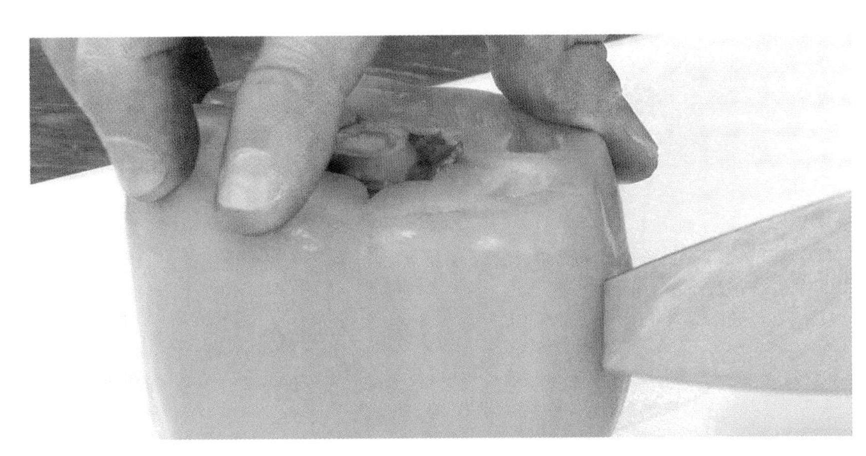

4. Lay the pepper on its side with the stem end facing away from the body. The vertical slice made in Step 3 should face the knife hand side and almost touch the cutting board. Insert the knife blade into the sliced section of the pepper and adjust the blade so that it is parallel with the cutting board.

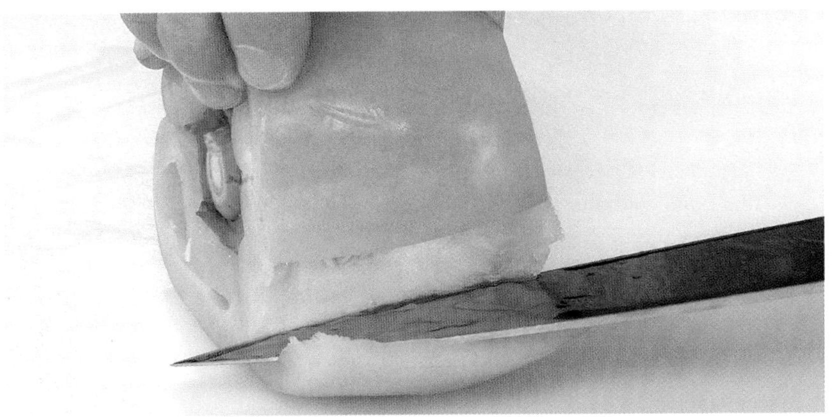

5. Keep the knife parallel to the cutting board and use a slight sawing motion to cut horizontally through the ribs of the pepper while rolling the pepper with the guiding hand in the same direction the knife is cutting.

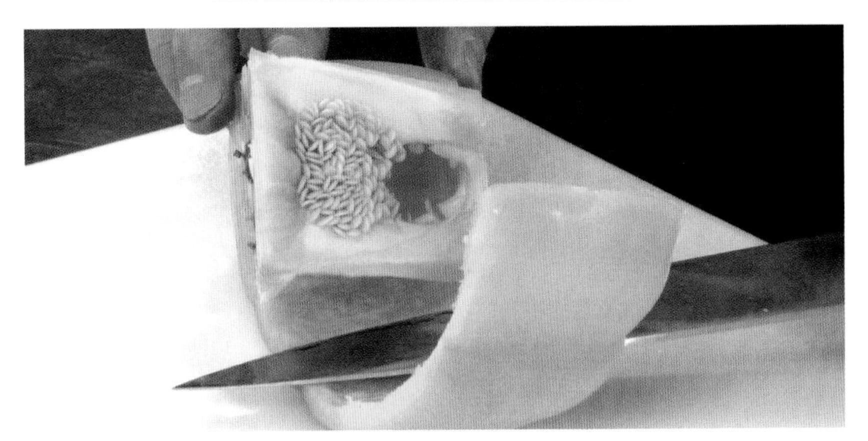

CORING SWEET PEPPERS USING A HORIZONTAL CUT (CONTINUED)

* This technique was executed by a left-handed chef.

6. Continue to cut and roll the pepper until the core that contains the ribs and seeds detaches from the sides of the pepper.

Finished cored sweet peppers using a horizontal cut produce one long rectangular piece of sweet pepper. The reserved end pieces can be cut as desired.

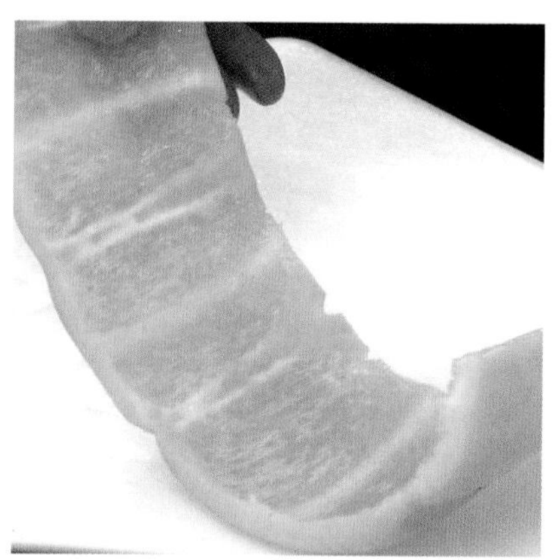

CORING SWEET PEPPERS USING A VERTICAL CUT

1. Place a sweet pepper on the cutting board with the stem end facing the knife hand side. Secure the pepper with a claw grip and use the tip-to-food method to slice off the top of the pepper just below the stem. Discard or compost the stem and reserve the top piece.

2. Rotate the pepper so that the bottom (blossom end) of the pepper faces the knife hand side. Secure the pepper with a claw grip and use the tip-to-food method to slice approximately ½ inch off the bottom of the pepper. Reserve this piece.

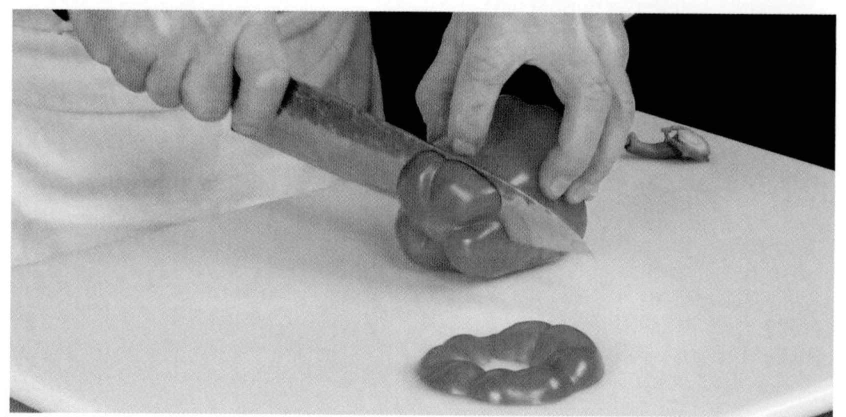

CORING SWEET PEPPERS USING A VERTICAL CUT (CONTINUED)

> **QUICK TIP**
> *Most sweet peppers have four distinct sections or "bumps" on the bottom (blossom end) of the pepper. These sections give the pepper a squarelike appearance that help emphasize the four sides of the pepper.*

3. Stand the pepper upright with the stem end against the cutting board and hold the top of the pepper with the guiding hand. Starting at the top of the pepper closest to the knife hand, cut off one of the pepper's four sides, making sure to avoid the ribs and the seeds.

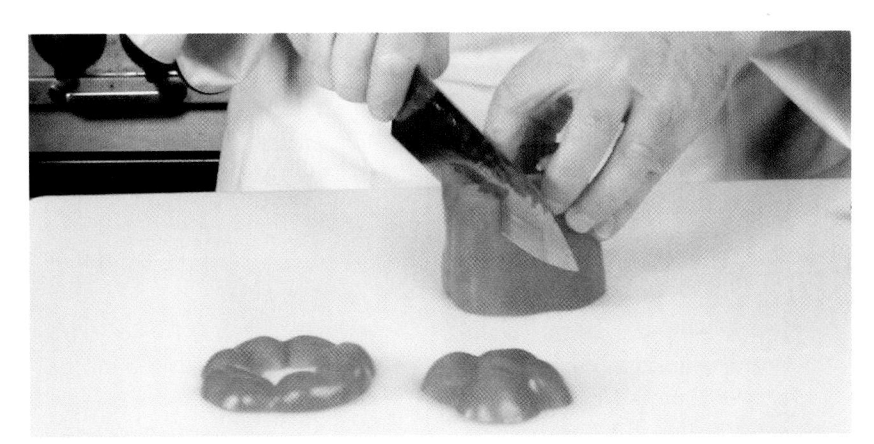

4. Rotate the pepper a quarter turn so that another side faces the knife hand and slice off that side. Repeat this step two more times until all four sides of the pepper are removed from the core that contains the ribs and seeds.

Finished cored sweet peppers using a vertical cut are four roughly shaped rectangular pieces of sweet pepper. The reserved end pieces can be cut as desired.

STICK-CUTTING SWEET PEPPERS

1. Place a cored sweet pepper flesh-side down on the cutting board with one of the removed ends facing away from the body.

2. To make a stick shape similar to a batonnet cut, secure the pepper with a claw grip and use the tip-to-board method to slice the pepper at ¼ inch intervals to produce consistently sized sticks.

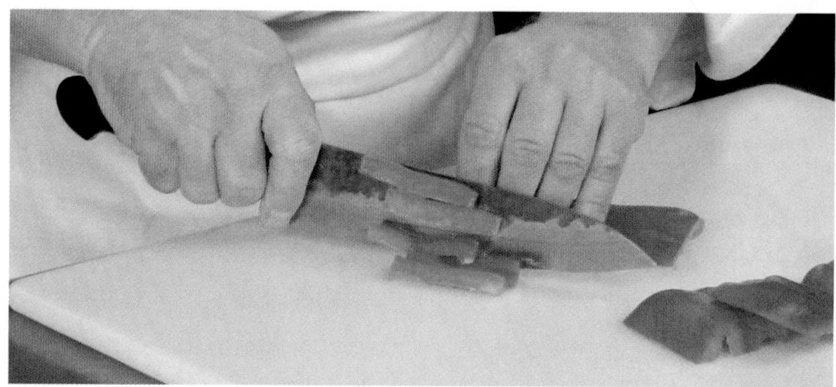

To make a stick shape similar to a julienne cut, slice through the pepper at ⅛ inch intervals.

Finished stick-cut sweet peppers have a consistent length and width.

DICING SWEET PEPPERS

* This technique was executed by a left-handed chef.

1. To make rough dice cuts, place a small bundle of stick-cut sweet peppers perpendicular to the knife blade. Use the side of the knife blade to align the ends of the stick bundle.

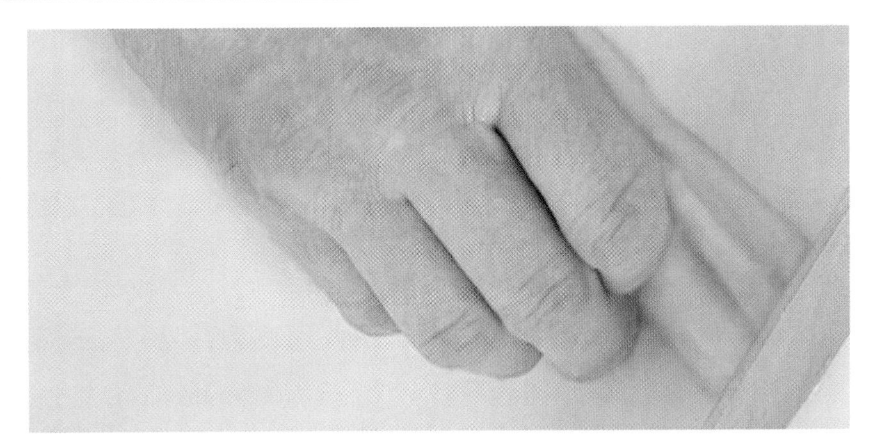

2. Secure the bundle with a claw grip and use the tip-to-board method to slice crosswise through the bundle at equally spaced intervals.

Finished dice-cut sweet peppers are regular and slightly irregular cube shapes that are close to uniform in size.

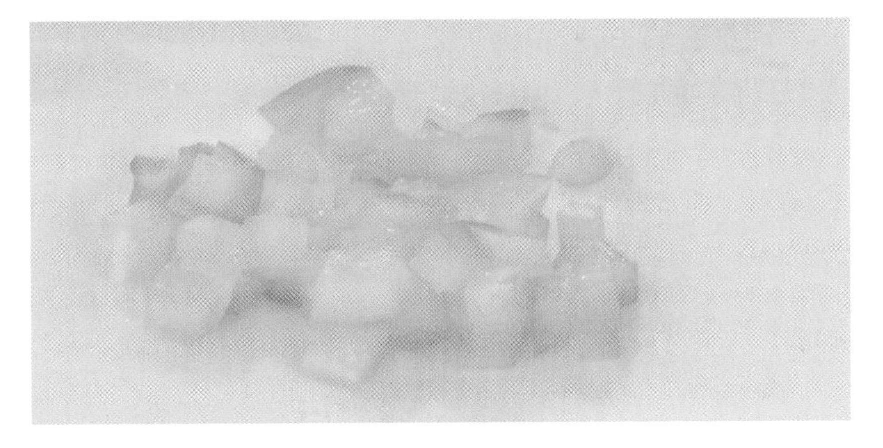

TECHNIQUE 21

PEELING AND CORING POMES

VIDEO 21

A pome is a fleshy fruit that contains a core of seeds and has an edible skin. Some of the most recognizable pomes include apples, pears, and quinces. Because the skin of pomes is more fibrous than the flesh and also tends to fall off when cooked, recipes may call for the skin of these fruits to be removed. The inedible core of pomes contains the seeds and should always be removed. Once pomes are peeled (optional) and cored, they can easily be sliced into thin wedges, sticks, or cubes and used raw or cooked in savory or sweet dishes.

Note: Disposable gloves are typically worn in the foodservice industry when preparing ready-to-eat food items.

Messermeister

Peeler

— or —

Paring Knife

Chef's Knife

— or —

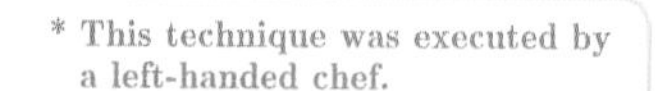

Santoku Knife

PEELING APPLES

* This technique was executed by a left-handed chef.

1. Hold an apple in the guiding hand with the stem end angled upward. Wrap the fingers of the dominant hand around a peeler or use a choke grip with a paring knife. Position the blade next to the apple's stem end and place the dominant hand thumb on the side of the apple.

QUICK TIP

Even if a fruit or vegetable, such as an apple, melon, or carrot, is going to be peeled before it is consumed, the item should be washed prior to cutting. This reduces the risk of transferring dirt and bacteria from the knife onto the cutting board and flesh of the fruit or vegetable being prepared.

PEELING APPLES (CONTINUED)

* This technique was executed by a left-handed chef.

2. Remove the skin by pulling the peeler or paring knife toward the thumb while using the guiding hand to rotate the apple into the blade.

   Note: To keep the flesh of the apple as intact as possible, peel only a thin layer of flesh with the skin.

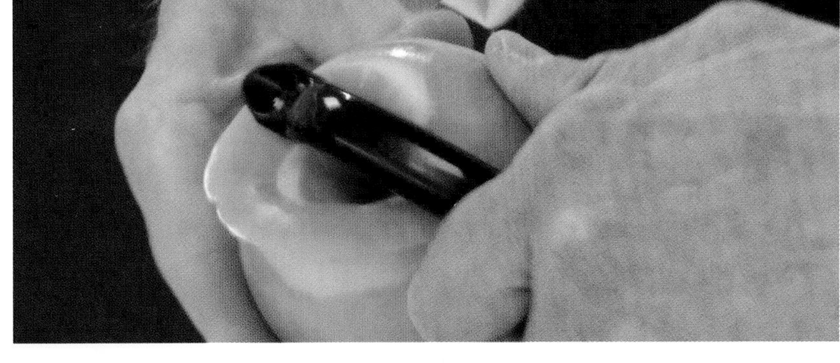

3. Reposition the thumb along the side of the apple and repeat Step 2, working down and around the apple until the skin is removed.

   Note: Inspect the apple and remove any remaining bits of skin.

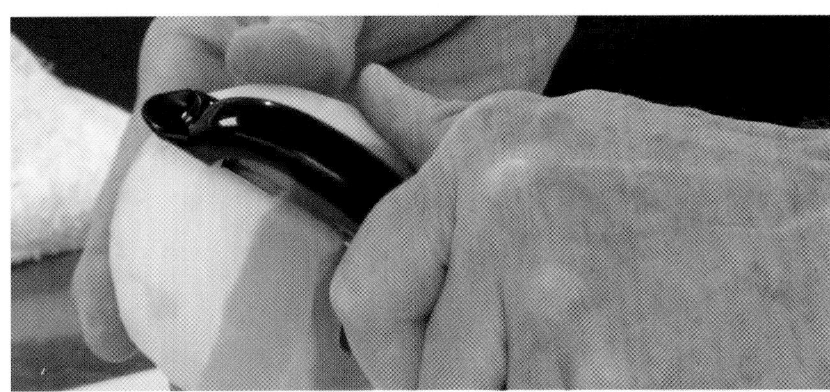

Finished peeled apples have had the skin removed but the apple flesh has been left intact.

CORING APPLES

* This technique was executed by a left-handed chef.

1. Place an apple stem-side up on the cutting board and secure with a claw grip. Slice the apple in half from stem end to blossom end with a chef's knife.

2. Place one of the apple halves cut-side down with the stem end facing away from the body and secure with a claw grip. Slice the apple in half from stem end to blossom end. Repeat this step with the remaining apple half to produce a quartered apple.

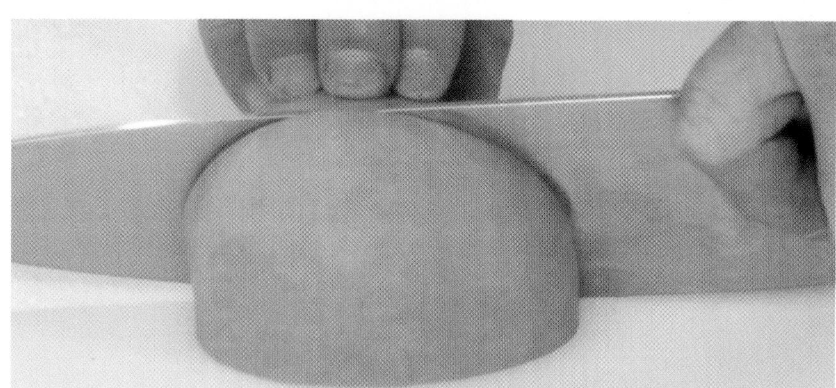

3. Position a quartered apple with one of the cut sides against the cutting board and the other cut side facing the knife hand side. Place the edge of the knife blade lengthwise just above the core and cut down at a slight angle to remove the core. Repeat this step with the remaining apple quarters.

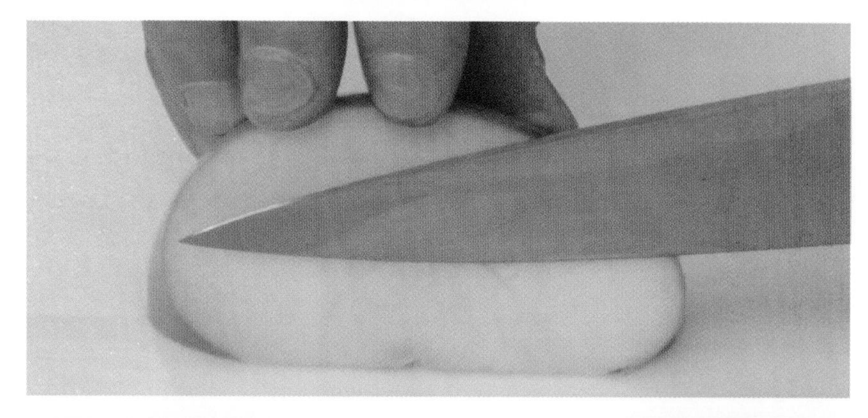

Finished cored apples are four equally sized apple wedges that are void of seeds and any rigid, fibrous pieces associated with the core.

QUICK TIP

Quartered apples can be served as thick wedges or sliced further into thinner wedges, stick cuts, or dice cuts based on the desired application.

VIDEO 22

CUTTING AVOCADOS

22 TECHNIQUE

Avocados have a rough green skin and a large pit, both of which are inedible and must be removed before the avocado can be eaten. The yellow-green flesh of an avocado has a buttery texture and mild flavor that is often mashed to a smooth or semismooth consistency and used in dips or spreads, such as guacamolé. Avocado flesh is also commonly sliced or diced and used raw in salsas, salads, and sandwiches and is a popular accompaniment with chilis, soups, and dishes that feature international cuisine.

Note: Disposable gloves are typically worn in the foodservice industry when preparing ready-to-eat food items.

Chef's Knife
— or —
Santoku Knife
Large Spoon

PITTING AVOCADOS

1. Lay an avocado on the cutting board and secure with the guiding hand. Hold the blade of a chef's knife parallel to the cutting board and cut into the middle of the avocado until the knife edge touches the pit.

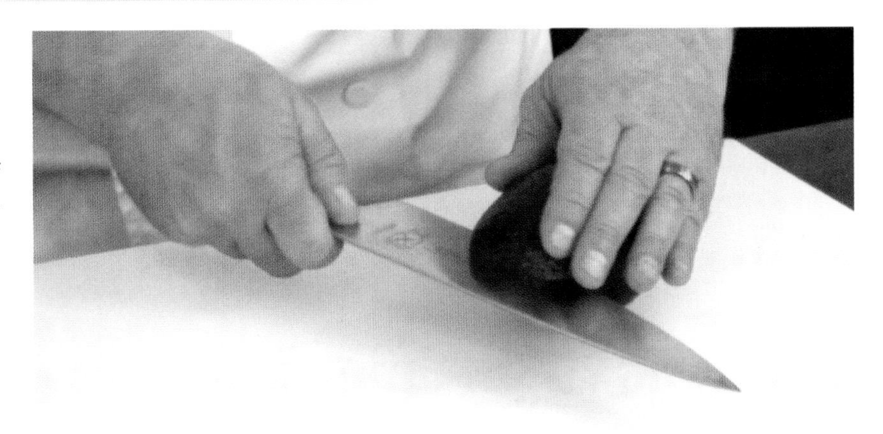

QUICK TIP

Ripe avocados should yield to gentle pressure. Firm, unripe avocados can be stored at room temperature in a brown paper bag to speed the ripening process. Placing unripe avocados in a bag exposes them to concentrated levels of ethylene gas, which promotes ripening.

2. Keep the knife blade edge in constant contact with the pit and rotate the avocado lengthwise around the blade. Continue rotating the avocado until a complete rotation (360°) has been made and the skin and flesh have been cut through.

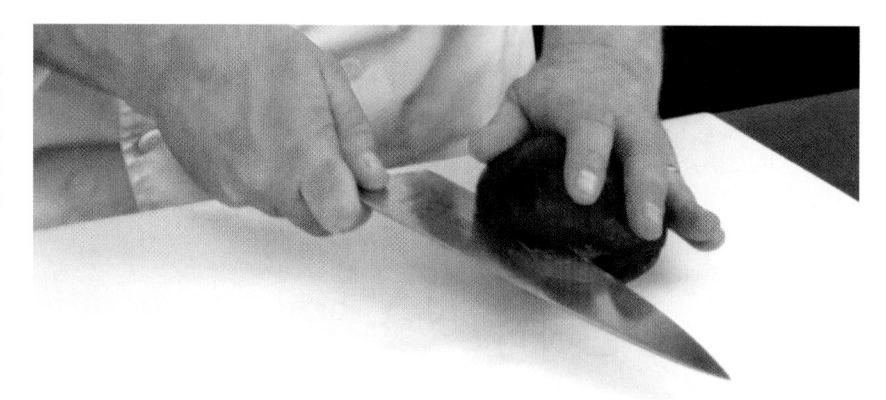

3. Use the hands to gently twist the avocado halves in opposite directions until the two halves can easily be pulled apart to reveal the pit.

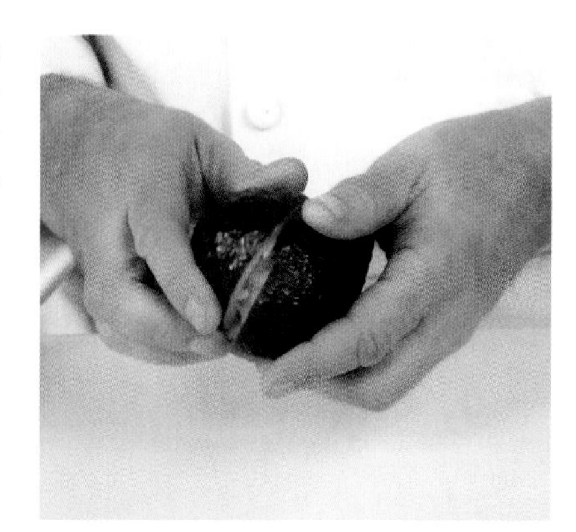

4. Hold the avocado half that contains the pit in the guiding hand. With the chef's knife, tap the heel of the blade into the pit. Twist the avocado and the knife in opposite directions until the pit loosens and can be lifted from the avocado while still attached to the knife.

   Note: Use a clean kitchen towel to pull the pit from the blade.

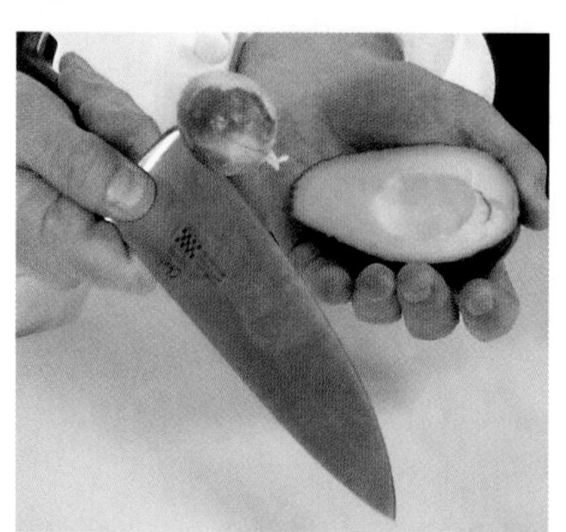

Finished pitted avocados are skin-on avocado halves that contain the flesh of the avocado but not the pit.

REMOVING AVOCADO FLESH

1. Start with a pitted avocado. Hold one of the avocado halves lengthwise in the guiding hand with the flesh-side up. Insert a large spoon where the skin meets the flesh.

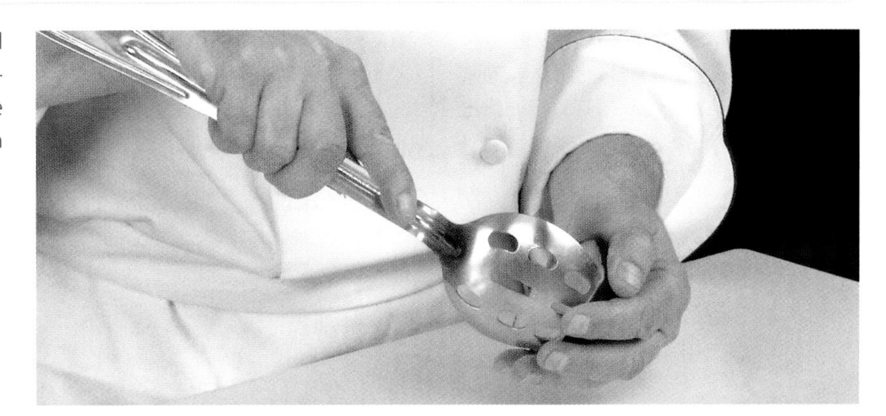

2. Keep the back of the spoon pressed against the skin and rotate the avocado and spoon in opposite directions until the flesh separates from the skin and can be lifted away in one piece. Repeat this step with the remaining avocado half.

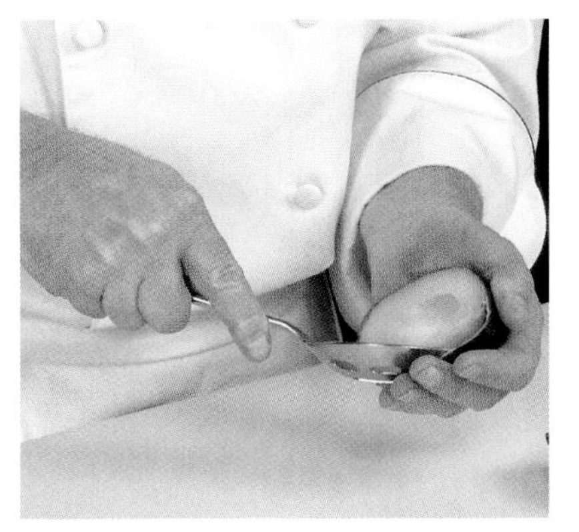

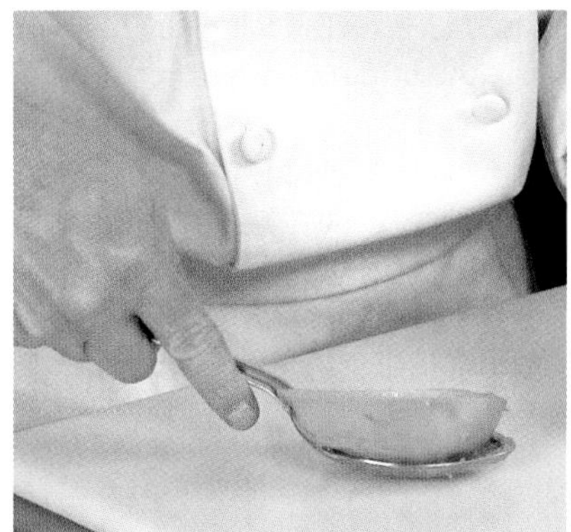

Removing avocado flesh from a pitted avocado produces two pieces of skinless flesh that are close to uniform in shape and size.

QUICK TIP

To prevent avocados from turning brown after the flesh is exposed to air, coat the flesh with an acid, such as lemon or lime juice, and store the avocados in an air-tight container.

SLICING AVOCADOS

1. Place the flesh from a halved avocado flat-side on the cutting board with an end facing the knife hand side. Hold the avocado with a claw grip and position the tip of a chef's knife on top of the avocado near the end.

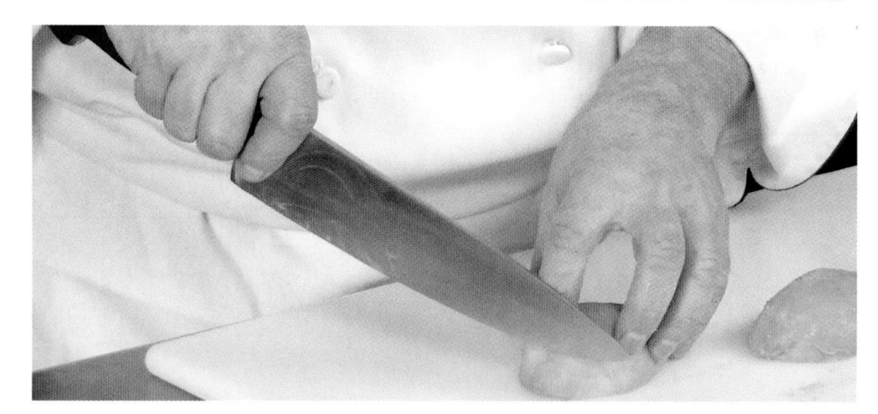

2. To make slices that do not stick to the blade, keep the knife handle angled approximately 45° above the cutting board and slice crosswise through the avocado. As the tip of the blade makes contact with the cutting board, pull the knife straight back toward the body while maintaining the handle's 45° angle with the cutting board.

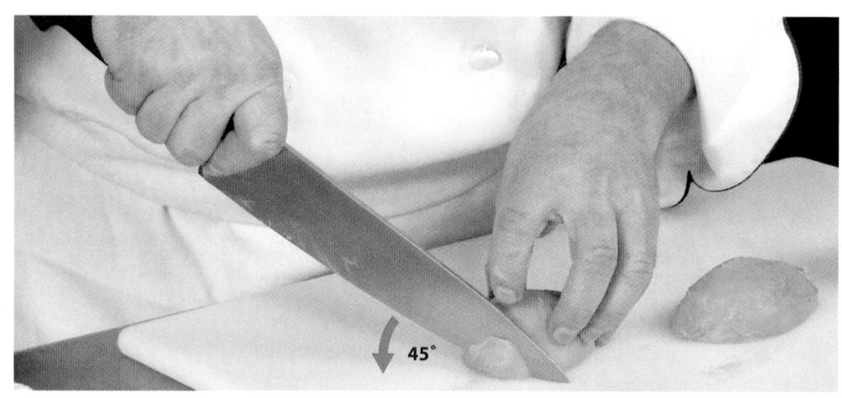

Repeat this procedure to make consistently sized slices that keep the shape of the avocado intact.

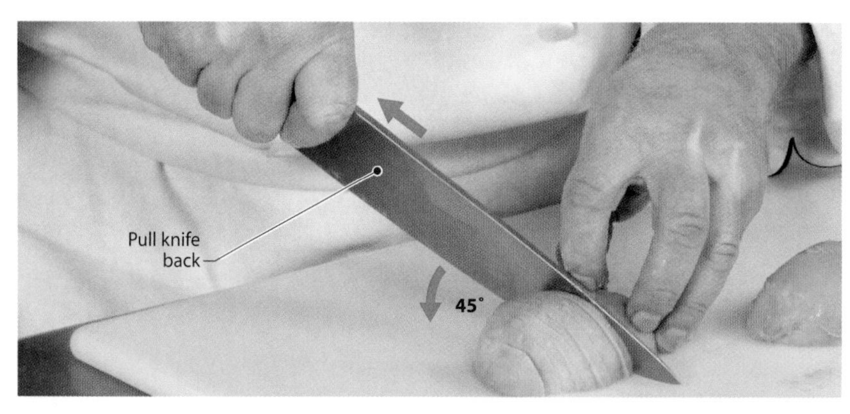

Finished sliced avocado halves produce semicircular cuts.

DICING AVOCADOS

1. Place the flesh from a halved avocado flat-side down on the cutting board with an end facing the knife hand side. Place the palm of the guiding hand on top of the avocado and make a horizontal cut through the middle of the avocado.

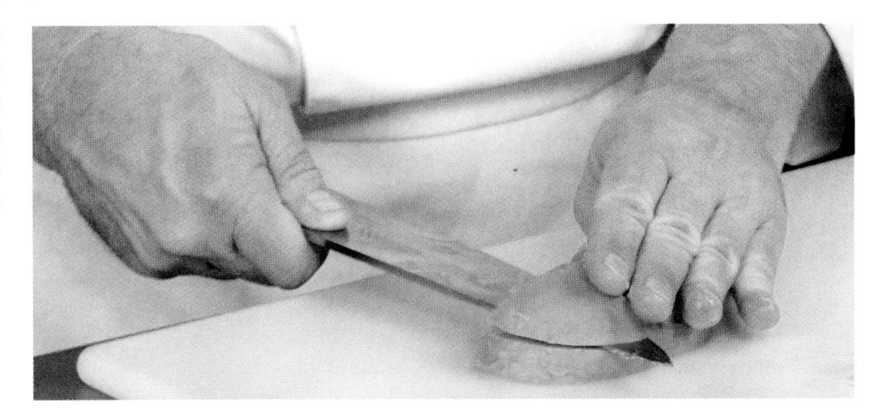

2. Rotate the avocado approximately 90° and secure the avocado with a claw grip. Starting at the side of the avocado closest to the knife hand, use the tip-to-food method to make vertical (lengthwise) slices at equally spaced intervals until reaching the other side of the avocado.

   Note: This step yields roughly shaped stick cuts, which may be called for in some applications.

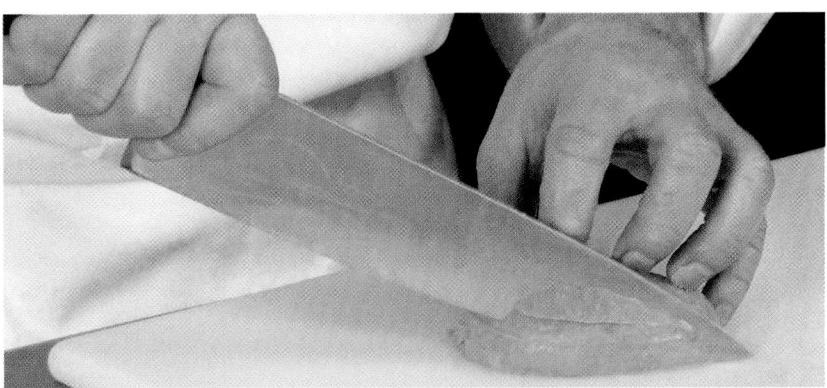

3. Rotate the avocado back to its previous position with an end facing the knife hand side. Secure the avocado with a claw grip and slice across the stick cuts at equally spaced intervals to produce rough dice cuts.

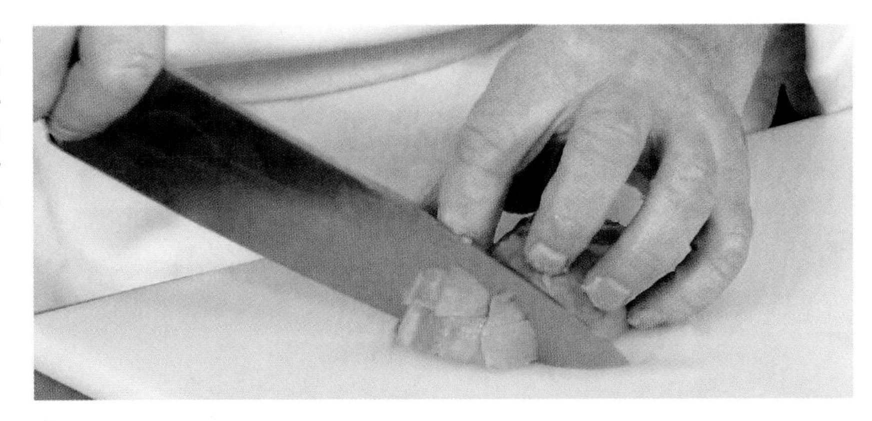

Finished diced avocados are regular and slightly irregular cube shapes that are close to uniform in size.

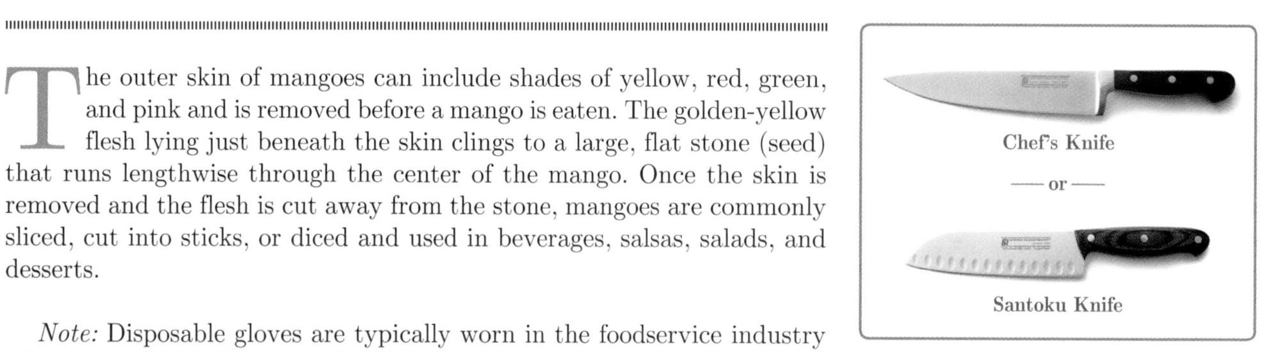

CUTTING MANGOES

VIDEO 23

The outer skin of mangoes can include shades of yellow, red, green, and pink and is removed before a mango is eaten. The golden-yellow flesh lying just beneath the skin clings to a large, flat stone (seed) that runs lengthwise through the center of the mango. Once the skin is removed and the flesh is cut away from the stone, mangoes are commonly sliced, cut into sticks, or diced and used in beverages, salsas, salads, and desserts.

Note: Disposable gloves are typically worn in the foodservice industry when preparing ready-to-eat food items.

Chef's Knife

— or —

Santoku Knife

REMOVING MANGO FLESH

* This technique was executed by a left-handed chef.

1. Place a mango on the cutting board so that an end (stem or blossom end) faces the knife hand side and secure the mango with a claw grip. Use a chef's knife to cut a small slice from the end of the mango to slightly expose the stone. Rotate the mango 180° and repeat this cut on the opposite end of the mango.

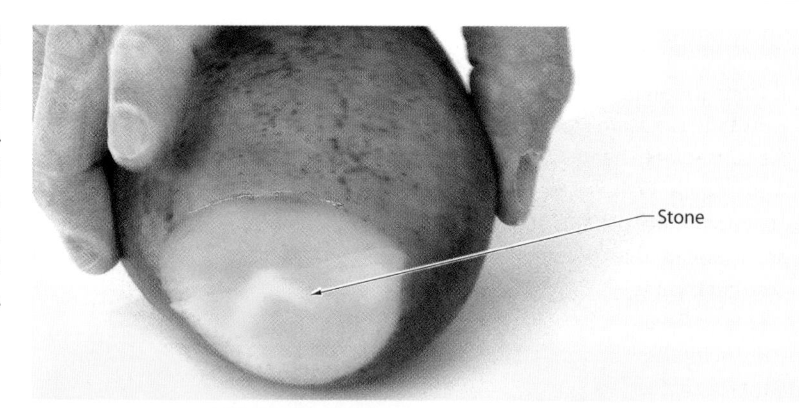

2. Stand the mango upright with the wider end of the mango against the cutting board. Hold the upper portion of the mango in a claw grip and position the edge of the knife tip on top of the mango where the flesh meets the skin. Following the contour of the mango, cut down the side using a slight sawing motion to remove a thin strip of the skin.

REMOVING MANGO FLESH (CONTINUED)

* This technique was executed by a left-handed chef.

3. Use the guiding hand to slightly rotate the mango toward the body. Return the edge of the knife tip to the top of the mango and use a slight sawing motion to cut away another thin strip of skin. Continue to rotate and cut down the sides of the mango until all of the skin is removed.

4. Secure the top of the mango with a claw grip and position the knife blade on top of the mango with the edge parallel to one side of the stone. Slice down the side of the mango, keeping the knife blade against the flat side of the stone. Rotate the mango 180° and repeat this cut to remove flesh from the flat sides of the stone.

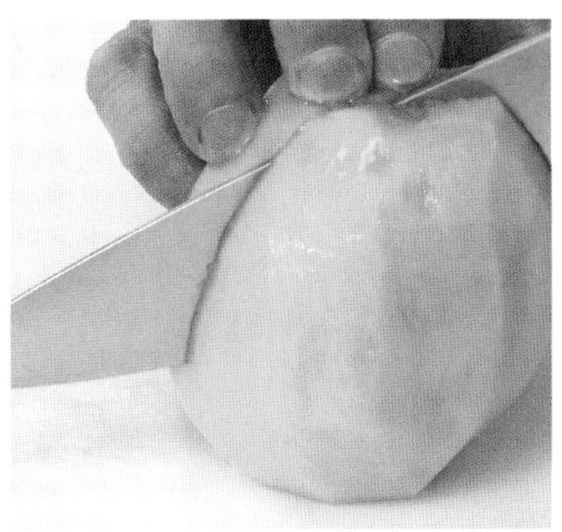

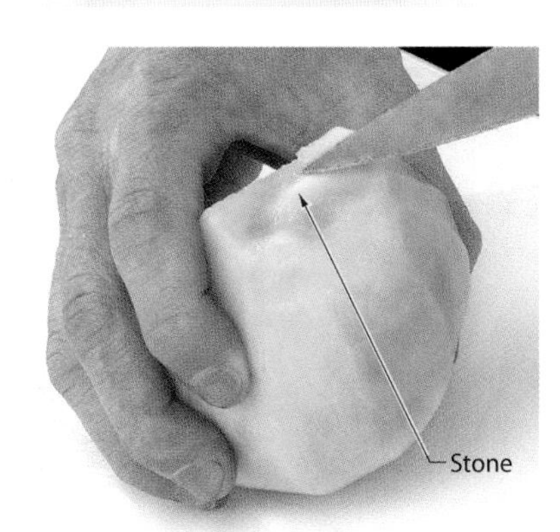

5. Place the mango piece that contains the stone flat-side down on the cutting board and secure with a claw grip. Cut the remaining flesh from the stone by slicing down one of the rounded sides, keeping the knife angled to avoid the stone. Rotate the mango 180°, and repeat this cut to remove the remaining flesh from the other side of the stone.

STICK-CUTTING MANGOES

* This technique was executed by a left-handed chef.

1. Place the flesh that has been cut from the side of a mango flat-side down on the cutting board with one of the ends (stem or blossom end) facing the knife hand side. With the palm of the guiding hand on top of the mango and the knife blade parallel to the cutting board, make horizontal cuts at approximately ¼ inch intervals through the mango.

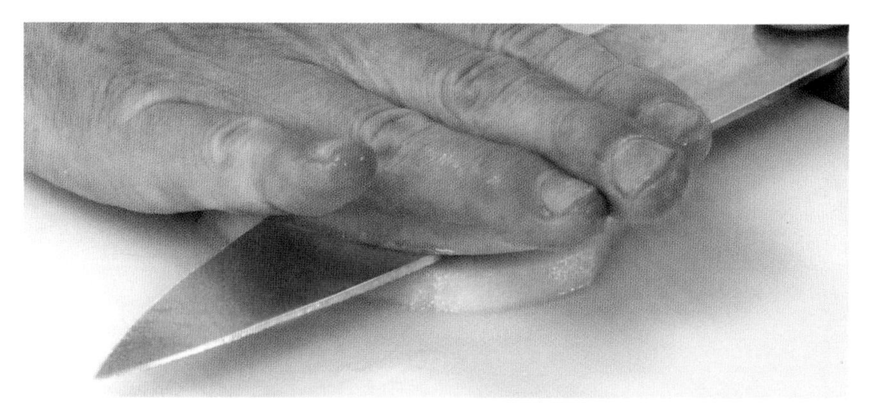

2. Rotate the mango approximately 90° and secure the mango with a claw grip. Starting at the side of the mango closest to the knife hand, use the tip-to-food method to make vertical (lengthwise) slices at approximately ¼ inch intervals until reaching the other side of the mango.

Finished mango stick cuts that have been cut horizontally and vertically at ¼ inch intervals are similar to batonnets but vary in length due to the natural contour of the mango.

DICING MANGOES

* This technique was executed by a left-handed chef.

1. Start with a stick-cut mango that has been sliced at ¼ inch intervals. To make dice cuts, place the stick cuts perpendicular to the knife blade. Use the side of the knife blade to align the ends of the mango.

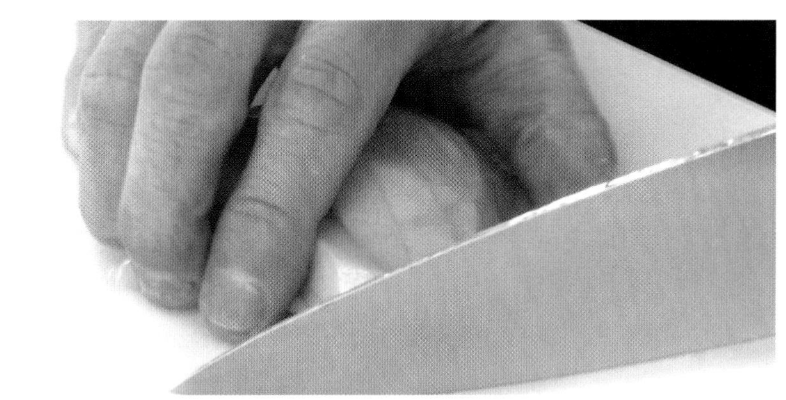

2. Hold the stick cuts in a claw grip and use the tip-to-board method to slice the mango crosswise at approximately ¼ inch intervals to produce rough dice cuts.

Finished diced mangoes that resemble a small dice cut have sides measuring approximately ¼ inch each.

QUICK TIP

If larger dice cuts are desired, use ½ – 1 inch intervals when making the horizontal, vertical, and crosswise cuts.

TECHNIQUE 24

CUTTING MELONS

VIDEO 24

Some of the most common types of melons include cantaloupes, honeydew melons, muskmelons, and watermelons. The outermost skin, or rind, on melons is often removed during preparation. Seeds found clustered at the center of melons, such as cantaloupes, are scooped out with a large spoon. In contrast, seeds scattered throughout the flesh of melons, such as watermelons, are typically left intact.

After removing the rind and seeds (when necessary), melons are commonly cut into slices or diced to produce cubelike shapes. Cut melons are often used as a garnish or in fruit trays, hors d'oeuvres, cold soups, salads, sides, and desserts.

Chef's Knife
— or —
Santoku Knife
Large Spoon

PEELING AND DESEEDING CANTALOUPES

1. Place a cantaloupe on the cutting board with an end facing the knife hand side and secure with a claw grip. Using a chef's knife, cut a thin slice from the end to slightly expose the flesh. Rotate the cantaloupe 180° and repeat this cut on the opposite end of the cantaloupe.

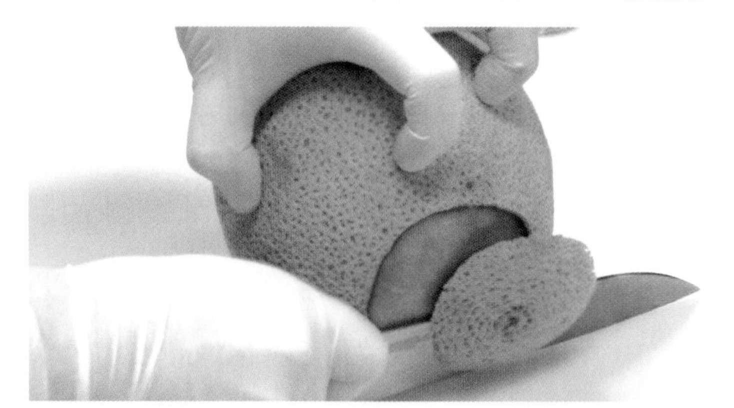

2. Stand the cantaloupe upright with one of the cut sides against the cutting board. Hold the top of the cantaloupe in a claw grip and place the knife blade edge just behind the rind where the orange flesh begins. Follow the contour of the cantaloupe and cut down the side using a slight sawing motion to remove a strip of rind.

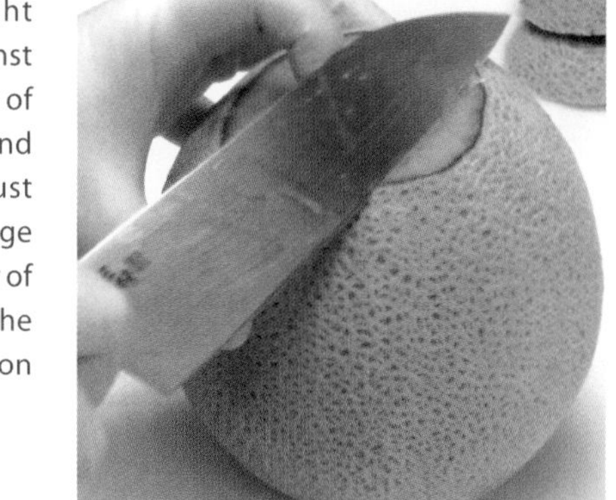

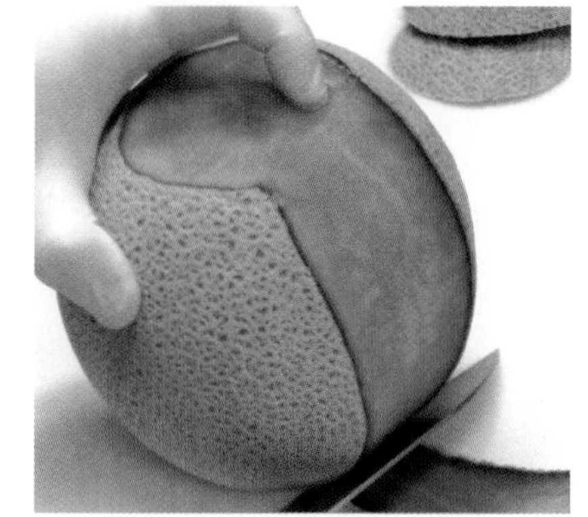

PEELING AND DESEEDING CANTALOUPES (CONTINUED)

3. Use the guiding hand to slightly rotate the cantaloupe toward the body. Return the knife blade edge to the top of the cantaloupe and use a slight sawing motion to cut away another thin strip of rind. Continue to rotate and cut down the sides of the cantaloupe until the entire rind is removed.

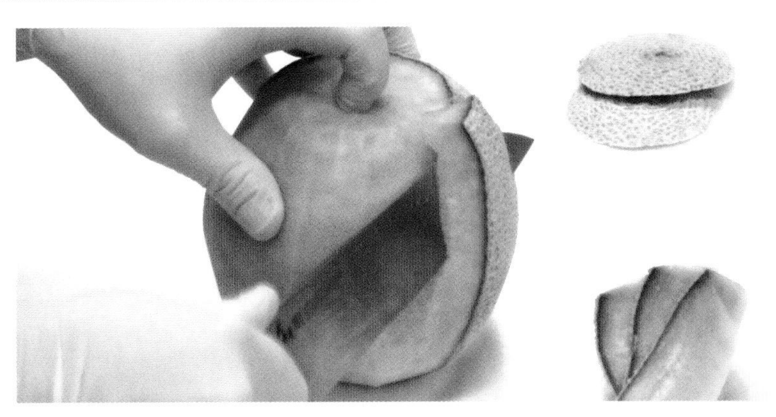

4. With the cantaloupe standing upright and secured with the guiding hand, use the tip-to-food method to slice the cantaloupe in half.

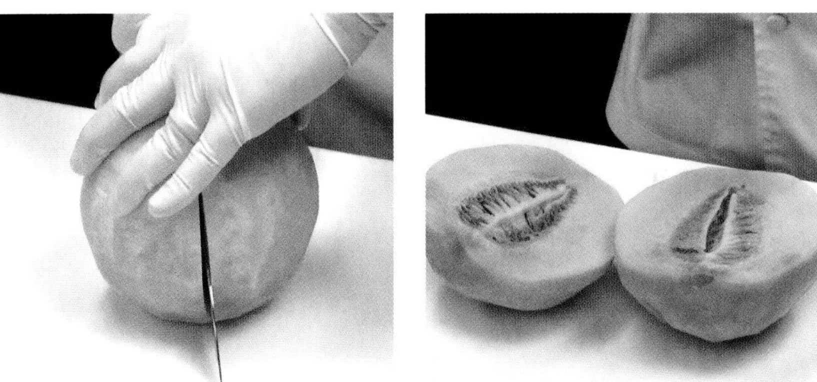

5. Use a large spoon to scoop out the seeds from the center of each cantaloupe half.

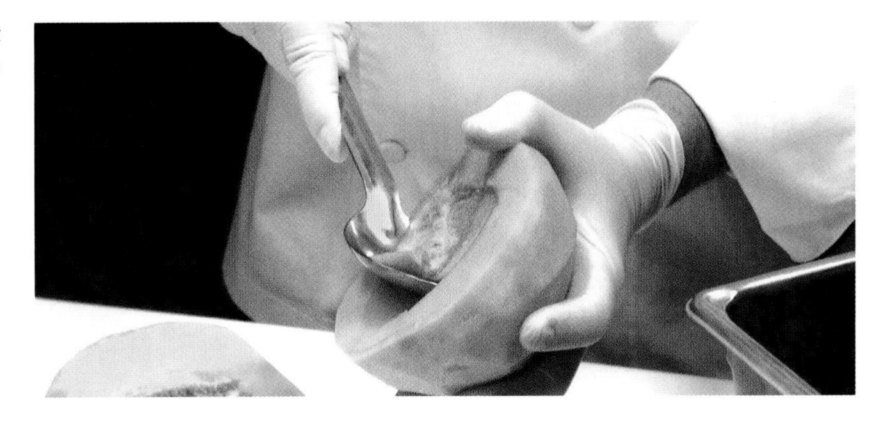

Finished peeled and deseeded cantaloupes yield two cantaloupe pieces that are void of both rind and seeds and close to uniform in shape and size.

SLICING CANTALOUPES

1. Place a peeled and deseeded cantaloupe half rounded-side up on the cutting board with an end facing the knife hand side. Secure the cantaloupe in a claw grip and position the edge of a chef's knife on top of the cantaloupe near the end.

2. Use the tip-to-food method to cut the cantaloupe into consistently sized slices.

   Note: As the knife tip makes contact with the cutting board, pull the knife straight back toward the body rather than rocking the blade up and down. This will prevent cantaloupe slices from sticking to the blade.

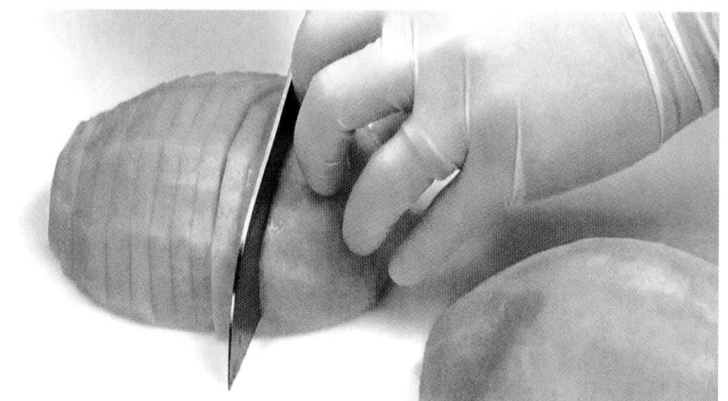

Finished sliced cantaloupe pieces are arc-shaped and vary in length due to the contour of the cantaloupe.

QUICK TIP

Most fruits, including melons, are the sweetest when served at room temperature.

DICING CANTALOUPES

1. Place a peeled and deseeded cantaloupe half rounded-side up on the cutting board with an end facing the knife hand side. Set the palm of the guiding hand on top of the cantaloupe and hold a chef's knife parallel with the cutting board at the midpoint of the cantaloupe.

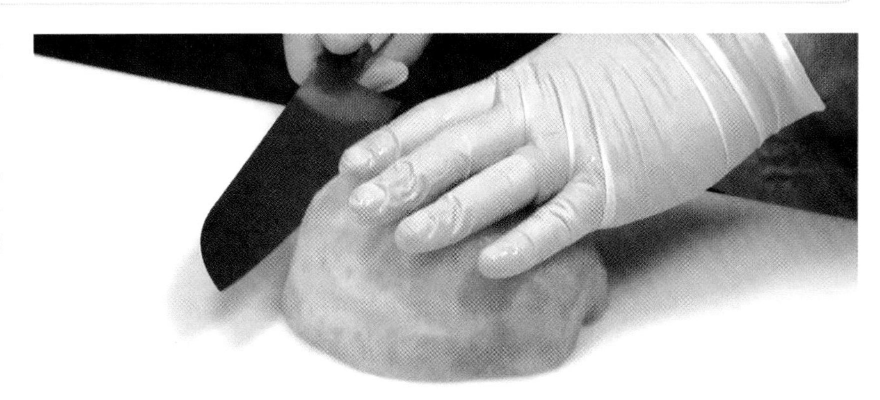

2. To produce extra-large dice cuts, start by making a horizontal cut through the center of the cantaloupe.

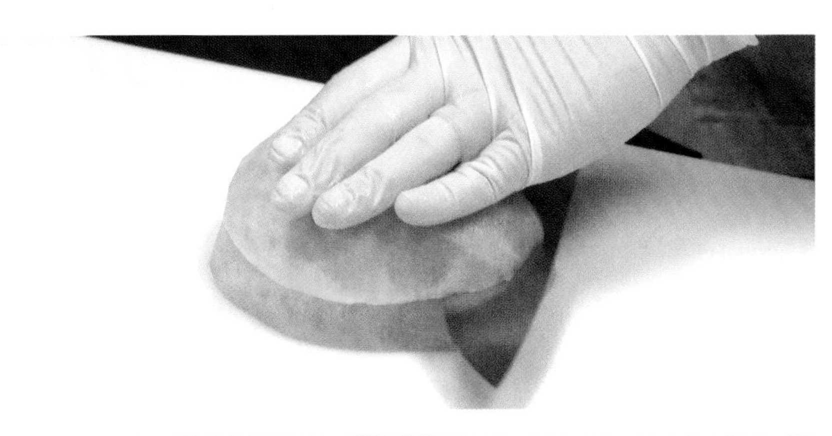

3. Rotate the cantaloupe 90° and secure with a claw grip. Starting at the side of the cantaloupe closest to the knife hand, use the tip-to-food method to make lengthwise slices at 1–1½ inch intervals until reaching the other side of the cantaloupe.

   Note: This step produces roughly shaped stick cuts.

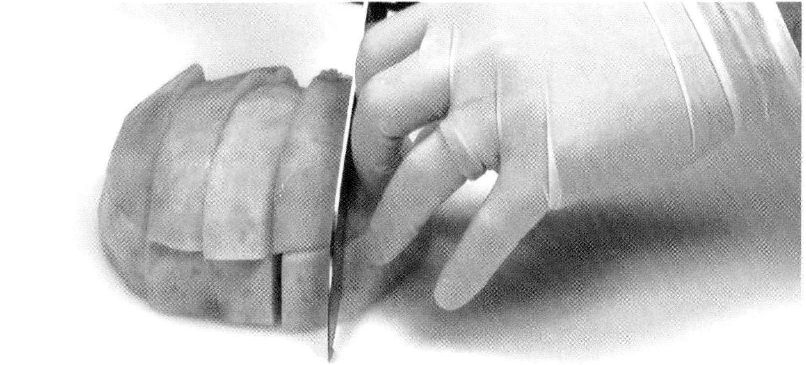

4. Rotate the cantaloupe 90°, back to its previous position. Secure the cantaloupe with a claw grip and slice the cantaloupe across the stick cuts at 1–1½ inch intervals.

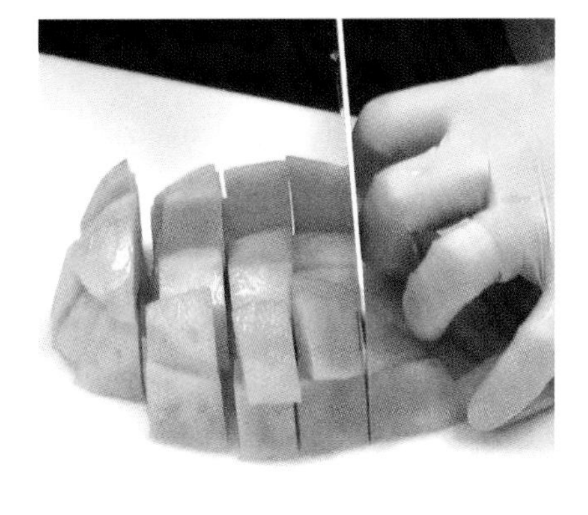

Finished diced cantaloupes are regular and slightly irregular cube shapes that are close to uniform in size.

TECHNIQUE 25

CUTTING PINEAPPLES

VIDEO 25

Pineapples are a tropical fruit recognized for their broad cylindrical shape, crown of elongated, pointy leaves, and spiny brown, green, or yellow skin. Just beneath the skin, pineapples have rough circular markings known as eyes. The eyes are the many flowers that wrap around a centralized core and develop into the fruit. When preparing pineapples, the eyes, crown, base, skin, and core are removed and the flesh is commonly sliced or cut into wedges. After pineapples are cut, they are often served raw or grilled in salsas, salads, sides, stir-fry dishes, and desserts.

Note: Disposable gloves are typically worn in the foodservice industry when preparing ready-to-eat food items.

PEELING, QUARTERING, AND SLICING PINEAPPLES

* This technique was executed by a left-handed chef.

1. Lay a pineapple on the cutting board with the crown of leaves angled toward the knife hand side and secure the pineapple with the guiding hand. Position a chef's knife approximately 1 inch from the crown and slice downward to remove the crown. Rotate the pineapple 180° and repeat this cut to remove the base.

2. Stand the pineapple upright with the base side on the cutting board and locate the pineapple eyes.

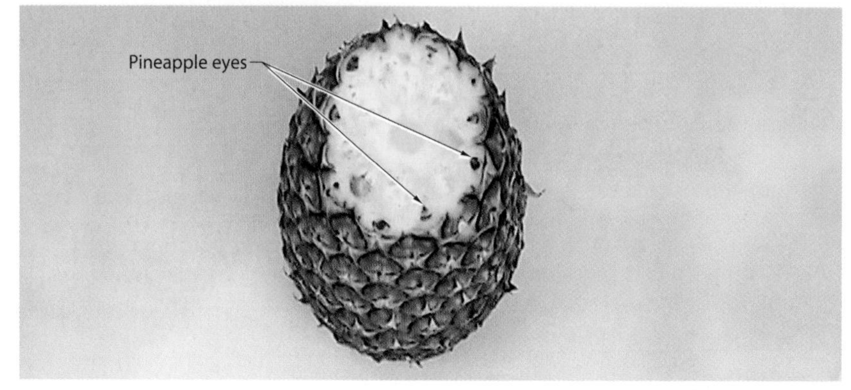

PEELING, QUARTERING, AND SLICING PINEAPPLES (CONTINUED)

* This technique was executed by a left-handed chef.

3. Secure the pineapple with the guiding hand and position the edge of the knife blade just behind several of the pineapple eyes. Follow the contour of the pineapple and cut down the side using a sawing motion to remove a thin strip of pineapple skin that includes the eyes but not too much flesh.

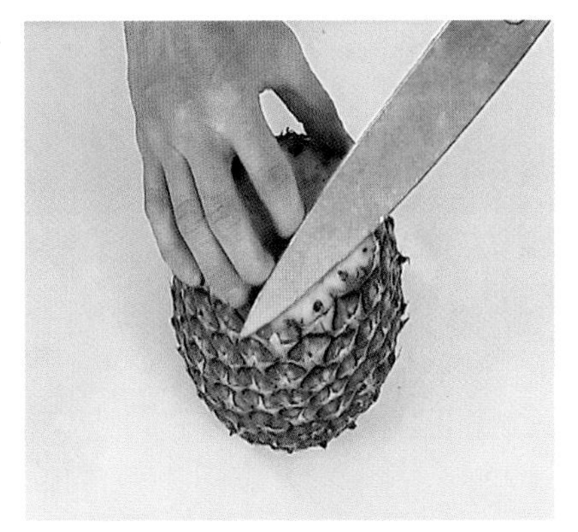

4. Slightly rotate the pineapple toward the body and repeat Step 3 until all the skin and eyes are removed from the pineapple. Trim any eyes that may have remained on the flesh.

5. Lay the pineapple on the cutting board so that an end faces the body. Hold the pineapple securely with the guiding hand and position the edge of the knife lengthwise on top of the pineapple. Cut down to slice the pineapple in half.

6. Place a pineapple half cut-side down on the cutting board and secure with the guiding hand. To quarter the pineapple half, make another lengthwise slice through the pineapple. Repeat this cut on the other pineapple half.

PEELING, QUARTERING, AND SLICING PINEAPPLES (CONTINUED)

* This technique was executed by a left-handed chef.

7. Place a pineapple quarter on its side so that part of the core is on the cutting board and part of the core is facing the knife hand side. Secure the pineapple with a claw grip and position the edge of the knife blade at a slight angle along the upper portion of the core. Cut downward at an angle to remove the core. Repeat this step on the remaining pineapple quarters.

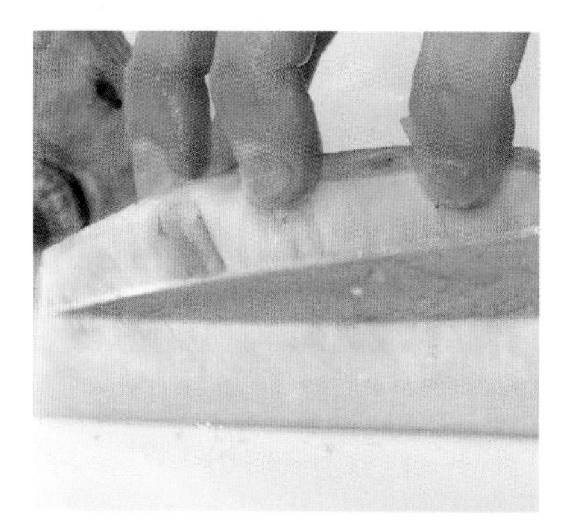

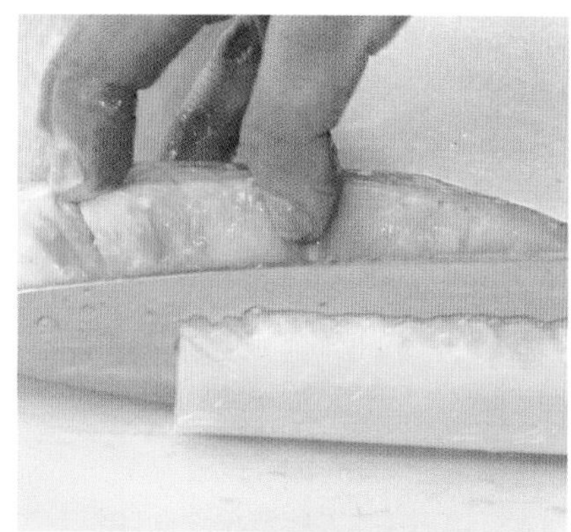

8. To slice the pineapple, place a pineapple quarter cored-side down on the cutting board and secure with a claw grip. Start at the end closest to the knife hand and use the tip-to-food method to cut the pineapple crosswise into consistently sized slices. Repeat this step on the remaining pineapple quarters.

Finished peeled, quartered, and sliced pineapples have also had the core remove to produce wedge-shaped pieces.

QUICK TIP

Pineapples do not continue to ripen after being harvested, and the outside color does not indicate ripeness. Therefore, a green pineapple and a golden brown pineapple can both be at the peak of ripeness. The ripest pineapples tend to have a crown of fresh-looking green leaves, a firm exterior, and a sweet, pineapple-like scent.

CUTTING CITRUS SUPREMES

TECHNIQUE 26

The flesh of citrus fruits, such as lemons, limes, oranges, tangerines, and grapefruits, lies beneath the skin and white, spongy pith of the fruit. Citrus flesh is naturally divided into wedge-shaped segments that are separated by a thin, white membrane. These segments are often cut into supremes. A supreme is the flesh from a segment of citrus fruit that has been cut away from the membrane. Supreming citrus fruits gives the segments an attractive appearance with a more tender texture and sweeter flavor. Citrus supremes are often used as a garnish or to elevate the presentation and flavor of salads, sides, entrées, and desserts.

Note: Disposable gloves are typically worn in the foodservice industry when preparing ready-to-eat food items.

CUTTING ORANGE SUPREMES

1. Place an orange on the cutting board with the stem end facing the knife hand side and secure the orange with a claw grip. Use a chef's knife to cut a thin slice from the stem end to slightly expose the flesh. Rotate the orange 180° and repeat this cut on the opposite end (blossom end) of the orange.

2. Stand the orange upright with one of the cut sides against the cutting board. Hold the upper portion of the orange in a claw grip and position the edge of the knife tip on top of the orange where the flesh meets the pith. Follow the contour of the orange and cut down the side with a slight sawing motion to remove a strip of the rind.

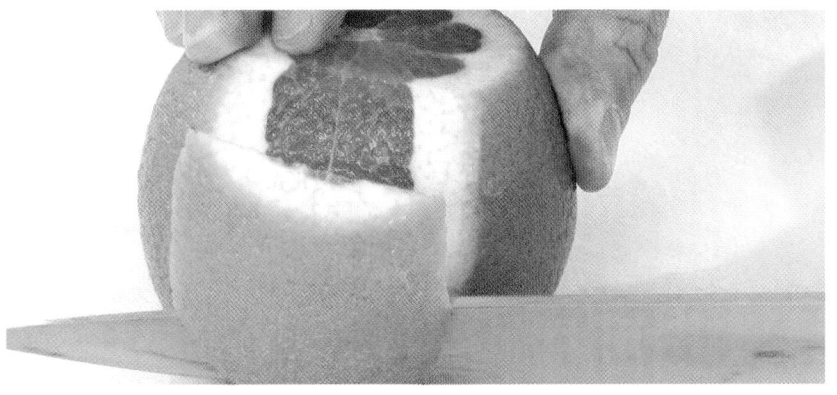

CUTTING ORANGE SUPREMES (CONTINUED)

3. Use the guiding hand to slightly rotate the orange toward the body. Return the edge of the knife tip to the top of the orange and use a slight sawing motion to cut away another thin strip of the rind. Continue to rotate and cut down the sides of the orange until the entire rind is removed.

4. If necessary, trim any remaining rind or white pith from the flesh.

5. Place the peeled orange on the cutting board with the stem end facing away from the body and secure the orange with a claw grip. Use the knife tip to cut along the white membrane on one side of an orange segment. To release the segment, cut along the membrane on the other side of the segment.

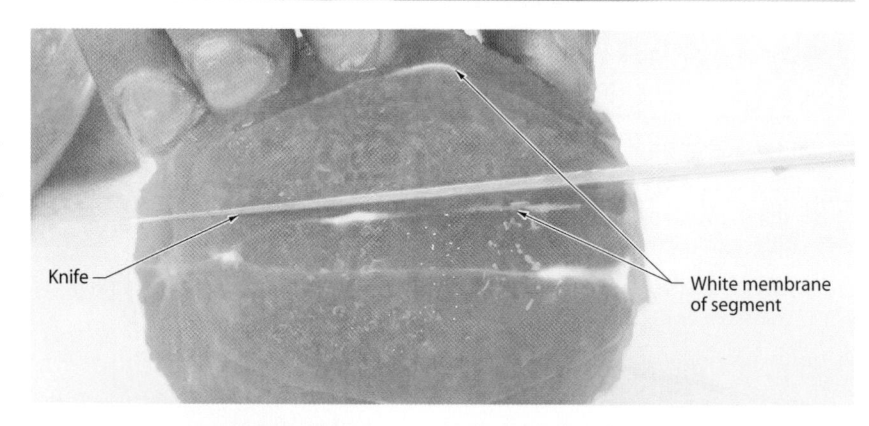

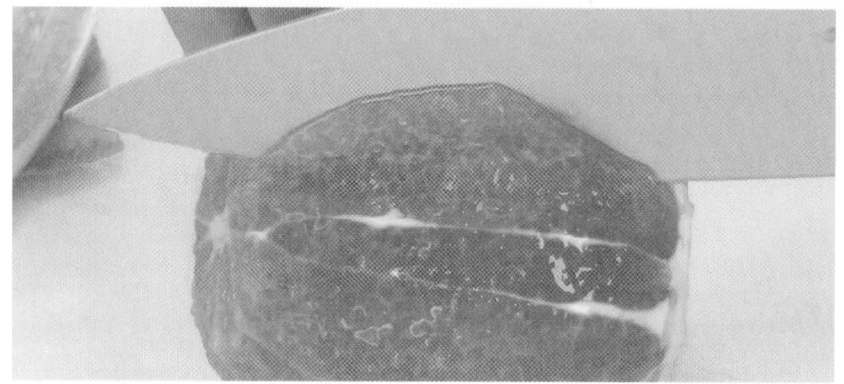

CUTTING ORANGE SUPREMES (CONTINUED)

6. Roll the orange slightly toward the guiding hand and cut along the membranes of the next orange segment to release this segment. Continue rolling and cutting along the membranes of each segment until all segments have been separated.

   Note: If seeds are present, remove with the tip of the knife blade.

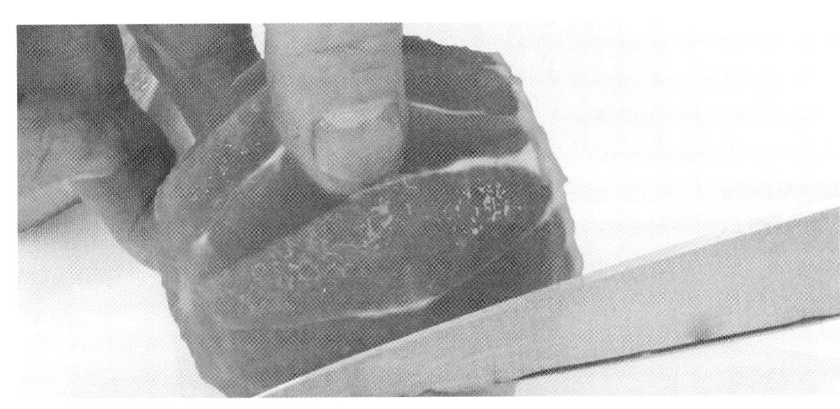

Finished orange supremes are orange segments that are void of a peel, pith, membranes, and seeds.

TECHNIQUE 27

CUTTING ARTICHOKES

VIDEO 27

Artichokes consist of rigid, pointy leaves that can be sharp and must be trimmed during preparation. A fibrous choke near the center is also removed to reveal the artichoke heart, the area where the leaves attach to the stem. Only the heart and the base of each leaf are edible.

Artichokes are often cut in half or broken down further to use only the heart. Artichoke halves are commonly covered with a topping, baked, and served as an appetizer. Artichoke hearts are commonly used in dips, salads, and pasta dishes.

Note: When cutting artichokes, they should periodically be rubbed with a cut lemon to prevent them from oxidizing and turning brown.

Chef's Knife

— or —

Santoku Knife

Kitchen Shears

Spoon

Paring Knife

CUTTING AND TRIMMING ARTICHOKE HALVES

1. Lay an artichoke on the cutting board with the leaves facing the knife hand side and secure the base with a claw grip. To expose the choke, cut off the top half of the artichoke with a chef's knife.

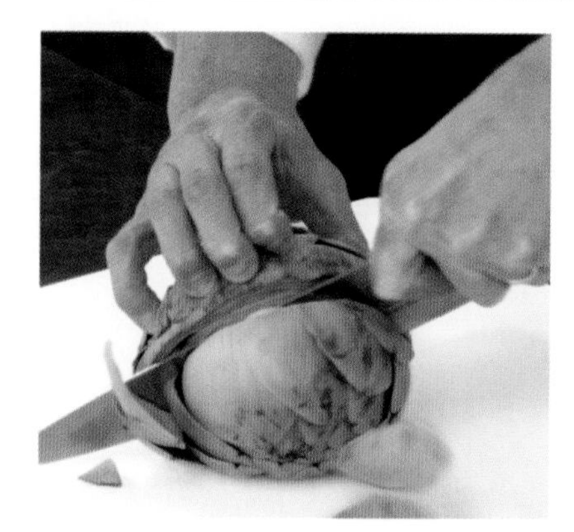

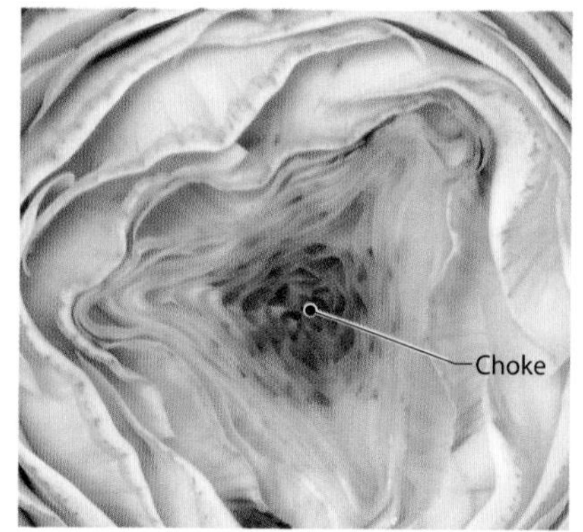

CUTTING AND TRIMMING ARTICHOKE HALVES (CONTINUED)

2. Hold the artichoke stem with the guiding hand and use kitchen shears to snip off the sharp tips of the exposed leaves.

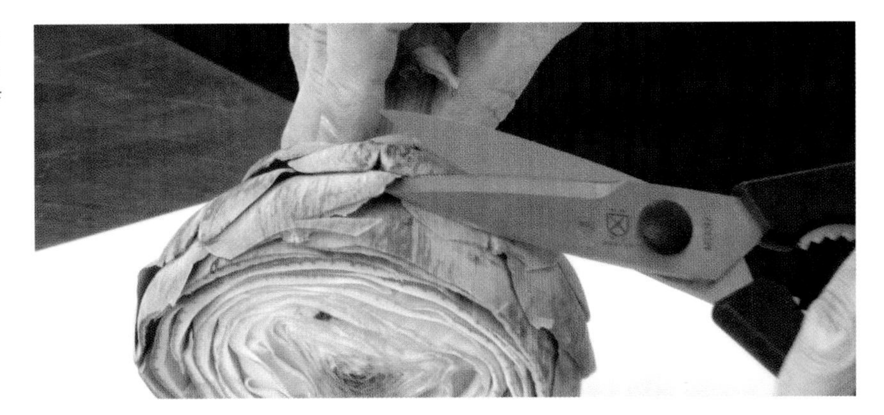

QUICK TIP

If the desired application calls for just the artichoke leaves to be eaten, remove only 1½ inches from the top of the artichoke, trim the sharp tips of the leaves, and cook as required.

3. Stand the artichoke upright on its stem. Grasp the bottom of the artichoke between both hands and use the thumbs to spread open the center and expose the choke.

4. Lay the artichoke on the cutting board with the choke facing the dominant hand. Use a spoon to scrape out the fibrous, purple-colored choke in the center.

   Note: Remove only the choke so as not to damage the artichoke heart that lies beneath.

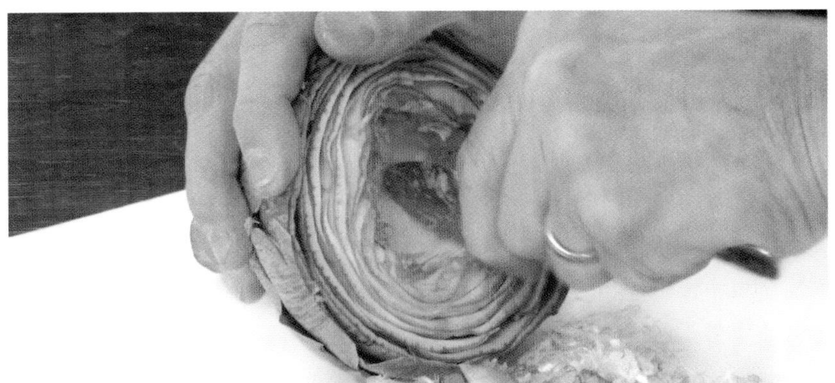

CUTTING AND TRIMMING ARTICHOKE HALVES (CONTINUED)

5. Rotate the artichoke so that the stem faces the knife hand side. Secure the leaves with the guiding hand and cut off the stem at the base of the artichoke.

Finished artichoke halves consist of the bottom half of an artichoke that has the sharp tips of the leaves trimmed away and both the choke and stem removed.

CUTTING ARTICHOKE HEARTS

1. Hold an artichoke in the guiding hand with the stem facing the body. Peel the outer layer of tough, dark-green leaves down and away from the artichoke with the dominant hand.

QUICK TIP

Reserve the leaves and trimmed sections of the artichoke to make stock or essence (a concentrated stock similar to an extract that is used in small amounts to enhance flavor).

CUTTING ARTICHOKE HEARTS (CONTINUED)

2. Continue peeling away the leaves until a defined edge is revealed at the base of the artichoke and the exposed leaves become more tender and lighter in color. This is the location of the artichoke heart.

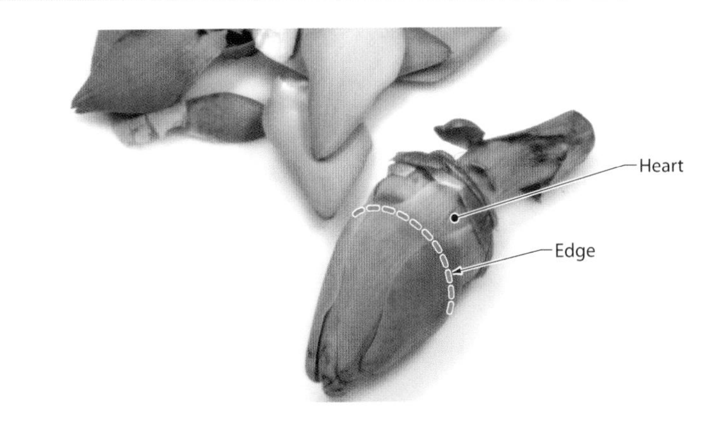

3. Place the artichoke on the cutting board with the stem angled toward the guiding hand side and secure with a claw grip. Position a chef's knife along the rim of the edge and slice through the artichoke to remove the top portion just above the heart.

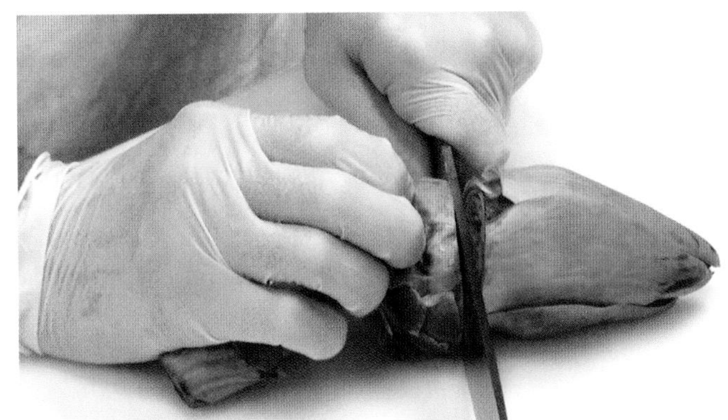

4. Hold the artichoke in the guiding hand and grasp a paring knife in a choke grip. Position the edge of the blade and knife hand thumb along the side of the artichoke.

5. Trim a layer of leaves from all sides of the artichoke by pulling the knife blade toward the thumb and simultaneously rotating the artichoke toward the knife blade.

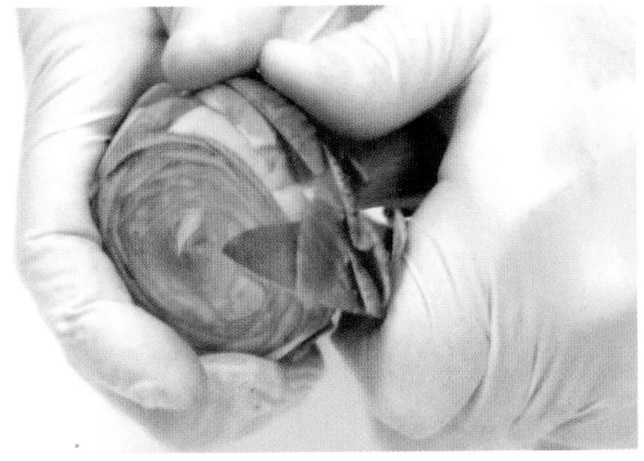

CUTTING ARTICHOKE HEARTS (CONTINUED)

6. Place the artichoke on the cutting board with the stem end facing the knife hand side. Secure the artichoke with the guiding hand and use the paring knife to remove approximately 1 inch from the bottom of the stem.

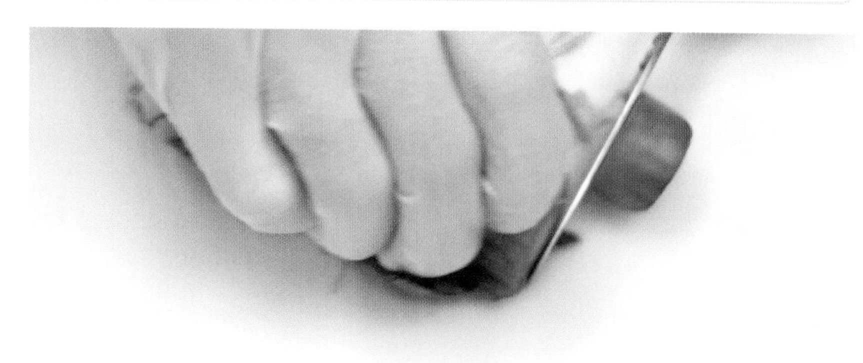

7. Hold the artichoke in the guiding hand with the stem facing away from the body. Grasp the paring knife in a choke grip and position the edge of the blade at the end of the stem and the knife hand thumb near the top of the artichoke.

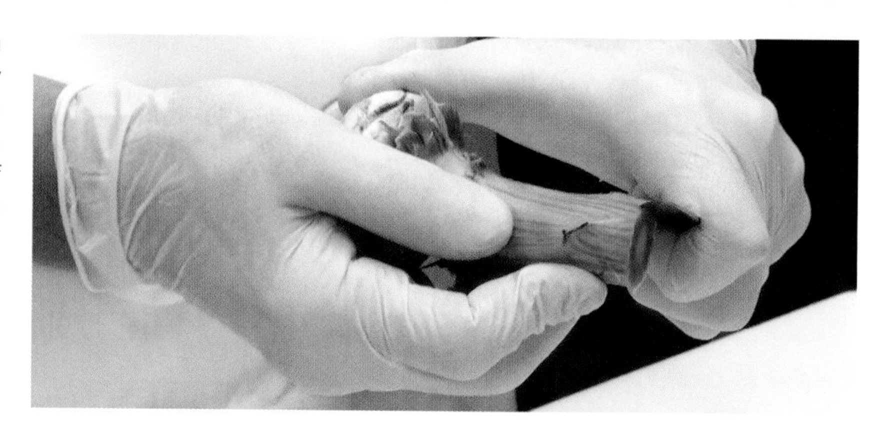

QUICK TIP

Artichokes are at the peak of ripeness from March through May. For best quality, choose artichokes that feel heavy for their size and have a deep green color with a tight formation of leaves. Artichokes range from baby to jumbo size. Baby artichokes are often sautéed, roasted, or marinated, while jumbo artichokes are generally stuffed.

8. Remove a fibrous strip of skin from the stem by pulling the knife blade toward the knife hand thumb. Slightly rotate the artichoke and repeat this step until the entire stem is peeled.

   Note: Inspect the artichoke and, if necessary, continue to trim the stem and edge to remove any tough, fibrous outer layers.

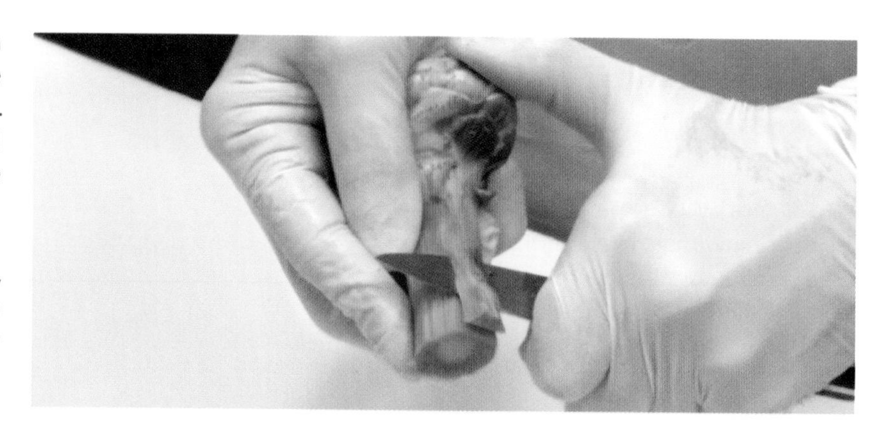

CUTTING ARTICHOKE HEARTS (CONTINUED)

9. To reveal the artichoke heart, use a spoon to scrape out the stringy choke.

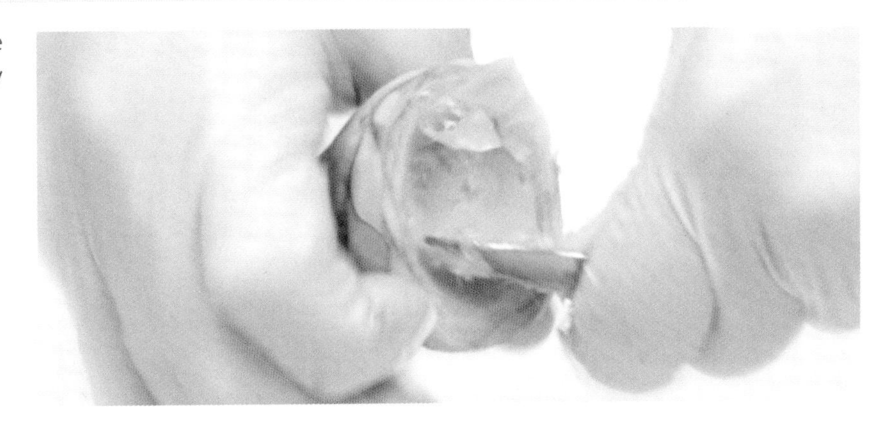

Finished artichoke hearts can be kept whole or sliced lengthwise into halves or quarters and used as desired.

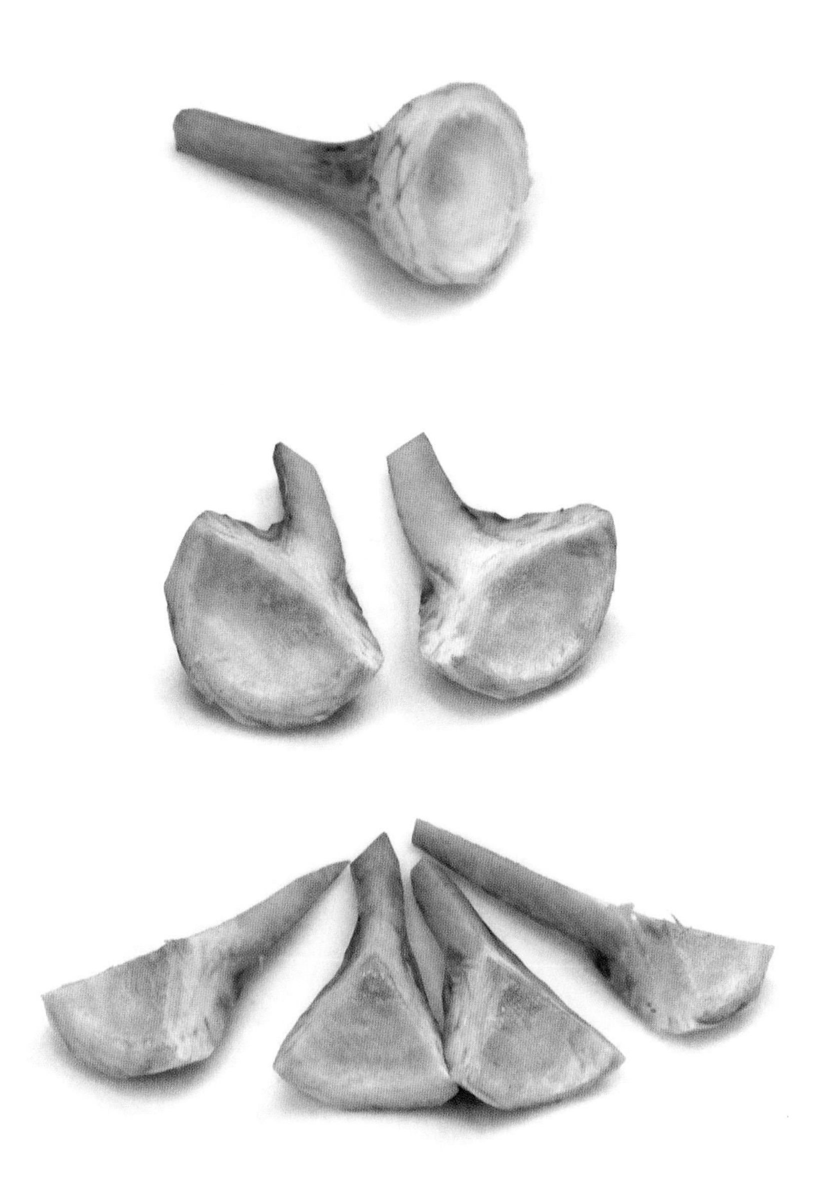

SECTION IV
CUTTING AND DEBONING POULTRY

- CUTTING POULTRY INTO HALVES, QUARTERS, AND EIGHTHS
- FABRICATING WHOLE POULTRY
- DEBONING POULTRY BREASTS
- CUTTING AIRLINE POULTRY BREASTS
- DEBONING POULTRY THIGHS AND LEGS
- DEBONING WHOLE POULTRY

LEARNER RESOURCES
ATPeResources.com/QuickLinks™
Access Code: 583241

TECHNIQUE 28

CUTTING POULTRY INTO HALVES, QUARTERS, AND EIGHTHS

VIDEO 28

Stiff Boning Knife

Poultry is a term used to describe chickens, turkeys, ducks, geese, guinea fowls, and squabs. Because all types of poultry have similar muscular and skeletal structures, they can be fabricated or cut in a similar manner. Whole poultry that has been properly fabricated minimizes waste, increases utilization of the entire carcass, and is more economical than purchasing prefabricated cuts.

Whole poultry is commonly cut into halves, quarters, or eighths, depending on the intended use. When whole poultry is cut in half, each half includes a leg (drumstick and thigh), wing, and breast. Quartered poultry yields two leg portions and two breast and wing portions. To produce eight pieces from quartered poultry, the drumstick and thigh quarters are split and the breast and wing quarters are split. Cutting poultry into halves, quarters, or eighths allows for greater control over cooking times and ease of serving.

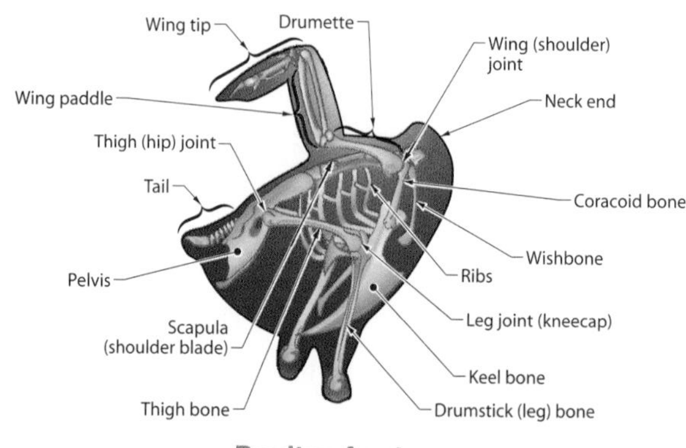

Poultry Anatomy

CUTTING CHICKENS INTO HALVES

* This technique was executed by a left-handed chef.

1. Place a chicken carcass back-side down on the cutting board with the tail facing the knife hand side. Use a stiff boning knife to remove the tail where it connects to the carcass.

* This technique was executed by a left-handed chef.

QUICK TIP

According to the United States Department of Agriculture (USDA), poultry should not be washed prior to cooking as this can increase the risk of spreading bacteria from the raw poultry to other surfaces, foods, and utensils.

2. Turn the carcass back-side up with the tail end near the side of the cutting board closest to the body and secure with the guiding hand. Make a shallow cut from the neck end to the tail end along one side of the backbone to score the carcass. Repeat this cut on the other side of the backbone.

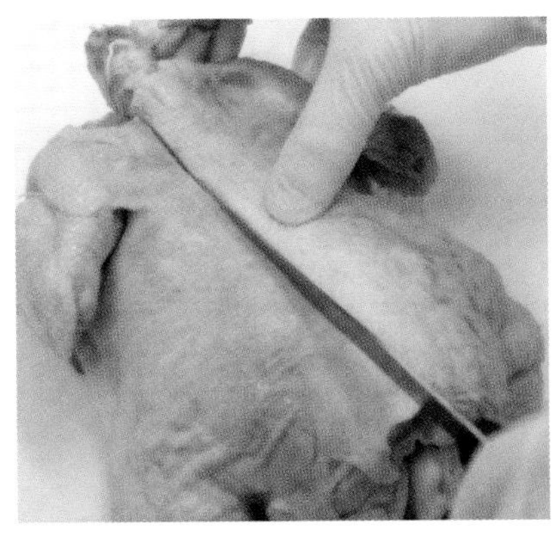

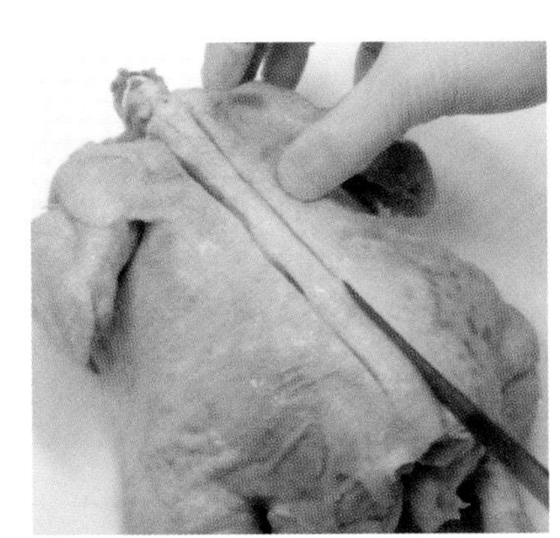

3. Stand the carcass upright with the tail end against the cutting board and secure with the guiding hand. Insert the knife tip at the score mark on the knife hand side and use a firm sawing motion to cut through the small bones along the side of the backbone. Stop the cut 3 inches from the tail end to help keep the carcass stable for further cutting.

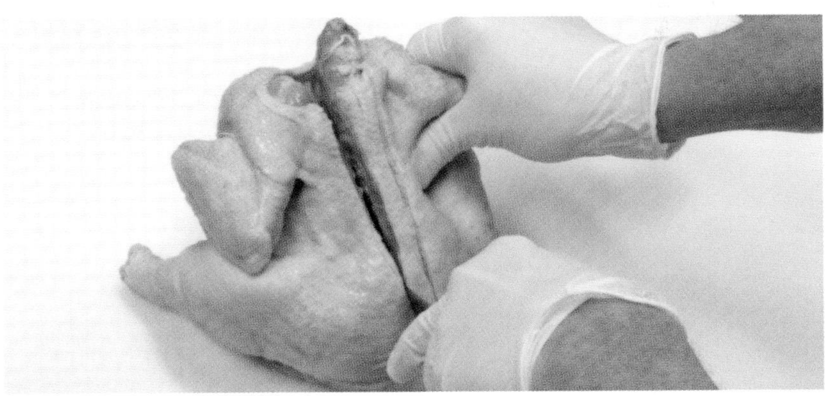

4. Insert the knife tip where the other score mark begins on the opposite side of the backbone. Cut through the small bones along this side of the backbone all the way down to the cutting board.

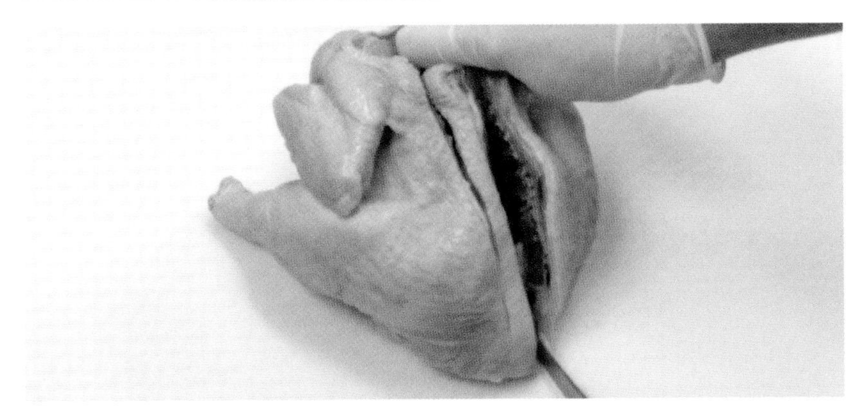

QUICK TIP

In order to cut through the small bones along each side of the backbone, it is important to keep the knife blade close to the backbone. If the knife blade is too far from the backbone, it will reach large bones that are too difficult to cut through.

* This technique was executed by a left-handed chef.

5. Hold the top of the backbone and breast with the guiding hand. Return to the cut that stopped 3 inches from the tail end and cut along the score mark until the backbone separates from the carcass.

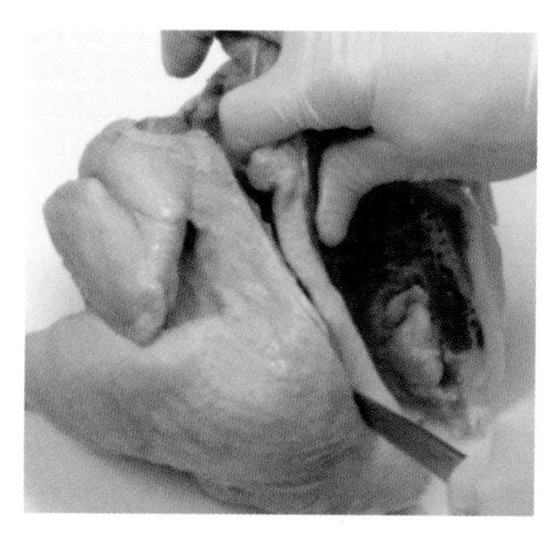

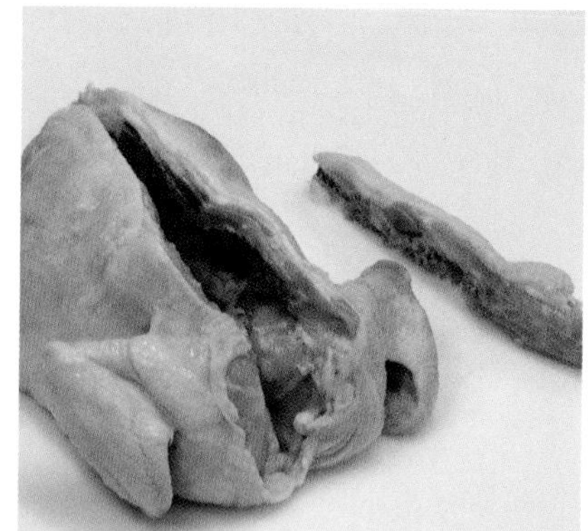

6. Use both hands to open the carcass and expose the cavity.

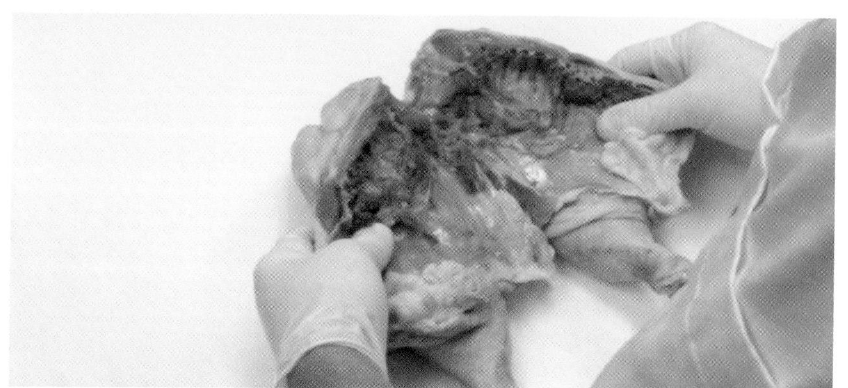

7. To split the chicken into halves, cut lengthwise down the center of the carcass between the two breasts.

   Note: The knife will cut through the wishbone, keel bone, and cartilage at the end of the keel bone.

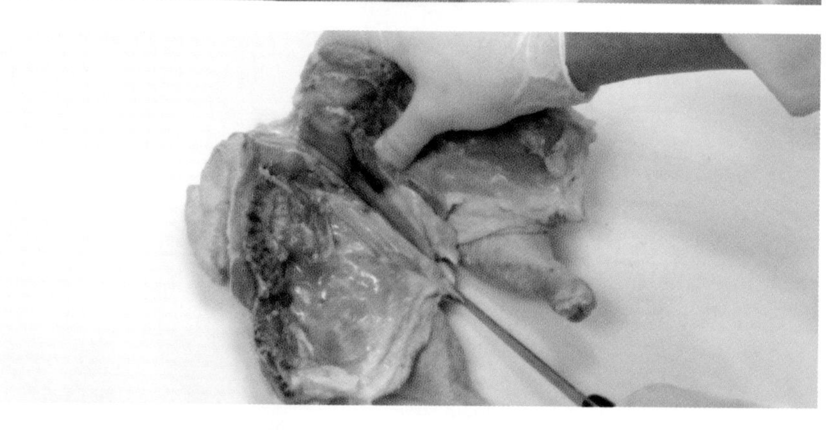

Each side of a halved chicken yields a leg (drumstick and thigh), wing, and breast.

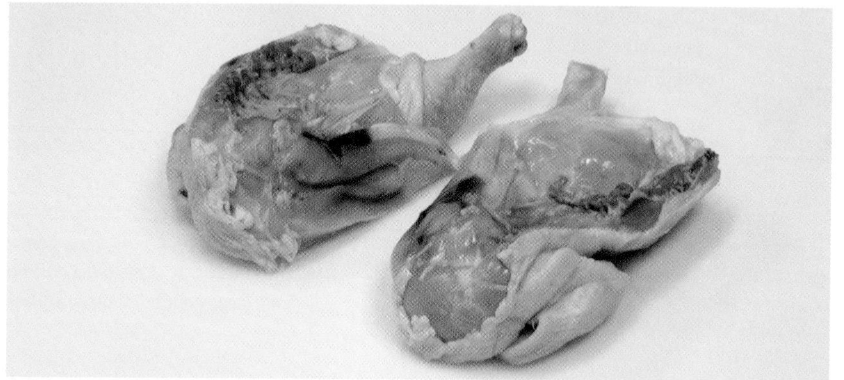

CUTTING CHICKENS INTO QUARTERS

* This technique was executed by a left-handed chef.

1. Place one side of a halved chicken skin-side up across the cutting board. Use the side of the guiding hand to press into the natural indentation that divides the leg (drumstick and thigh) from the breast and wing.

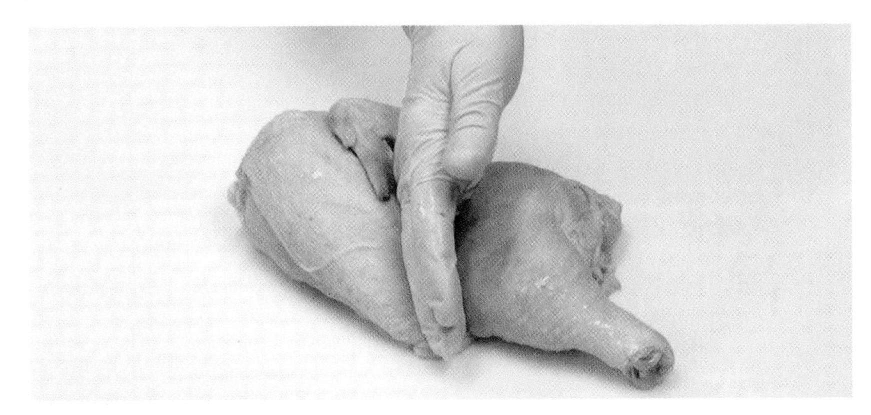

2. Use a stiff boning knife to cut along the indented line made in Step 1 to separate the leg from the breast and wing. Repeat Steps 1–2 on the other chicken half to produce a quartered chicken.

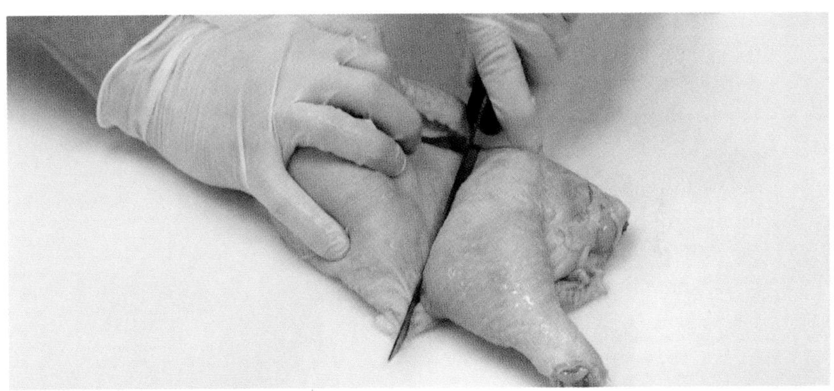

A chicken cut into quarters yields two leg (drumstick and thigh) portions and two breast and wing portions.

QUICK TIP

Chickens have both light and dark flesh. Dark flesh is the result of a substantial amount of myoglobin (red protein pigment) in the tissue. The more a muscle is worked, the more myoglobin that is present. Since chickens use their leg muscles more than their breast and wing muscles, the leg muscles have more myoglobin and darker flesh.

CUTTING CHICKENS INTO EIGHTHS

* This technique was executed by a left-handed chef.

1. Place a leg (drumstick and thigh) from a quartered chicken skin-side down with the thigh facing the knife hand side. Secure the drumstick with the guiding hand and position the edge of a stiff boning knife next to the drumstick side of the fat line.

   Note: The fat line runs directly over the leg joint and is used as a guide when separating the drumstick from the thigh.

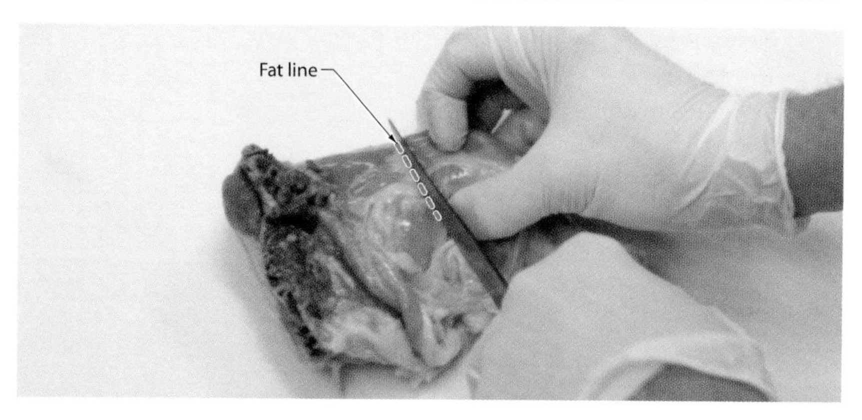

2. To separate the drumstick from the thigh, cut along the fat line and through the leg joint. Repeat Steps 1–2 on the other leg portion.

   Note: If the knife does not easily cut through the joint, move it slightly to the right or left to find the center of the joint.

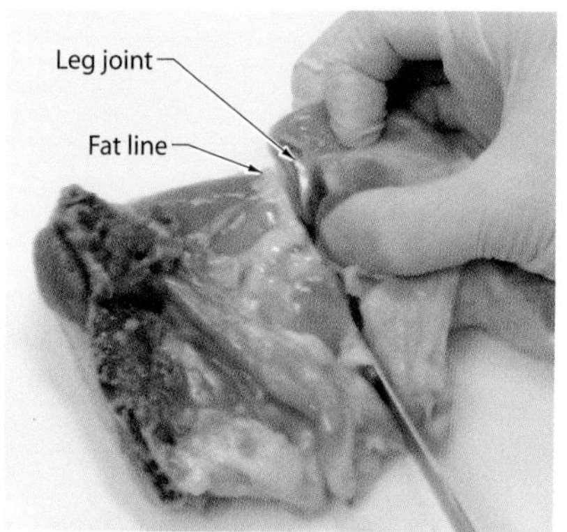

3. Place a breast and wing from a quartered chicken skin-side up across the cutting board. Pull the wing back with the guiding hand and position the knife blade above the wing (shoulder) joint.

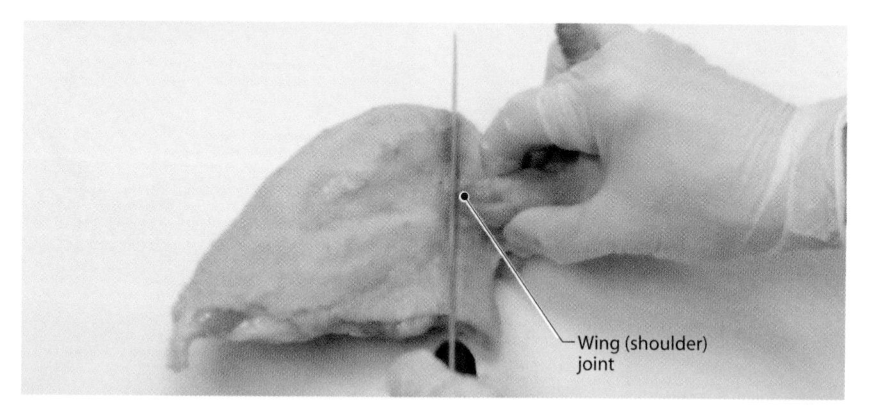

QUICK TIP

Salmonella and Campylobacterjejuni are serous bacterial concerns when handling raw poultry. Never use the same cutting boards or tools that were used with raw poultry for other products until they have been washed and sanitized. Hands should also be washed with soap and warm water for a minimum of 20 seconds before and after the poultry is handled.

CUTTING CHICKENS INTO EIGHTHS (CONTINUED)

* This technique was executed by a left-handed chef.

4. Make a shallow cut straight down to expose the joint, then angle the knife blade to cut under the joint. Continue cutting until the wing separates from the breast. Repeat Steps 3–4 on the other breast and wing portion.

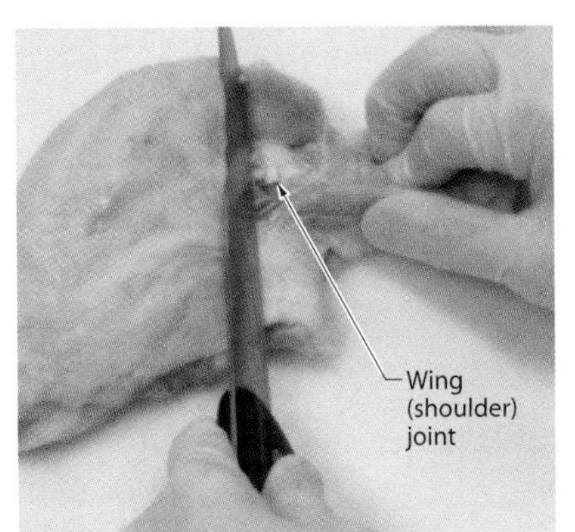

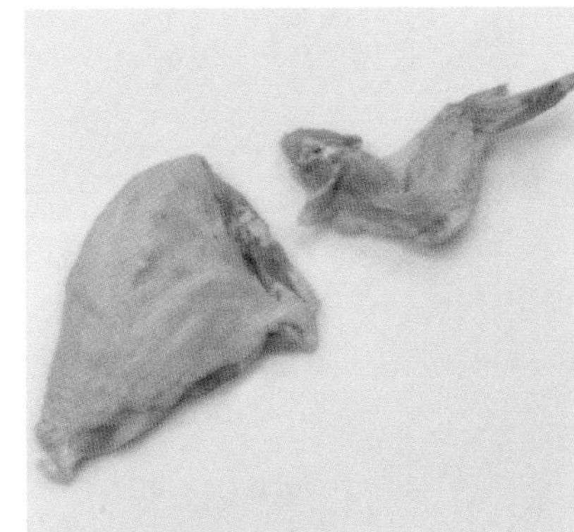

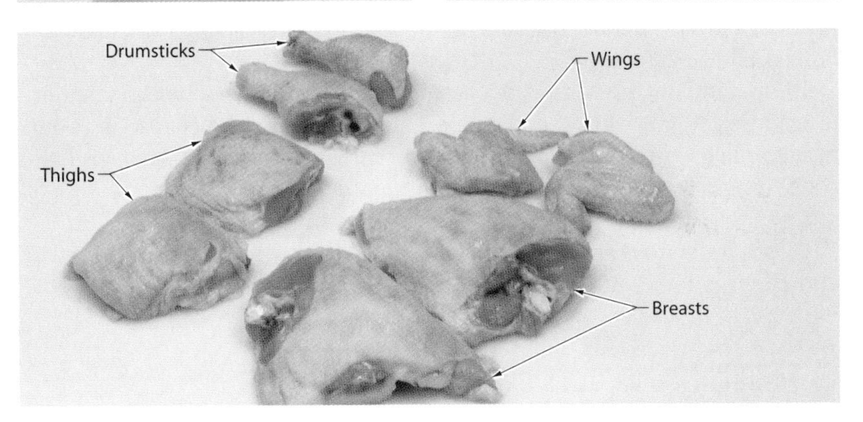

Chicken cut into eighths yields two drumsticks, two thighs, two breasts, and two wings.

TECHNIQUE 29

FABRICATING WHOLE POULTRY

VIDEO 29

Whole poultry is fabricated with this technique to produce two leg (drumstick and thigh) portions and two breast and wing portions that have been cut away from the carcass. As a result, the fabricated portions contain fewer bones than poultry that has been halved and quartered.

Many culinary professionals consider this technique to be the standard procedure for fabricating and preparing whole poultry because it results in high yields with little flesh left on the carcass. Therefore, this standard procedure is a cost-effective way to fabricate whole poultry.

Note: Refer to Technique 28: Cutting Poultry into Halves, Quarters, and Eighths for Poultry Anatomy diagram.

Stiff Boning Knife

FABRICATING WHOLE CHICKENS

* This technique was executed by a left-handed chef.

1. Stand a chicken carcass upright with the tail end on the cutting board and hold the neck area with the guiding hand. Use the tip of a boning knife to scrape the flesh from both sides of the V-shaped wishbone located just beneath the flesh at the top edge of each breast.

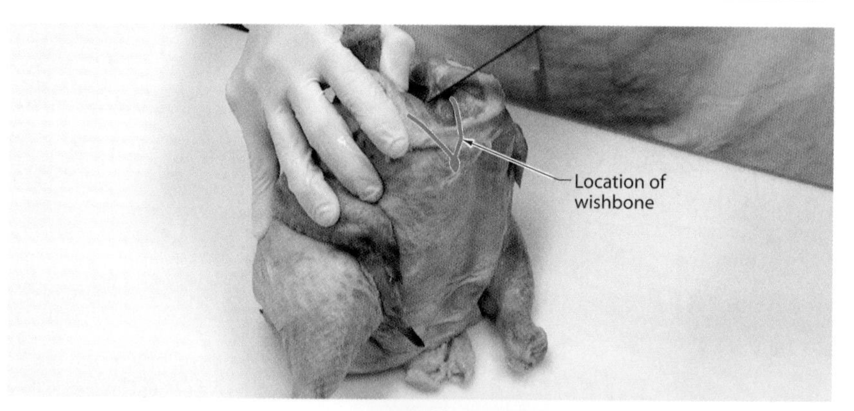

2. Use the fingers to remove flesh from the wishbone and loosen it from the breasts. Grasp the wishbone where both sides of the bone meet and pull back toward the neck opening to remove it from the carcass.

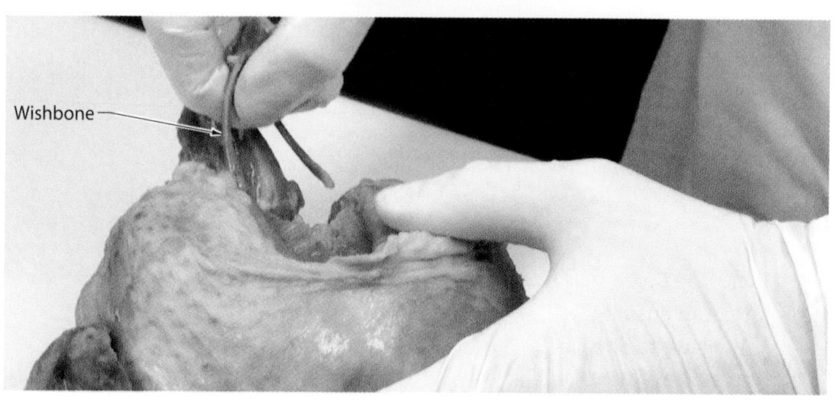

3. Lay the carcass back-side down with the tail facing the knife hand side and cut the tail from the carcass.

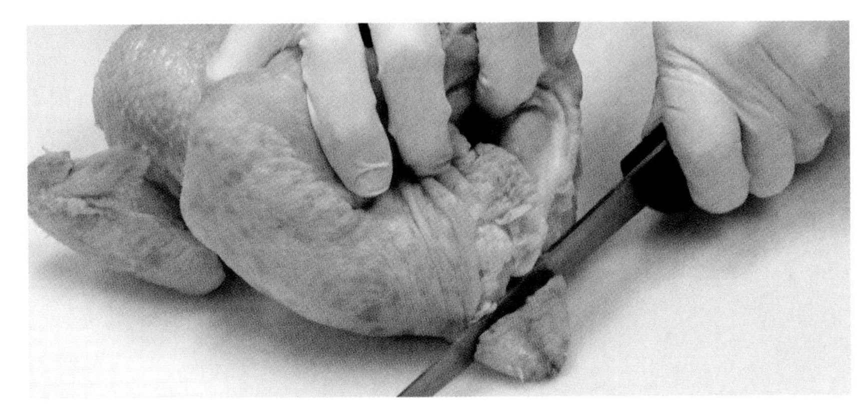

4. Turn the carcass back-side up with the tail end near the side of the cutting board closest to the body. Use the knife tip to make a shallow cut (score mark) along the center of the backbone from the neck end to the tail end.

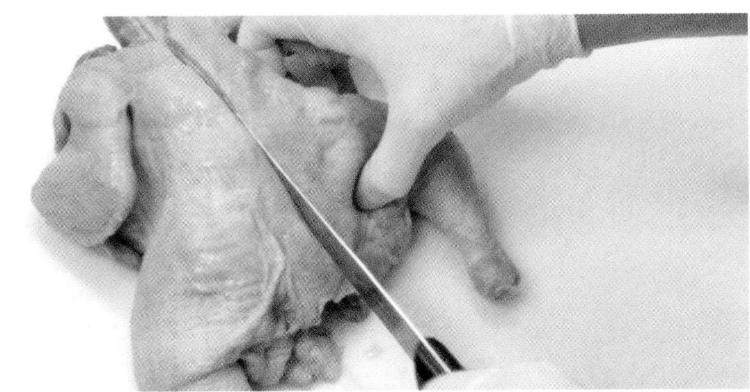

5. Find the two round bumps of flesh, known as oysters on either side of the backbone slightly above the hip area. With the edge of the knife in front of the oysters and perpendicular to the backbone, make a score mark that is 1 inch long.

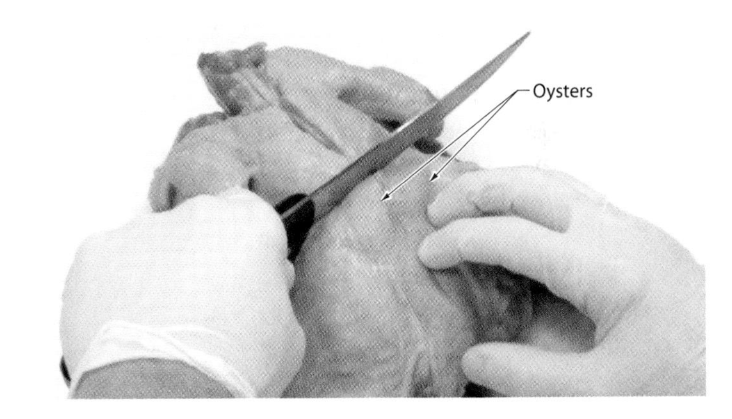

6. Turn the carcass back-side down and use the guiding hand thumb to pull the loose skin on the leg toward the breast. Cut through the taut skin on the thigh to make a 1½–2 inch score mark just above the kneecap. Repeat this step on the other thigh.

   Note: The skin is pulled toward the breast to ensure that enough skin is available to completely cover the top surface of the breast during cooking.

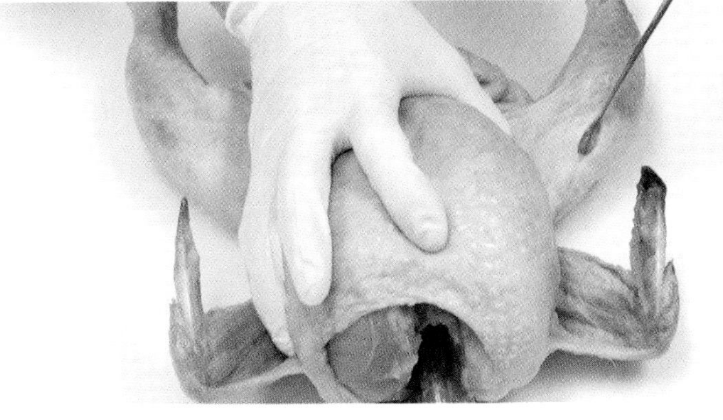

* This technique was executed by a left-handed chef.

7. Grasp one leg in each hand and insert the thumbs into the score marks made in Step 6. Bend each leg down and under the tail end to separate the thigh (hip) joint from the socket.

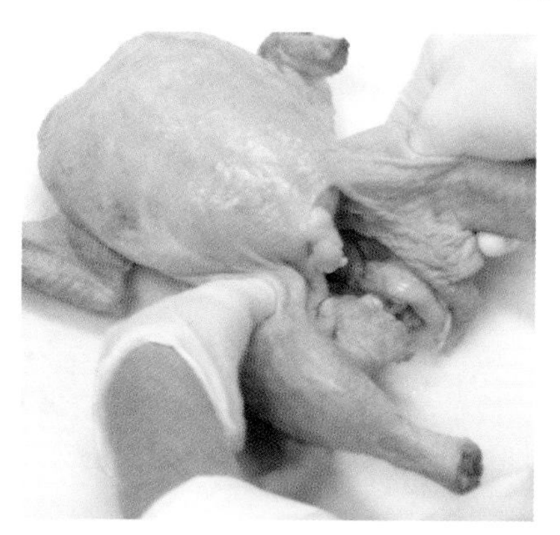

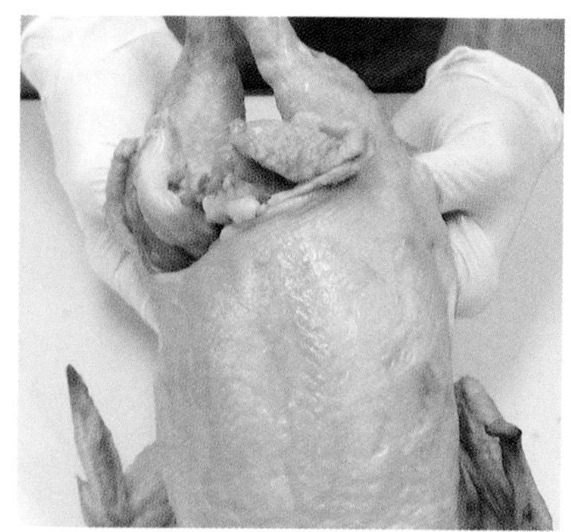

8. Grasp both legs in the guiding hand and flip the carcass backside up. Position the knife blade edge perpendicular to the bottom of the score mark on the thigh made in Step 6.

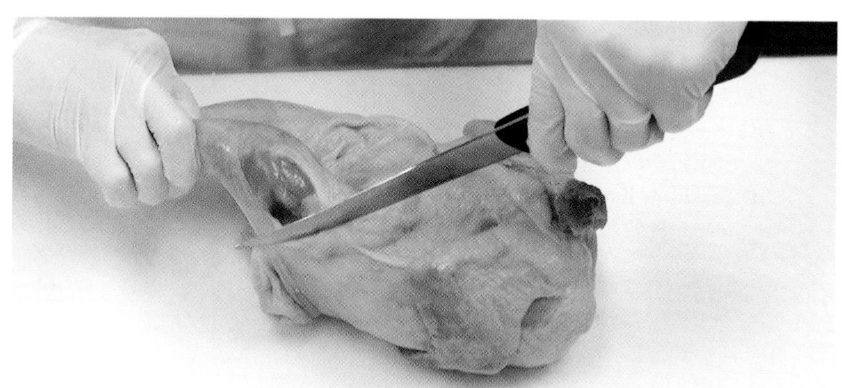

9. Using the score marks as a guide, make a cut ½ inch deep across the carcass that starts at the outside thigh, moves through the score mark at the oysters, and ends at the bottom of the score mark on the other thigh.

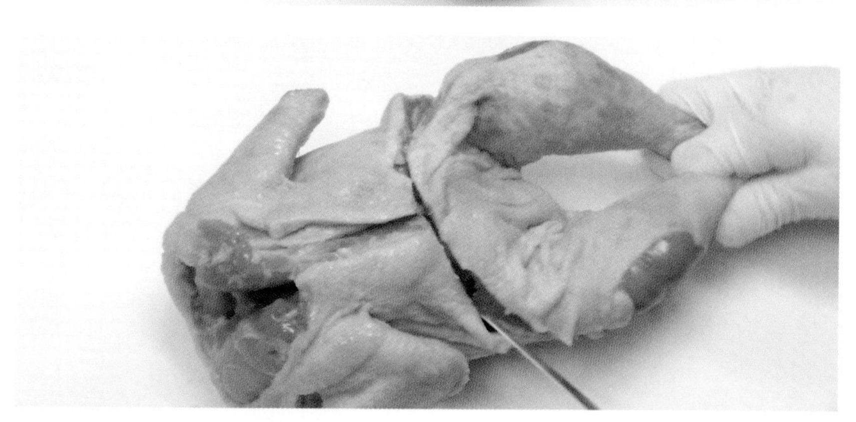

10. Make a small cut in front of the oyster located closest to the body and then scrape the oyster from its socket using the fingers.

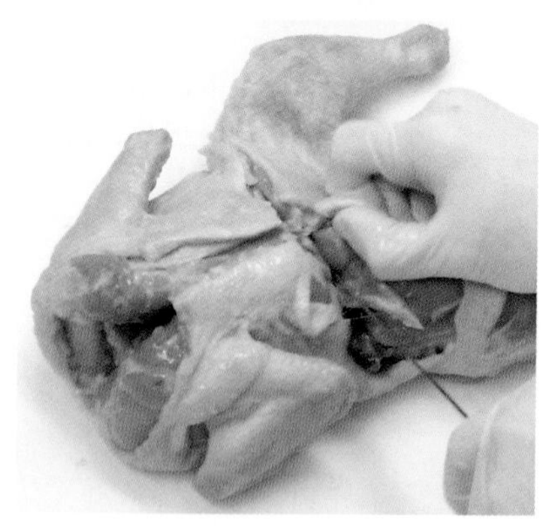

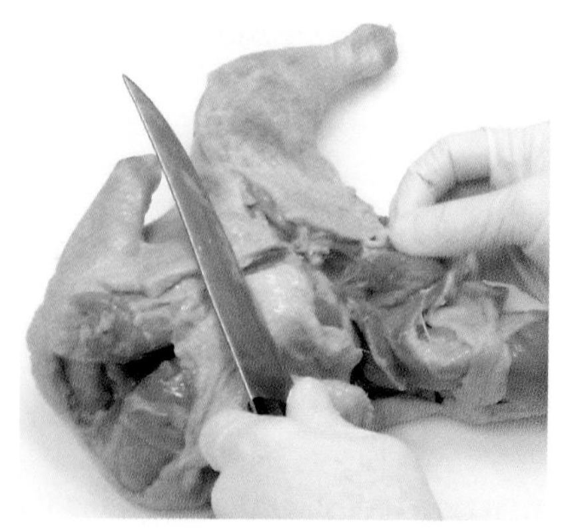

11. With the guiding hand, bend back the leg closest to the body to expose the thigh (hip) joint. Make a shallow incision beneath the joint to cut connective tissue.

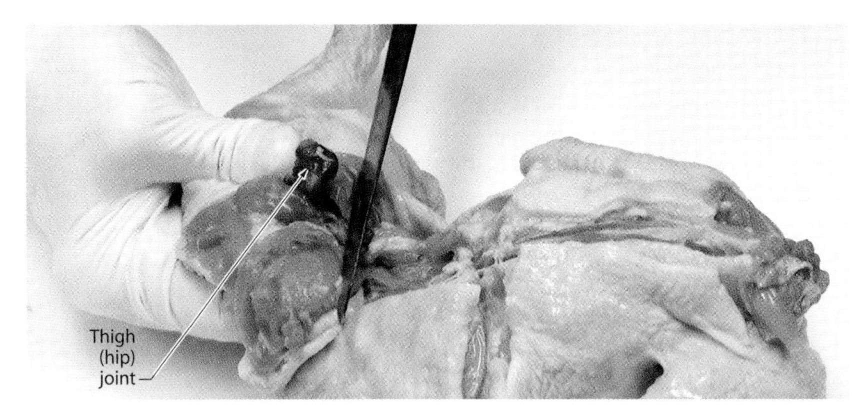

12. Grasp the thigh bone with the guiding hand. Place the side of the knife blade in the cut mark above the oyster to use for leverage and pull the leg back until it is connected to the carcass only by skin.

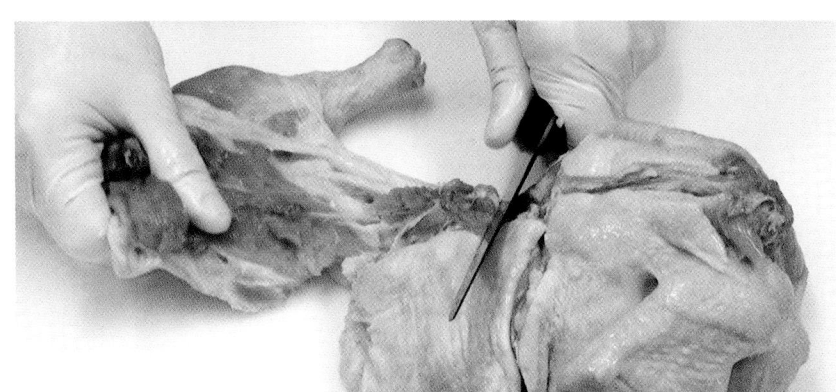

13. To remove the leg, cut through the skin that keeps the leg attached to the carcass. Repeat Steps 10–13 to remove the other leg.

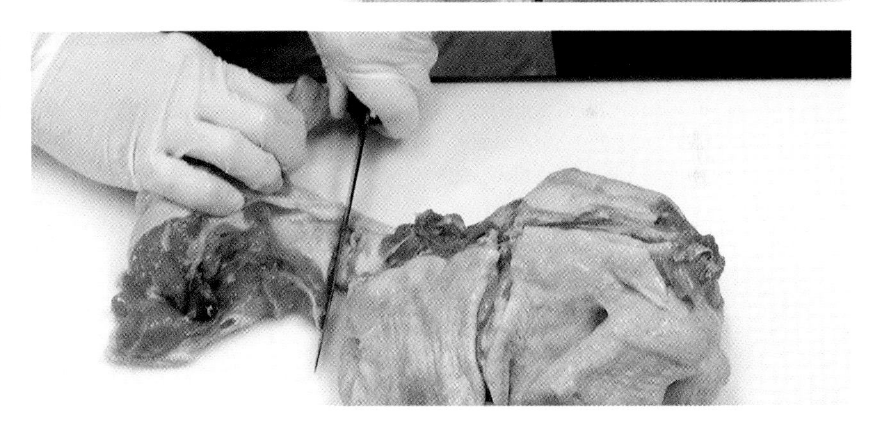

QUICK TIP

When fabricating whole poultry, allow for enough skin to completely cover each cut. Even skin coverage allows for more consistency when cooking and results in a more appealing finished product.

14. Turn the carcass breast-side up with the tail end near the side of the cutting board closest to the body. While holding the skin in place with the guiding hand, make a lengthwise cut down the center of the breasts that is deep enough to reach the keel bone.

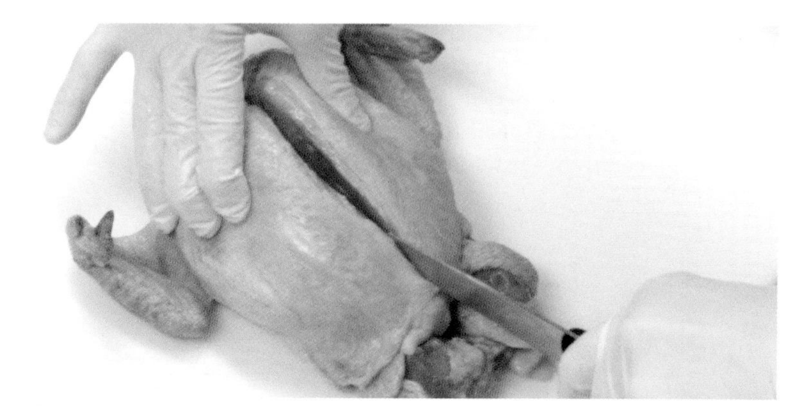

15. Cut along one side of the keel bone, pulling the knife back in smooth, even strokes until the breast remains attached to the carcass just beneath the wing (shoulder) joint. Repeat this step on the other side of the keel bone.

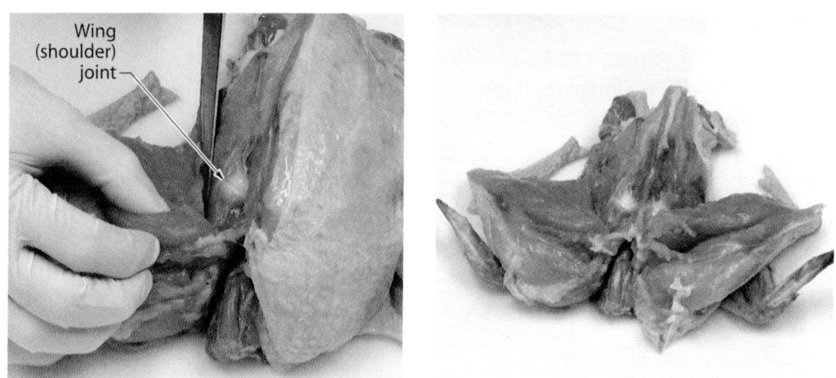

16. Rotate the carcass 180° so that the neck end is near the side of the cutting board closest to the body. Cut through the wing joint to separate the breast and wing from the carcass. Repeat this cut on the other side of the carcass to remove the remaining breast and wing.

Fabricating whole chickens with this standard technique yields two leg portions (drumstick and thigh) and two breast and wing portions that have been removed from the carcass.

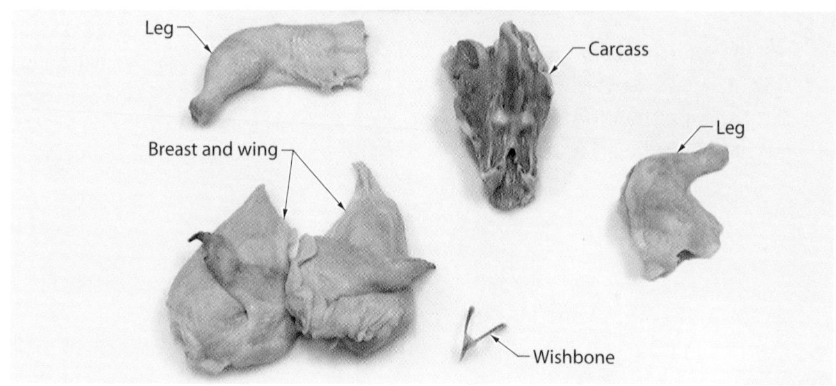

VIDEO 30

DEBONING POULTRY BREASTS

TECHNIQUE 30

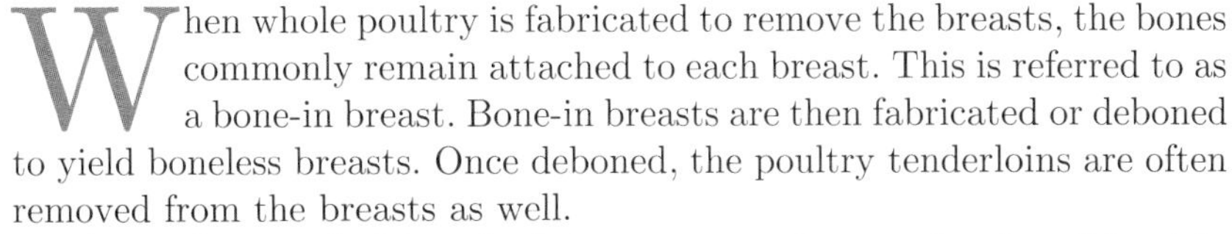

When whole poultry is fabricated to remove the breasts, the bones commonly remain attached to each breast. This is referred to as a bone-in breast. Bone-in breasts are then fabricated or deboned to yield boneless breasts. Once deboned, the poultry tenderloins are often removed from the breasts as well.

The poultry tenderloin, also known as a poultry tender, is a thin strip of muscle that runs along the inside, lower section of the breast and is situated close to the bone. Although poultry breasts may include skin, the tenderloin is skinless due to its location on the breast. Skinless poultry is often preferred because it is lower in calories and fat than equivalent cuts that include the skin. Poultry tenderloins and boneless, skinless poultry breasts are identical in terms of color, flavor, and texture, and both are well-suited for quick cooking methods such as grilling, pan-frying, sautéing, and broiling.

Note: Refer to Technique 28: Cutting Poultry into Halves, Quarters, and Eighths for Poultry Anatomy diagram.

Stiff Boning Knife

DEBONING AND SKINNING CHICKEN BREASTS

* This technique was executed by a left-handed chef.

1. Hold a bone-in chicken breast from an eight-cut chicken slightly upright on the cutting board with the bones facing the knife hand side. Use a boning knife to make a 1–2 inch cut behind the keel bone to start releasing it from the breast.

   Note: Refer to Technique 28: Cutting Poultry into Halves, Quarters, and Eighths for the procedure on fabricating a bone-in chicken breast.

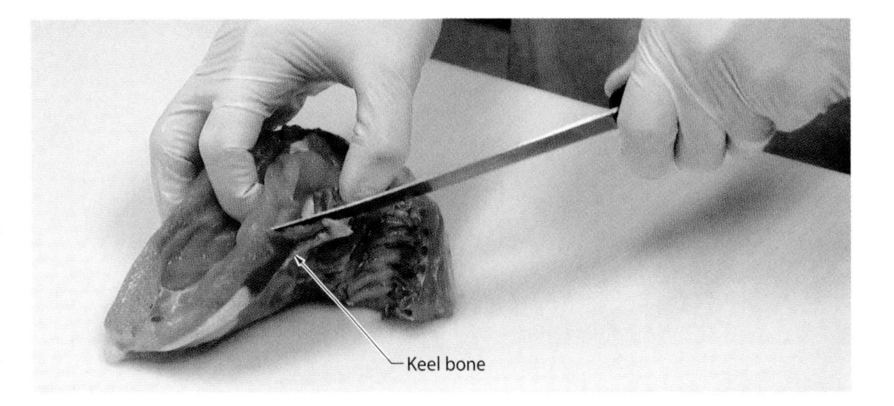

* This technique was executed by a left-handed chef.

2. Lay the breast bone-side down and place the side of the knife blade into the cut made behind the keel bone. Firmly press the keel bone against the cutting board with the side of the knife blade.

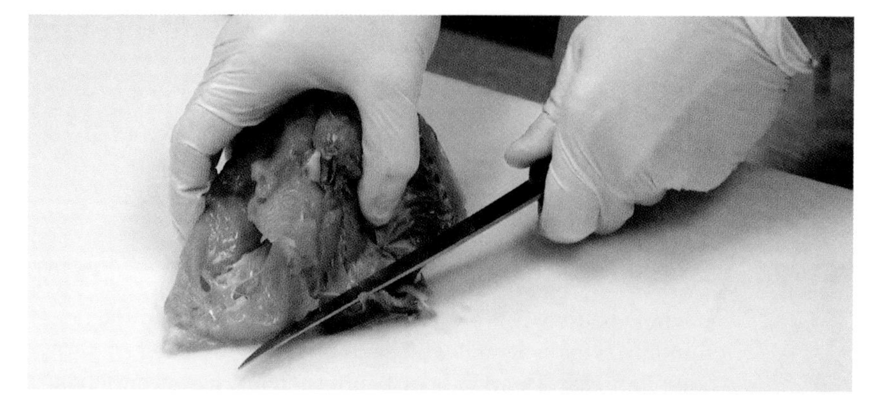

3. Keep the keel bone pressed against the cutting board and use the guiding hand to pull the breast back and away until the keel bone is removed from the breast.

4. Place the breast skin-side down and insert the knife blade behind the rib bones. Keeping the side of the blade against the rib bones, cut down at a slight angle to remove the rib bones from the breast.

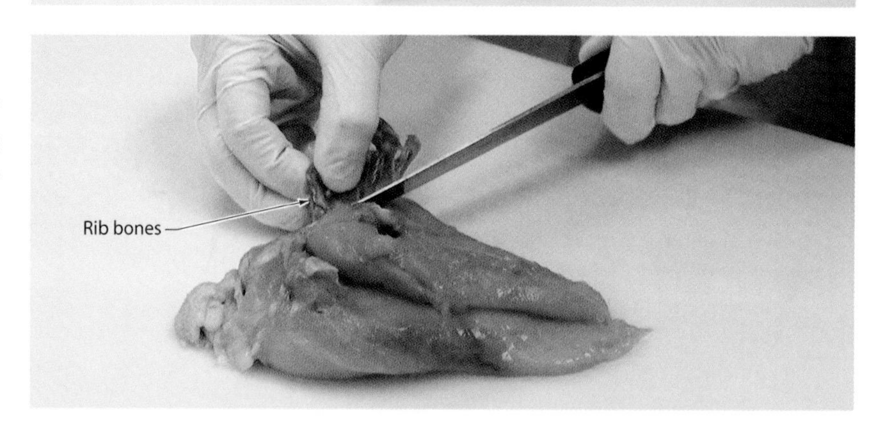

5. Pinch the end of the coracoid bone between the index finger and thumb of the guiding hand. Using the knife tip, cut along the bone until reaching the joint where the wing was attached.

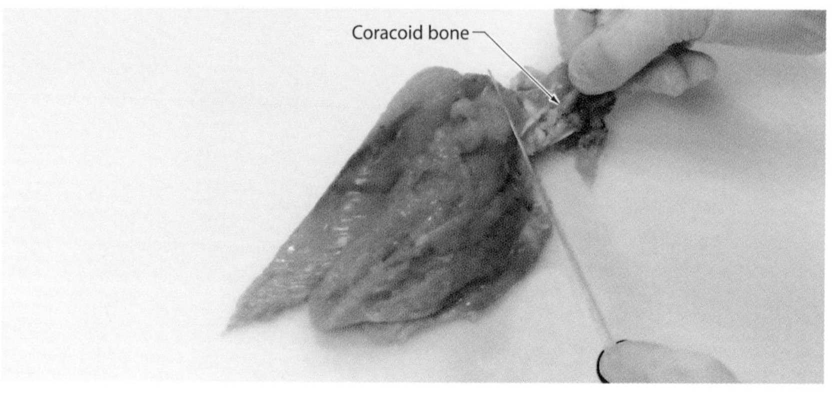

6. Scrape the flesh along the wishbone with the knife blade edge until it can be lifted and pulled away from the breast.

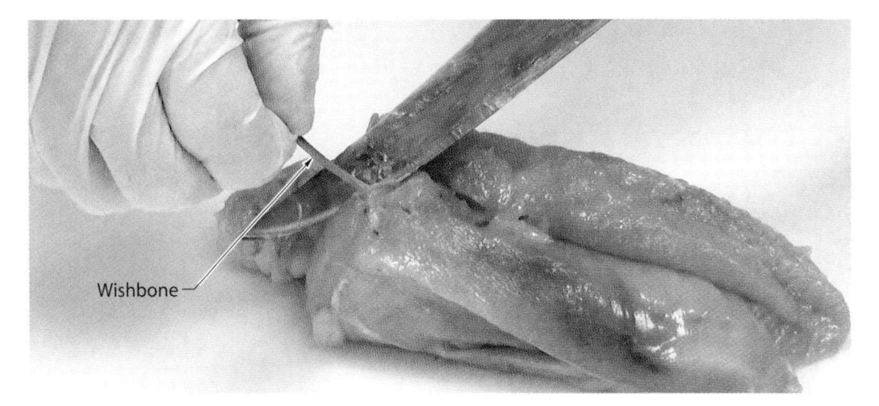

7. Hold the coracoid bone against the cutting board and cut around the joint at the end of the bone to remove it from the breast.

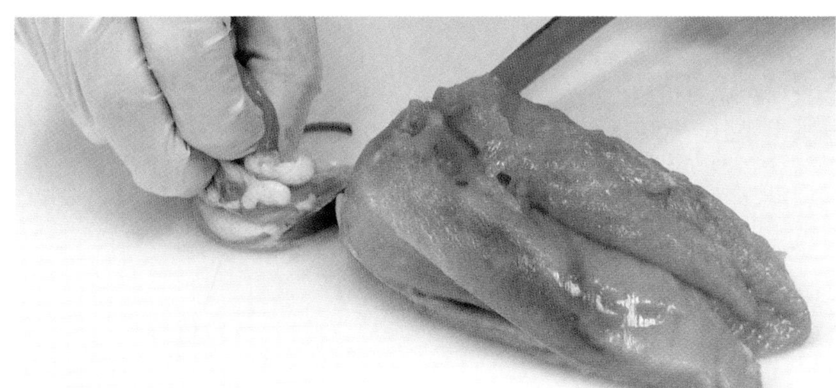

8. Turn the breast skin-side up and pull the skin away from the breast. Use the knife to trim any areas of skin or excess fat that remain attached to the breast.

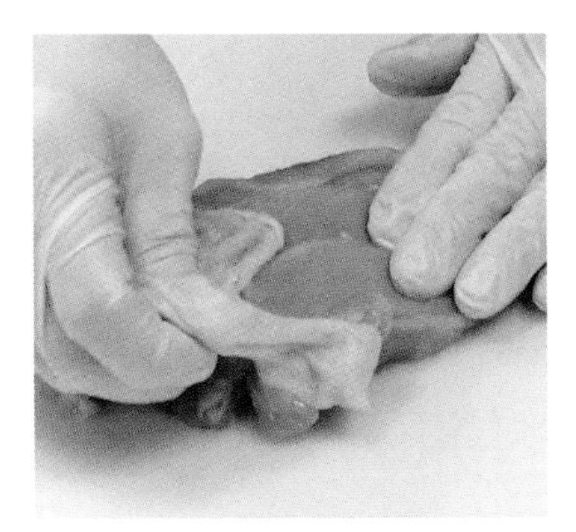

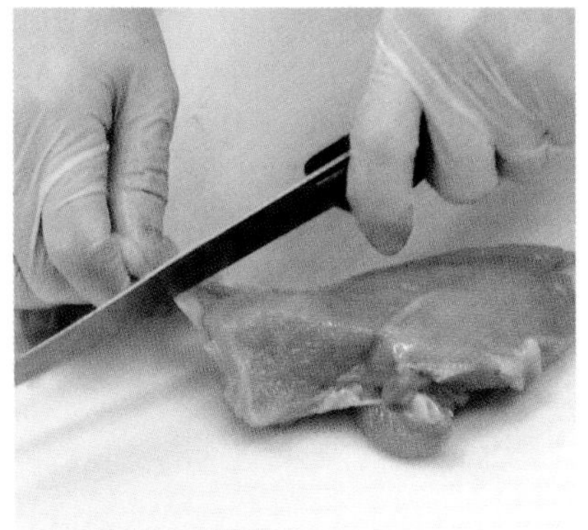

Finished deboned and skinned chicken breasts are free of bones and skin and are trimmed of any excess fat.

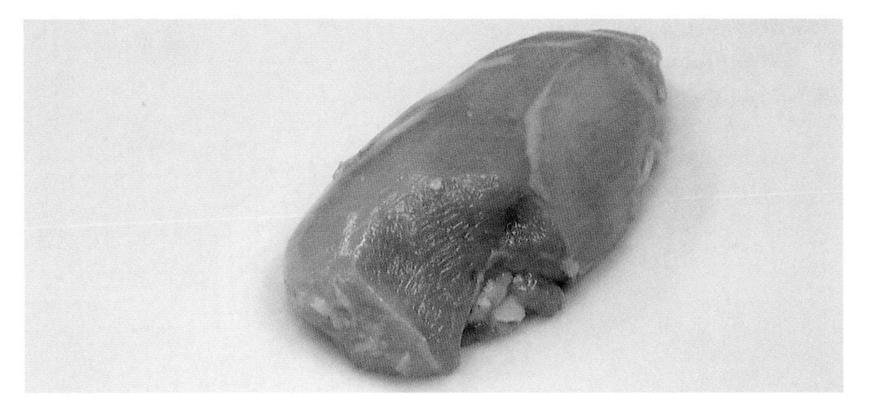

CUTTING AND TRIMMING CHICKEN TENDERLOINS

* This technique was executed by a left-handed chef.

1. Place a deboned chicken breast tenderloin-side up on the cutting board.

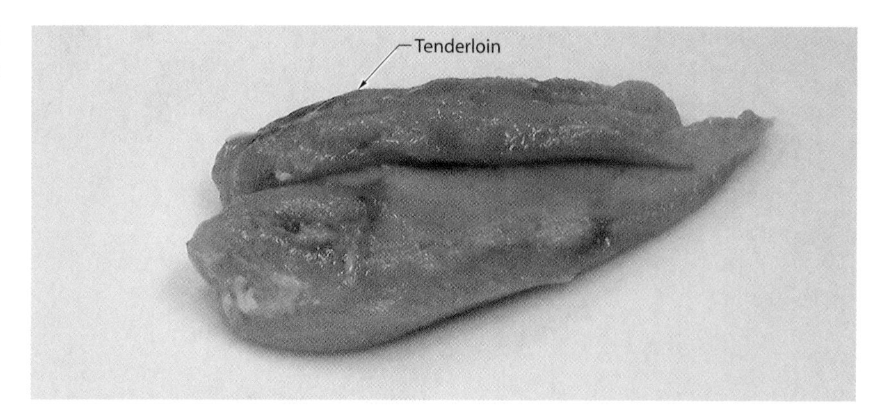

2. Lift the wider end of the tenderloin and pull it away from the breast. If the tenderloin does not release easily, use a boning knife to cut along the side of the tenderloin until it is removed or place the side of the knife blade on the breast for leverage and pull the tenderloin away.

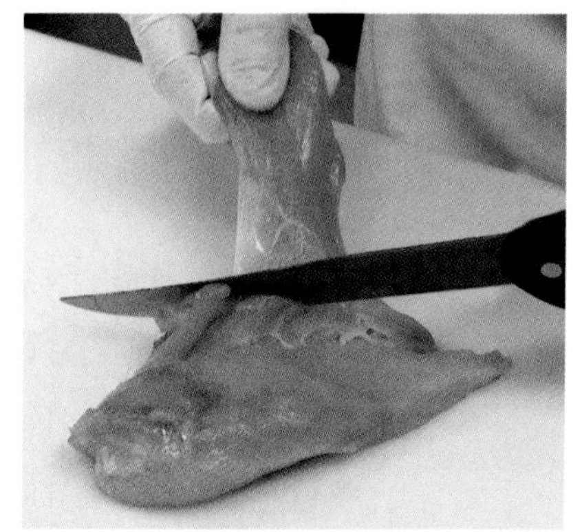

3. Locate the tough, white tendon that runs the length of the tenderloin. Secure the tenderloin with the guiding hand and scrape along the exposed end of the tendon with the knife tip.

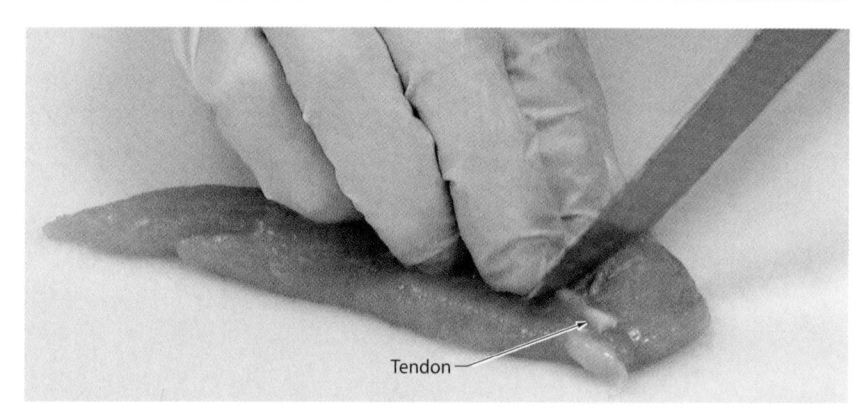

4. With the edge of the knife facing away from the guiding hand, insert the knife tip just beneath the exposed tendon. Cut toward the knife hand side to separate the end of the tendon from the flesh.

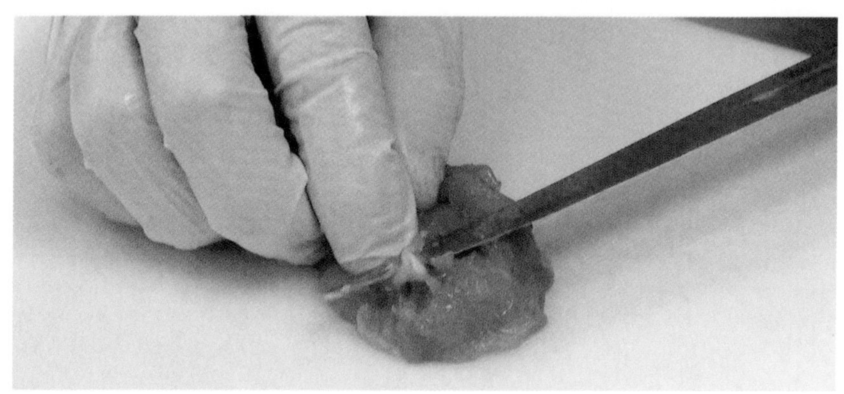

CUTTING AND TRIMMING CHICKEN TENDERLOINS (CONTINUED)

* This technique was executed by a left-handed chef.

5. Turn the tenderloin over and hold the tendon securely against the cutting board with the guiding hand. Use the edge and side of the knife blade to scrape and push the flesh back and away from the tendon until the tendon is removed.

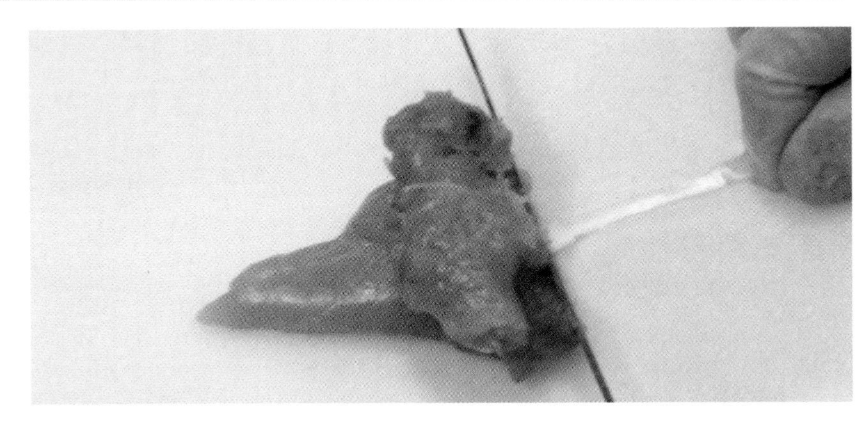

Finished chicken tenderloins are thin tapered strips of tendon-free flesh that have been removed from the breast.

QUICK TIP

Each 2-ounce chicken tenderloin has approximately 65 calories and 1 gram of fat. In comparison, a 6-ounce boneless chicken breast without skin has approximately 200 calories and 4 grams of fat, while a 6-ounce boneless chicken breast with skin has close to 300 calories and 16 grams of fat.

TECHNIQUE 31

CUTTING AIRLINE POULTRY BREASTS

VIDEO 31

An airline poultry breast is a skin-on, semiboneless breast with the first bone of the wing (drumette) still attached. The poultry breast is boneless, apart from the wing bone, and the tenderloin may be removed if desired. After the wing paddle and wing tip are removed, the clean drumette bone stands at an upright angle, adding a refined and elegant appearance when the cooked breast is served.

Note: Refer to Technique 28: Cutting Poultry into Halves, Quarters, and Eighths for Poultry Anatomy diagram.

Messermeister

Stiff Boning Knife

CUTTING AIRLINE CHICKEN BREASTS

* This technique was executed by a left-handed chef.

QUICK TIP *When preparing airline chicken breasts, separate the wing tips from the wing paddles and reserve both. The wing tips can be used for making stock, and the wing paddles are the ideal size for serving as appetizers or hors d'oeuvres.*

1. Place the breast and wing section from a fabricated whole chicken skin-side up and pull the wing with the guiding hand to elongate it. Position the blade of a stiff boning knife near the center of the drumette and cut down to the bone.

   Note: Refer to Technique 29: Fabricating Whole Poultry for the procedure on fabricating a chicken breast and wing.

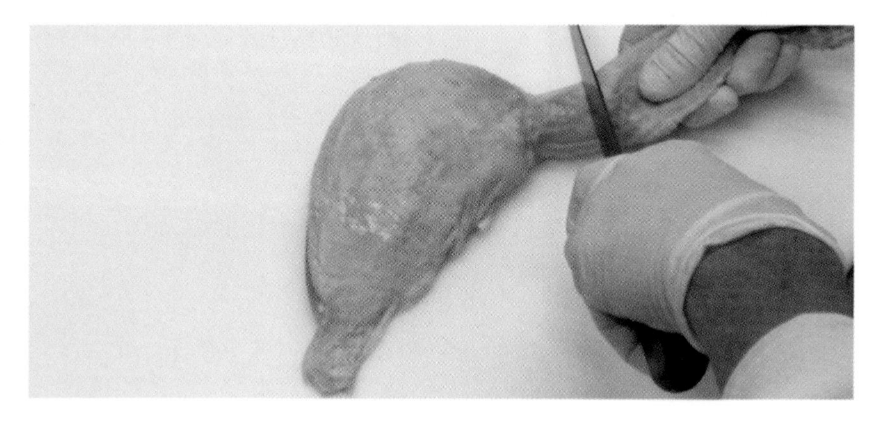

2. To expose the drumette bone, rotate the knife 360° around the bone, cutting through the skin and flesh that encircles it. Use the guiding hand to rotate the wing during this cut.

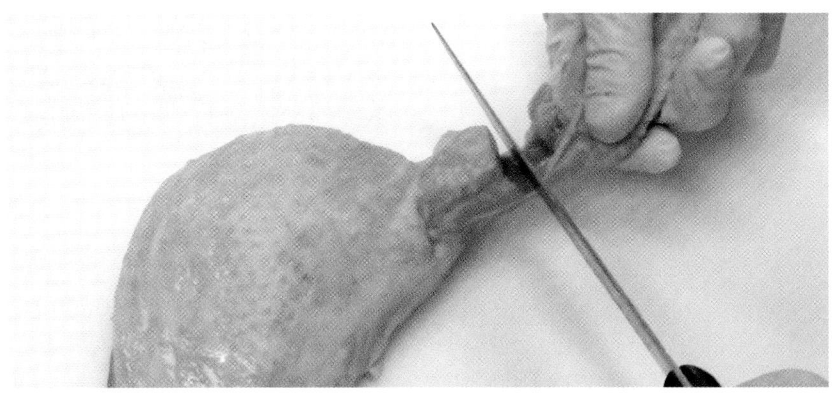

3. Grasp the wing paddle with one hand and the drumette with the other hand. To snap the joint and separate the wing paddle from the drumette, quickly and firmly bend the wing paddle backward in the opposite direction it is supposed to bend.

   Note: By quickly bending the wing paddle back to hyperextend it and not twisting it from the drumette, the cartilage from the wing joint stays with the paddle to leave the end of the drumette bone clean.

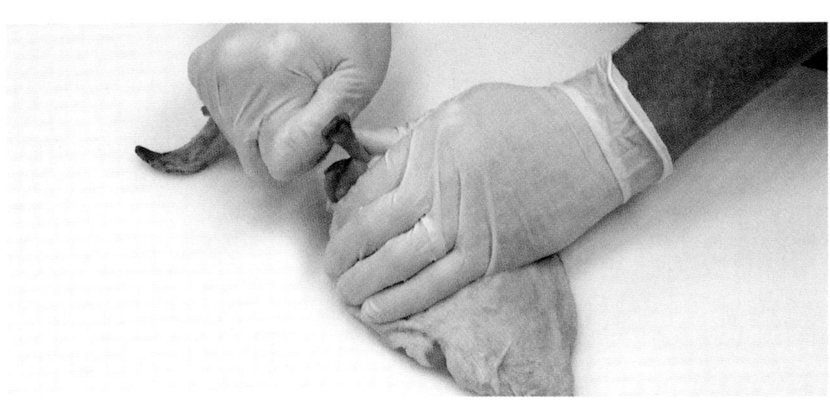

QUICK TIP

Reserve the wing paddle, cook as desired, and serve as an appetizer.

Finished airline chicken breasts consist of a clean drumette bone still attached to a skin-on boneless breast.

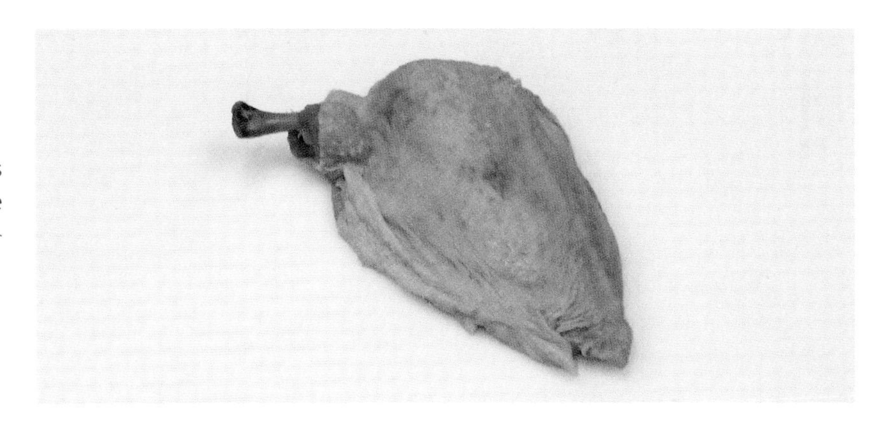

TECHNIQUE 32

DEBONING POULTRY THIGHS AND LEGS

VIDEO 32

The thigh bone is removed to produce a deboned poultry thigh while both the thigh and drumstick bones are removed to produce a deboned poultry leg. Deboning poultry thighs and legs is a relatively quick process and more cost-effective than purchasing these items already deboned.

Deboned poultry thighs and deboned poultry legs are also versatile. For example, they can be pounded flat and grilled, breaded and pan-fried, or prepared as a roulade. A roulade is a thin piece of meat or poultry that is stuffed (filled), rolled, and cooked. Once cooked, roulades are commonly sliced into elegant rounds. Deboned poultry thighs and deboned poultry legs can also be cut up for use in salads, kabobs, stir-fry dishes, and soups.

Stiff Boning Knife

Note: Refer to Technique 28: Cutting Poultry into Halves, Quarters, and Eighths for Poultry Anatomy diagram.

DEBONING CHICKEN THIGHS

* This technique was executed by a left-handed chef.

1. Place a leg (drumstick and thigh) from a fabricated whole chicken skin-side down on the cutting board with the thigh facing the knife hand side. Secure the drumstick with the guiding hand and cut next to the drumstick side of the fat line and down through the leg joint to separate the drumstick from the thigh.

   Note: Refer to Technique 29: Fabricating Whole Poultry for the procedure on fabricating a leg from a whole chicken.

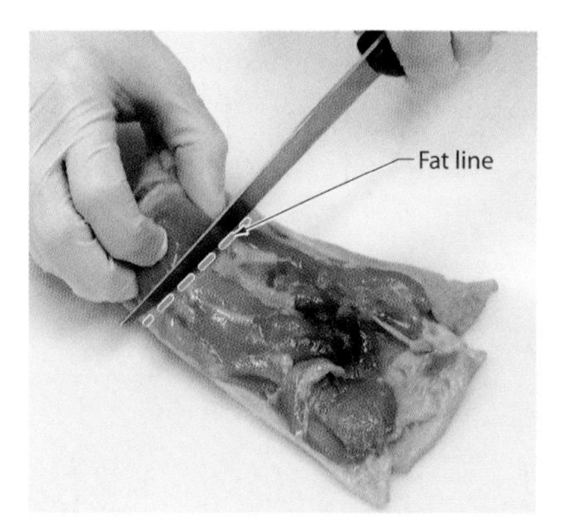

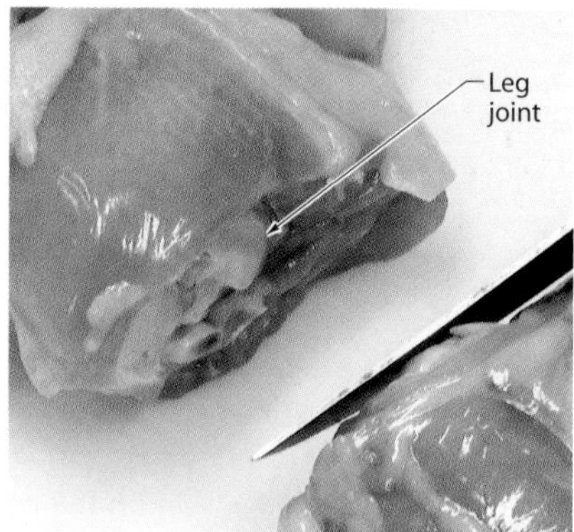

DEBONING CHICKEN THIGHS (CONTINUED)

* This technique was executed by a left-handed chef.

2. Secure the thigh with the guiding hand. Use the knife tip to make a shallow cut alongside the thigh bone.

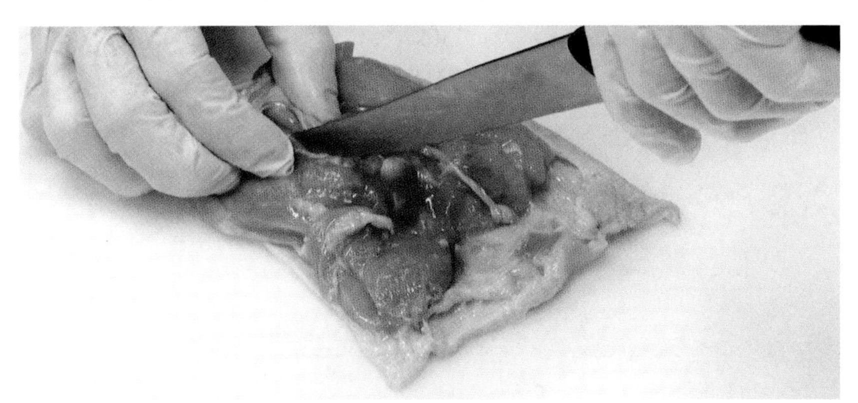

3. Scrape the thigh bone with the knife blade edge to remove flesh from the bone.

   Note: Scrape away from the guiding hand.

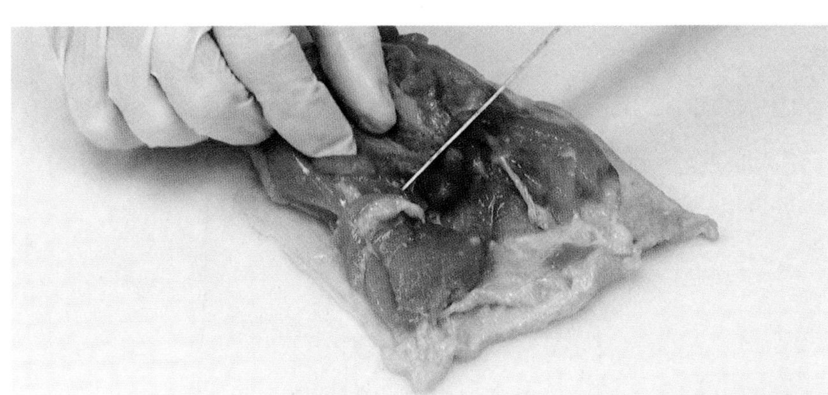

4. Pinch under the thigh bone with the index finger and thumb of the guiding hand to angle the end of the joint upward. Insert the knife tip under the joint with the edge facing the knife hand side and cut outward to separate the end of the bone from the flesh.

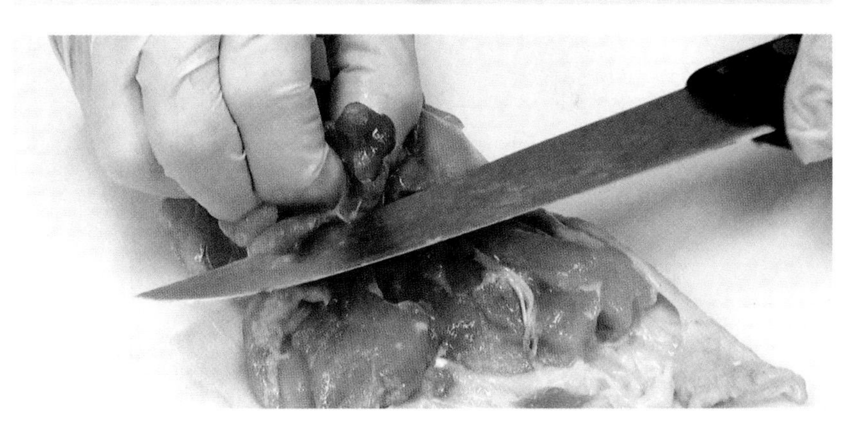

5. Hold the end of the thigh bone in the guiding hand and use the knife blade edge to scrape the flesh from the top of the bone down toward the leg joint (kneecap).

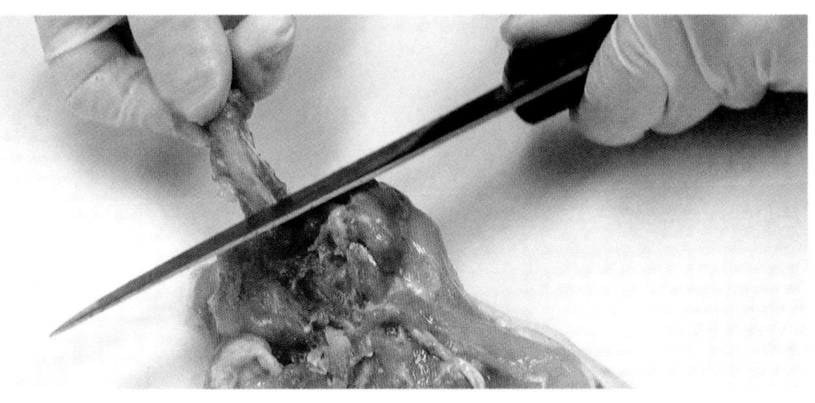

DEBONING CHICKEN THIGHS (CONTINUED)

* This technique was executed by a left-handed chef.

6. Cut around the leg joint and cartilage to separate the bone from the rest of the thigh.

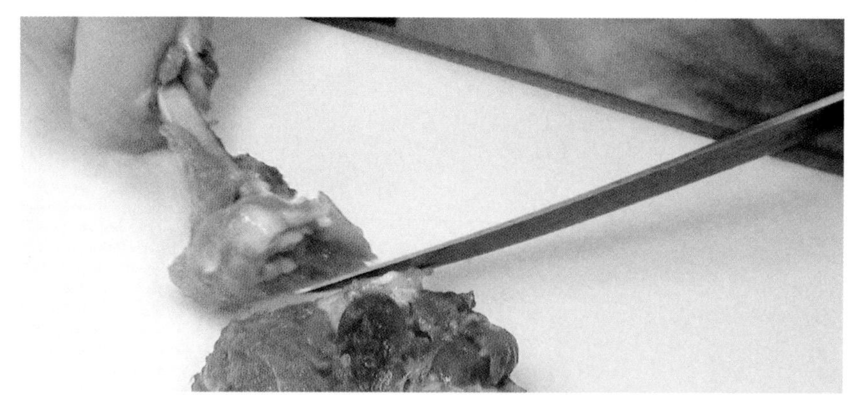

Finished deboned chicken thighs are the boneless flesh and skin of the thigh.

Note: The skin can be removed from the flesh for skinless, boneless thighs.

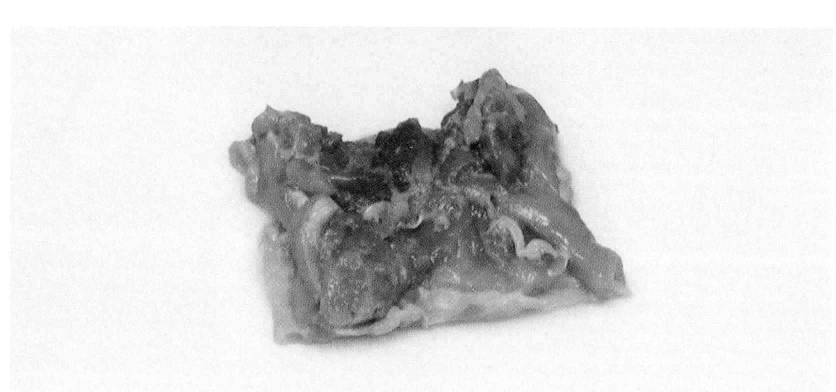

DEBONING CHICKEN LEGS

* This technique was executed by a left-handed chef.

1. Place a leg (drumstick and thigh) from a fabricated whole chicken skin-side down on the cutting board with the thigh facing the knife hand side. Secure the leg with the guiding hand and use the tip of a boning knife to make a shallow cut alongside the drumstick and thigh bones.

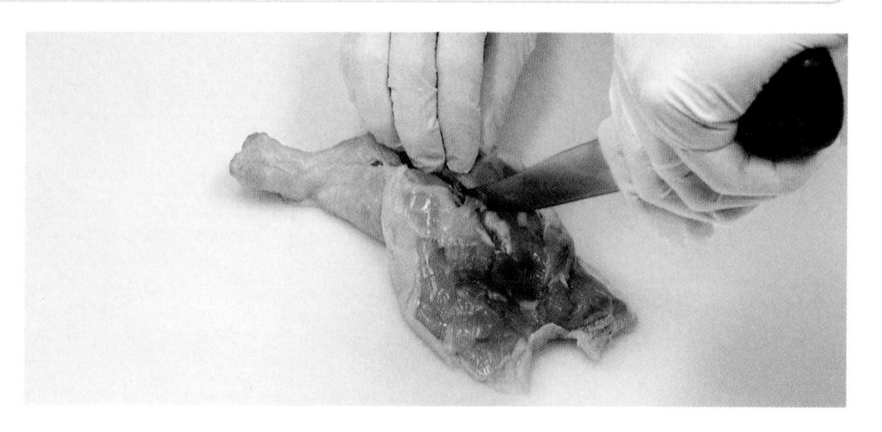

* This technique was executed by a left-handed chef.

2. At this point, leave the drumstick bone alone and refer to Deboning Chicken Thighs. Repeat the procedure outlined in Steps 3–5 on the thigh bone of the leg.

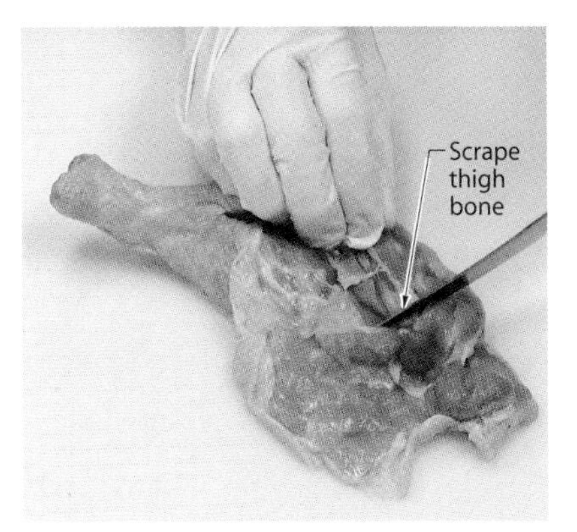

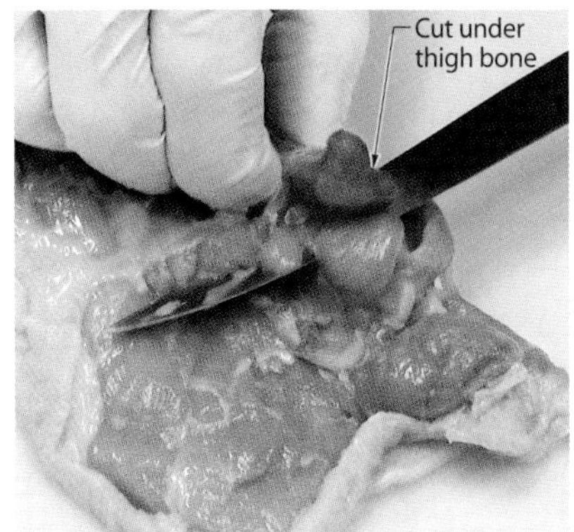

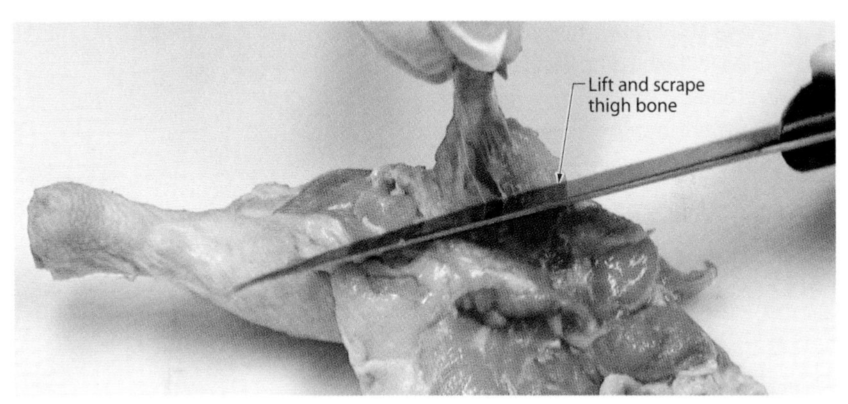

3. Rotate the leg so that the end of the drumstick faces the knife hand side. Secure the leg with the guiding hand and use the knife tip to slightly deepen the cut alongside the drumstick bone from Step 1.

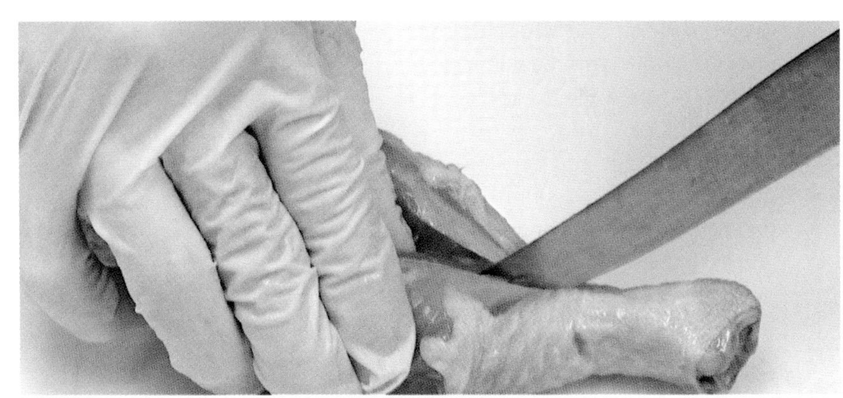

4. Pinch under the drumstick bone with the index finger and thumb to make a space for the knife.

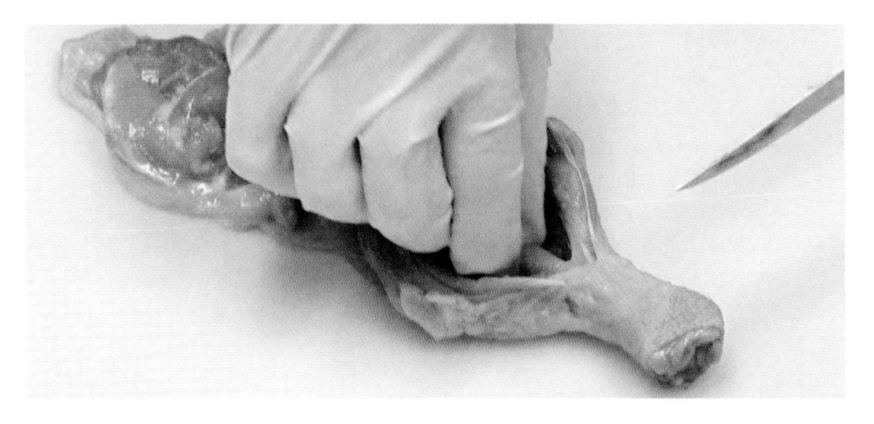

* This technique was executed by a left-handed chef.

5. Insert the knife tip under the drumstick bone with the edge facing the knife hand side and cut outward to separate the end of the bone from the flesh.

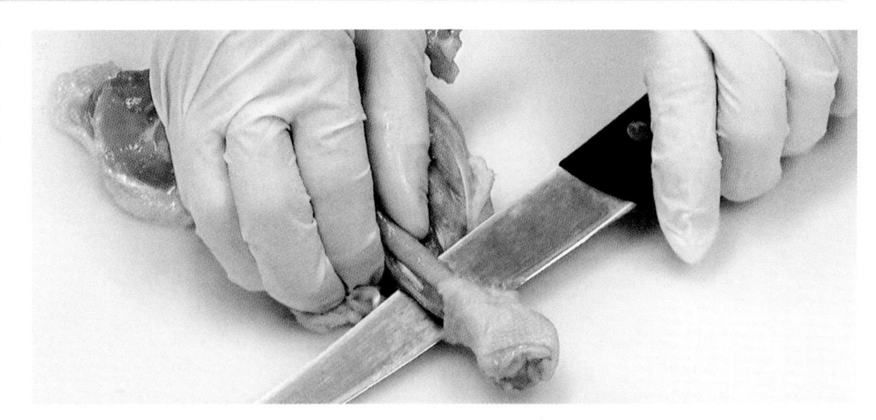

6. Grasp the end of the drumstick bone in the guiding hand and use the edge of the knife to scrape the flesh from the top of the bone down toward the leg joint (kneecap).

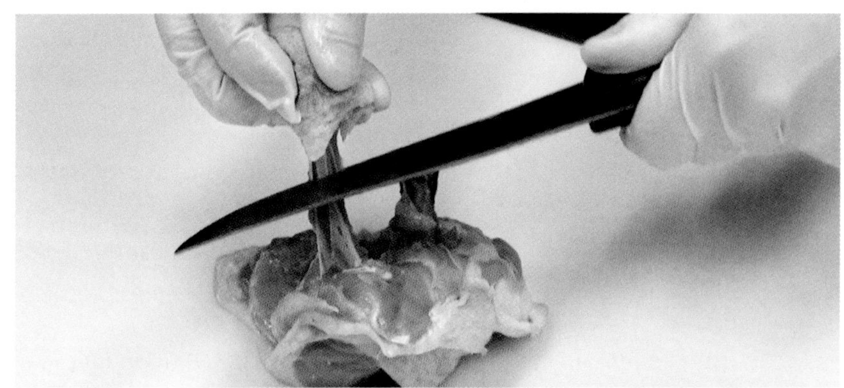

7. Grip the thigh and drumstick bones together with the guiding hand. Use the knife to scrape, push, and cut the flesh from around the joint and cartilage. Continue this process until the thigh and drumstick bones separate from the rest of the leg.

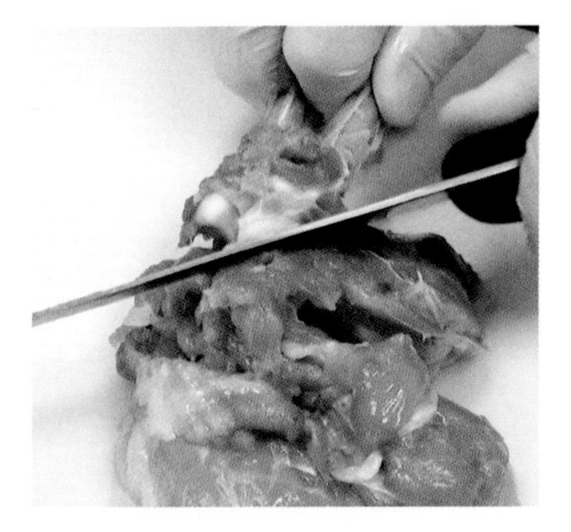

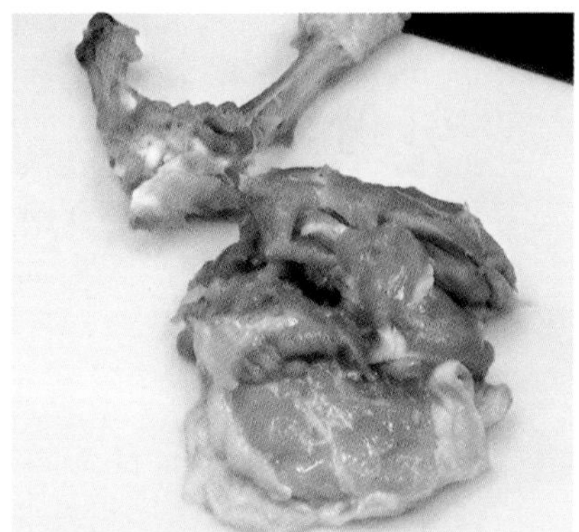

Finished deboned chicken legs are the boneless flesh and skin of the thigh and drumstick. They are commonly pounded to an even thickness and then rolled (often with a filling) prior to cooking.

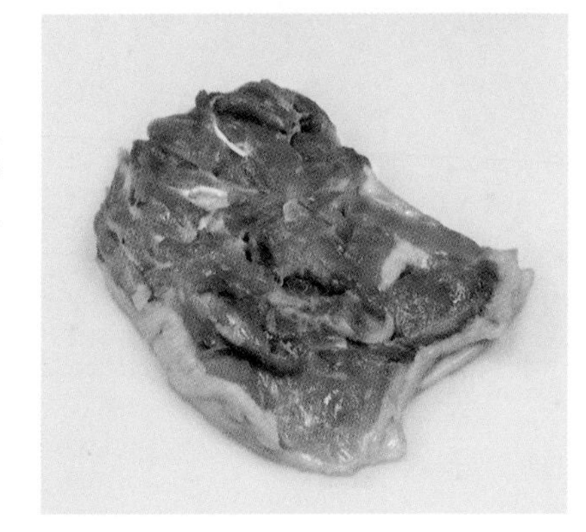

VIDEO 33

DEBONING WHOLE POULTRY

Stiff Boning Knife

Whole poultry is typically deboned in order to be stuffed (filled), cooked, and sliced without having to carve around bones. The boneless slices can be cut uniformly to enhance presentation and keep serving sizes consistent. In addition to removing all bones, it is equally important to keep the poultry flesh and skin in one piece. Keeping the flesh and skin intact allows the poultry to be neatly wrapped around a filling or seasoned, rolled, and trussed (tied with butcher's twine) before being cooked.

Although deboning whole poultry may take some practice, the results are impressive in terms of presentation and flavor development. Deboned whole poultry tends to cook evenly, yielding a tender, juicy product, and there are a wide variety of flavorful ingredients that can be used for fillings and seasonings.

Note: Refer to Technique 28: Cutting Poultry into Halves, Quarters, and Eighths for Poultry Anatomy diagram.

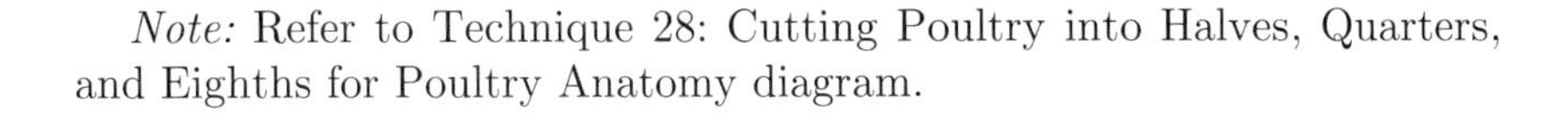

DEBONING WHOLE CHICKENS

* This technique was executed by a left-handed chef.

1. Start with a whole chicken that has had the wishbone and tail removed. Position the carcass back-side up on the cutting board with the tail end facing the body. Use the tip of a stiff boning knife to make a shallow cut (score mark) along the center of the backbone from the neck end to the tail end.

   Note: Refer to Technique 29: Fabricating Whole Poultry and follow Steps 1–3 from Fabricating Whole Chickens for the procedure on how to remove the wishbone and tail.

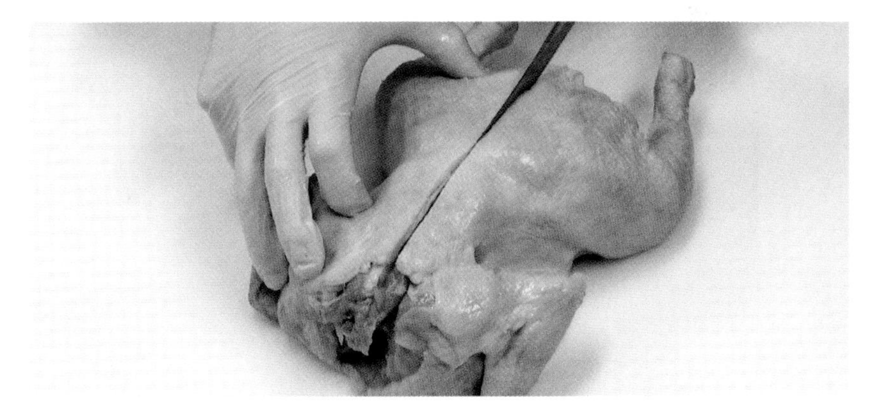

QUICK TIP

Poultry will shrink slightly when cooked. Therefore, to keep the stuffing intact when deboning and rolling whole poultry, do not overstuff the bird. When properly stuffed and rolled, the juices of the poultry seep into the stuffing as it cooks, adding an extra layer of flavor to the overall dish.

DEBONING WHOLE CHICKENS (CONTINUED)

* This technique was executed by a left-handed chef.

2. Lay the carcass on its side with the back facing the knife hand side and secure with the guiding hand. Position the knife tip at the wing (shoulder) joint and cut down at a slight angle until the joint is exposed.

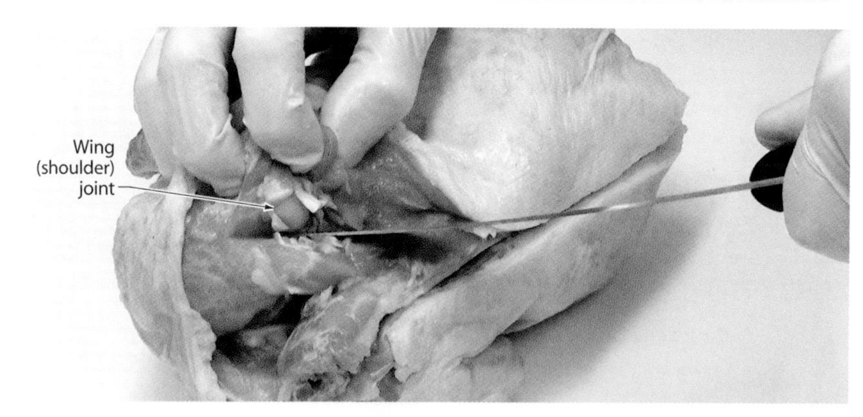

3. Place the side of the knife tip under the wing joint and pull the knife back to cut along the scapula (shoulder blade). Continue this cut until reaching the oyster located at the end of the scapula.

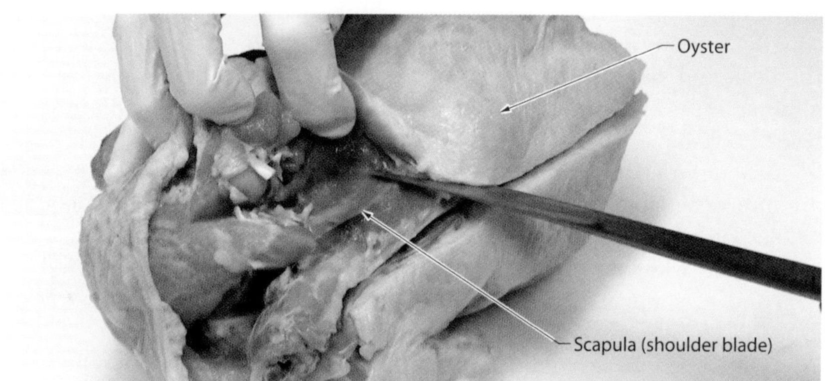

4. Use the knife tip to make a slight cut under the oyster and then use the fingers to scrape the oyster away from its socket.

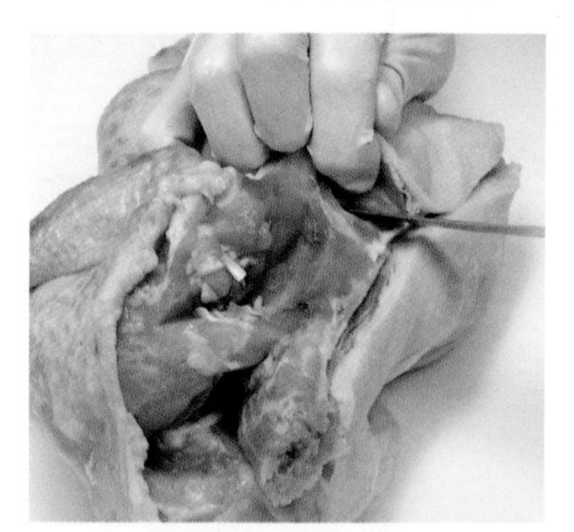

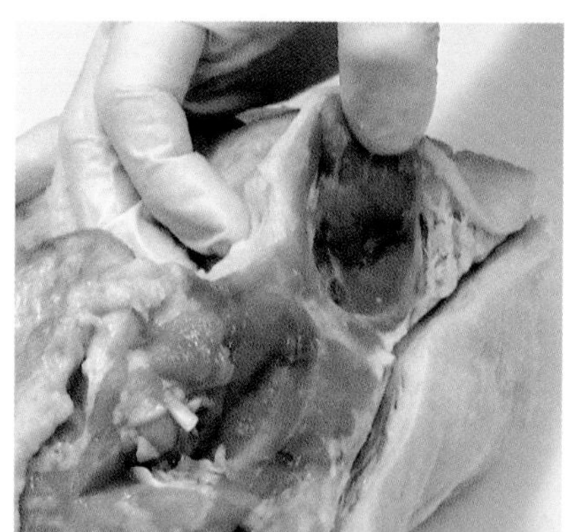

5. Start at the oyster and continue cutting alongside the bones until the thigh (hip) joint is exposed. To separate the ball joint from the socket, turn the carcass back-side up, grasp the leg, and bend it toward the backbone.

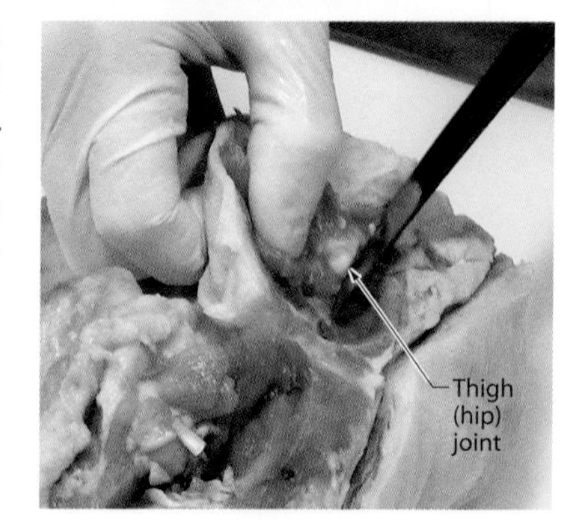

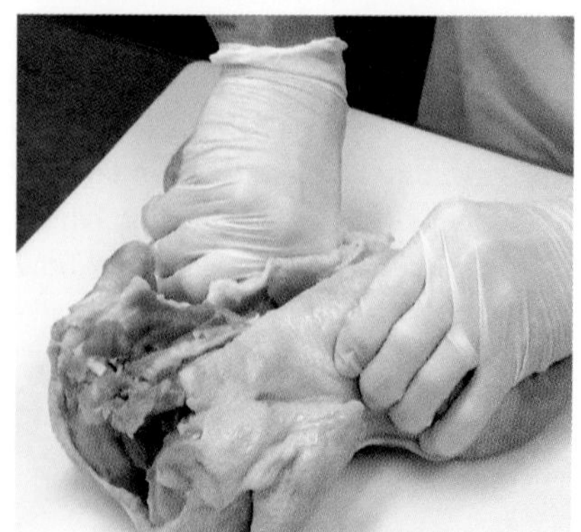

* This technique was executed by a left-handed chef.

6. Secure the carcass with the guiding hand and cut through the thigh (hip) joint. Continue to trim flesh from the carcass by pulling the knife blade along the pelvis area toward the tail end of the carcass.

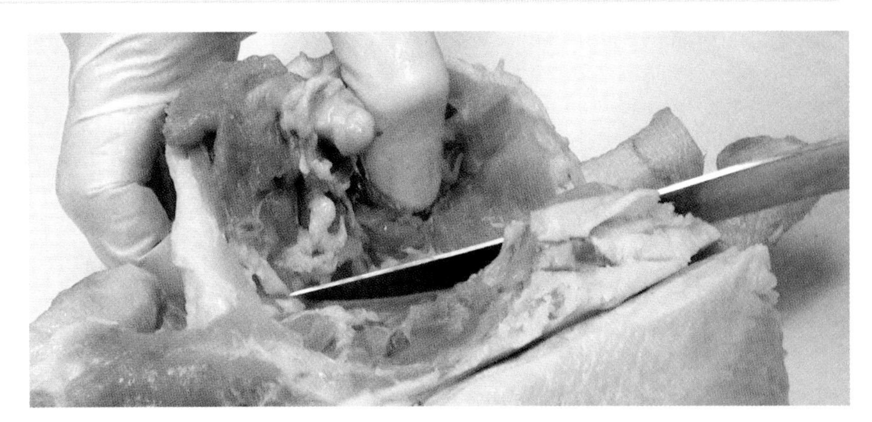

7. Reposition the knife behind the wing joint. Pull the knife back, making shallow cuts along the fat line of the breast.

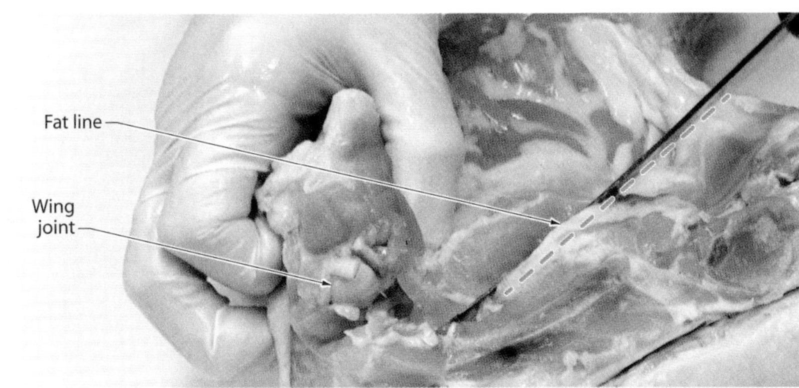

8. Continue making shallow cuts along the fat line, working the knife blade down the rib cage in smooth, even strokes until reaching the keel bone. Repeat Steps 2–8 to remove flesh from the other side of the carcass.

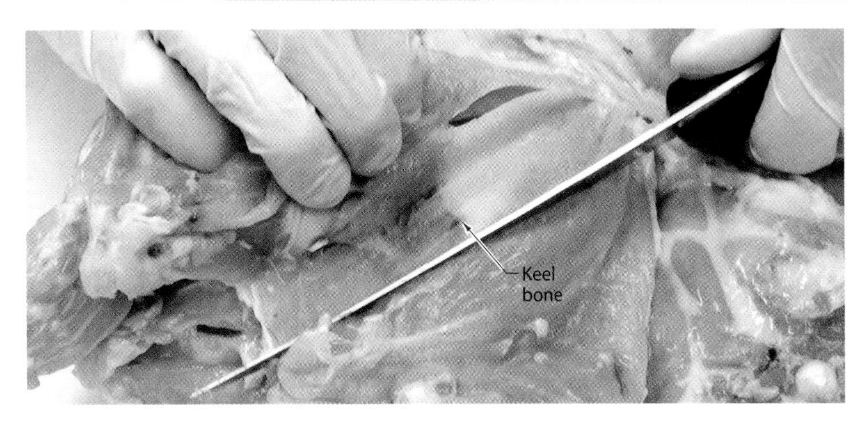

9. To remove the keel bone, keep the breasts pressed against the cutting board with one hand. Grasp around the rib cage with the other hand and pull back toward the tail until the keel bone separates from the breasts.

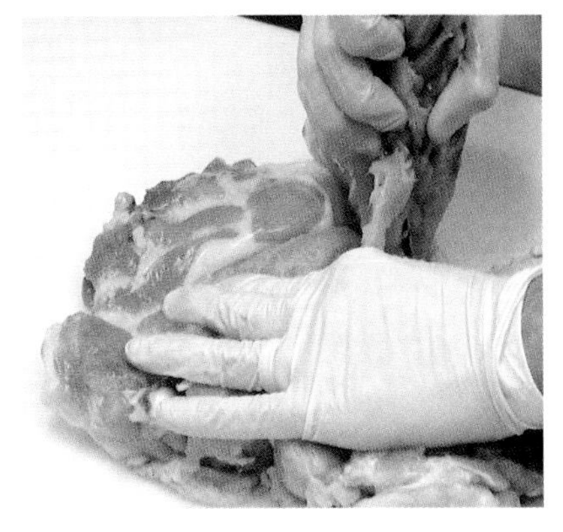

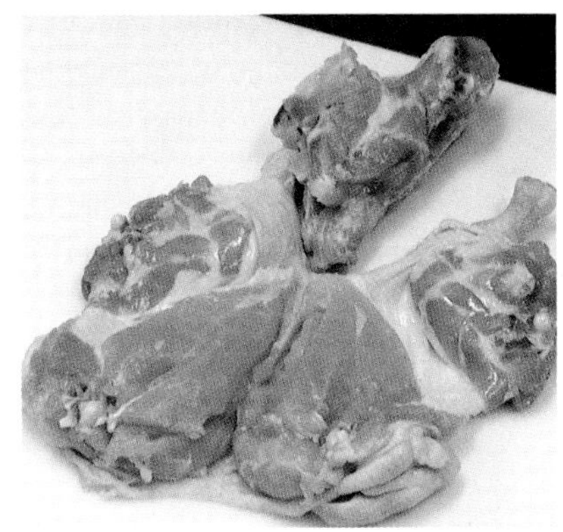

* This technique was executed by a left-handed chef.

10. To start removing the leg bones, use the knife tip to make a shallow cut alongside the thigh and drumstick bones.

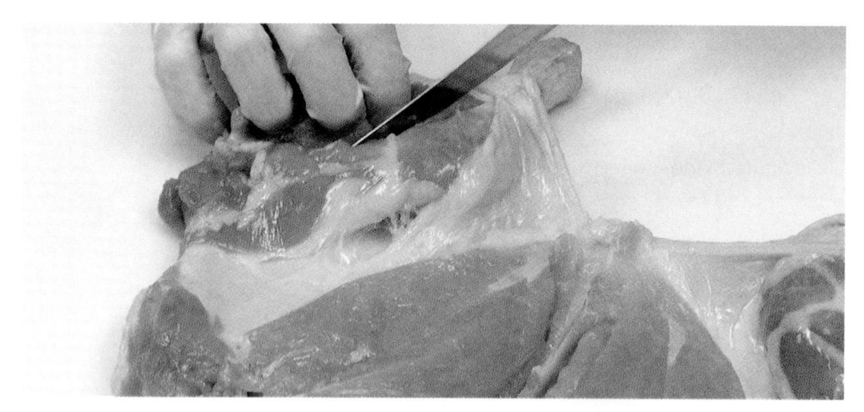

11. Pinch under the drumstick bone with the index finger and thumb. Insert the knife tip under the bone with the edge facing the knife hand side and cut outward to separate the end of the bone from the flesh.

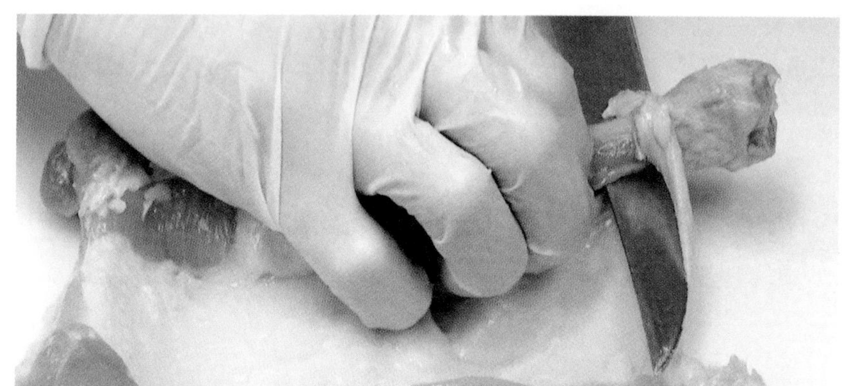

12. Grasp the end of the drumstick bone in the guiding hand and use the edge of the knife to scrape the flesh from the top of the bone down toward the leg joint (kneecap).

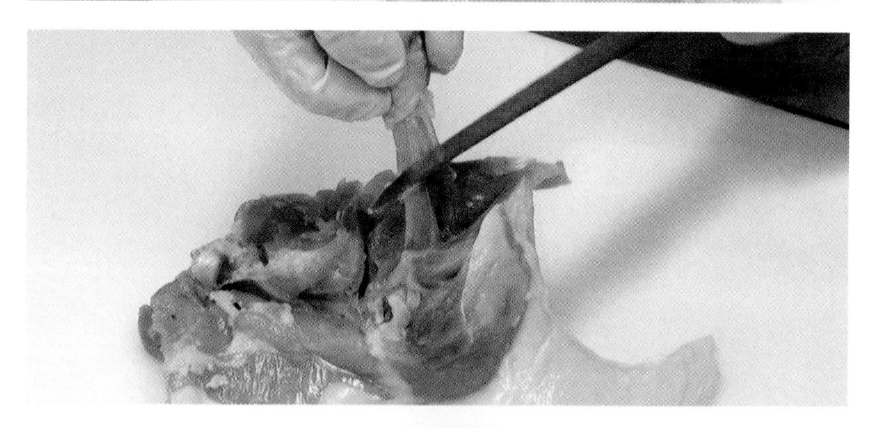

13. Pinch the thigh bone with the index finger and thumb of the guiding hand to angle the end of the joint upward. Insert the knife tip under the joint with the edge facing the knife hand side and cut outward to separate the end of the bone from the flesh.

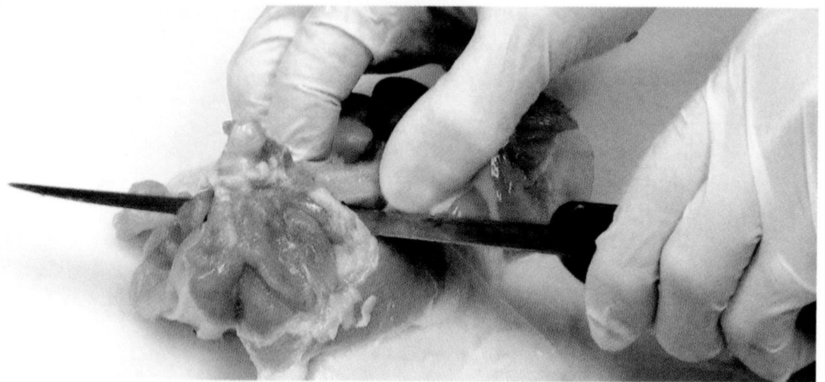

DEBONING WHOLE CHICKENS (CONTINUED)

* This technique was executed by a left-handed chef.

14. Grasp the end of the thigh bone in the guiding hand and use the knife blade edge to scrape the flesh from the top of the bone down toward the leg joint.

15. Grip the thigh and drumstick bones together with the guiding hand. Use the knife to scrape, push, and cut the flesh from around the joint and cartilage. Continue this process until the thigh and drumstick bones separate from the rest of the leg. Repeat Steps 10–15 to debone the other leg.

16. Hold a wing with the guiding hand. Cut under and around the wing joint to separate the wing from the carcass. Repeat this step to remove the other wing.

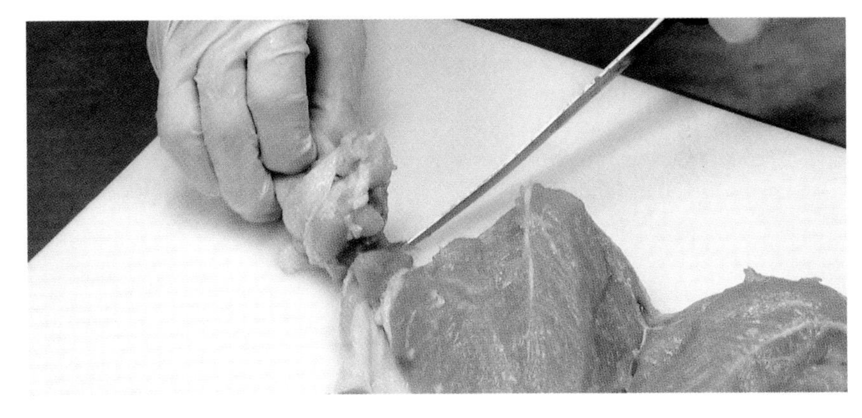

Deboned whole chickens include the boneless flesh and skin from the breasts and legs.

TRIMMING AND ROLLING DEBONED WHOLE CHICKENS

* This technique was executed by a left-handed chef.

1. Place a deboned whole chicken skin-side down across the cutting board with the legs positioned alongside the breasts. Locate the tenderloins.

   Note: There will be an area of skin between the breasts and the legs.

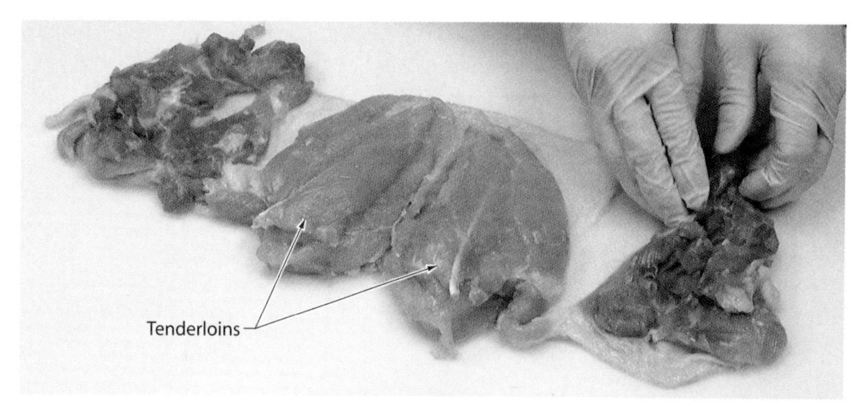

2. Lift the wider end of the tenderloin and pull it away from the breast. If the tenderloin does not release easily, use a boning knife to cut along the side of the tenderloin until it is removed or place the side of the knife blade on the breast for leverage and pull the tenderloin away.

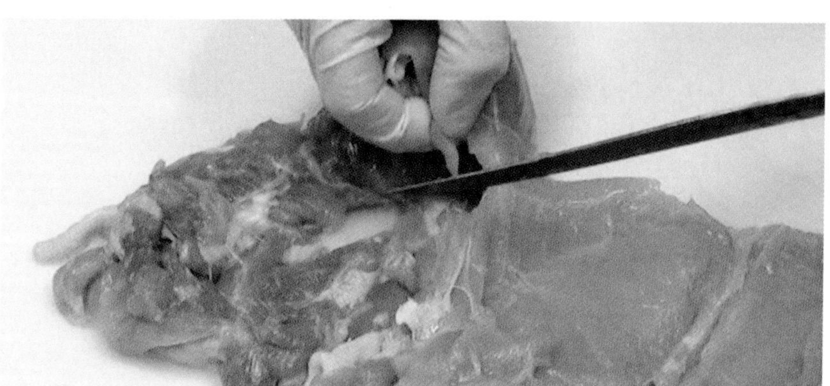

3. Remove the tendon from the tenderloin and then place the tenderloin in the area of skin between the breast and the leg. Repeat Steps 2–3 to remove and place the other tenderloin.

   Note: Refer to Technique 30: Deboning Poultry Breasts and follow Steps 3–5 from Cutting and Trimming Chicken Tenderloins for the procedure on how to remove the tendon.

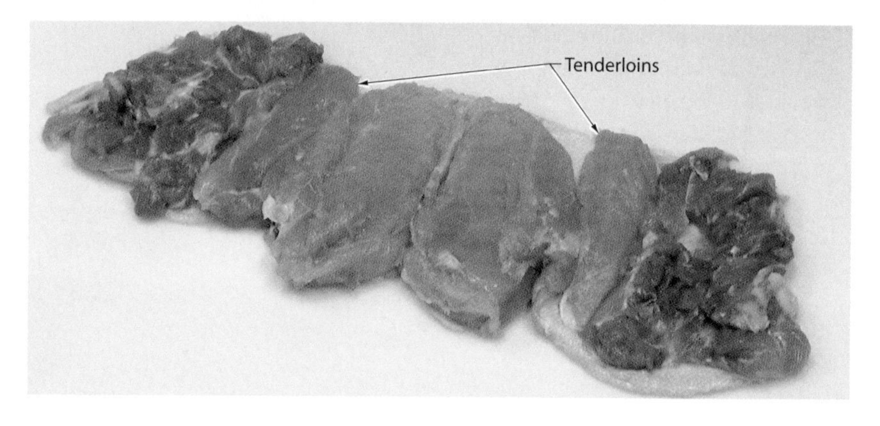

QUICK TIP

To help ensure even cooking, cover deboned poultry with a sheet of plastic wrap and lightly pound to a consistent thickness with a meat mallet. The poultry can also be seasoned, filled, and rolled at this point.

TRIMMING AND ROLLING DEBONED WHOLE CHICKENS (CONTINUED)

* This technique was executed by a left-handed chef.

4. Starting at one end and working toward the other end, roll and fold the flesh away from the body while tucking the skin under the flesh. If necessary, trim excess skin along the ends of the roll.

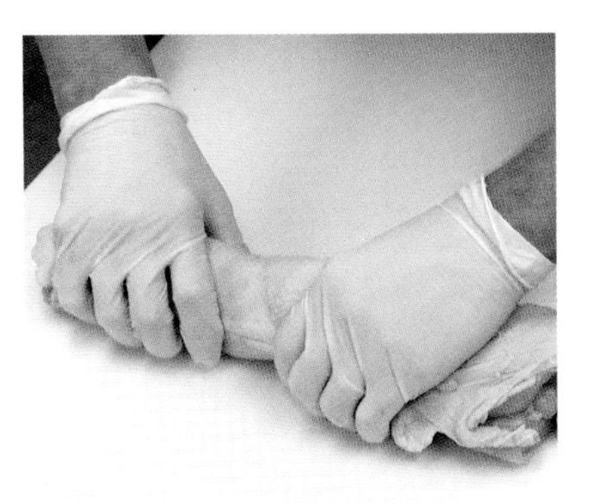

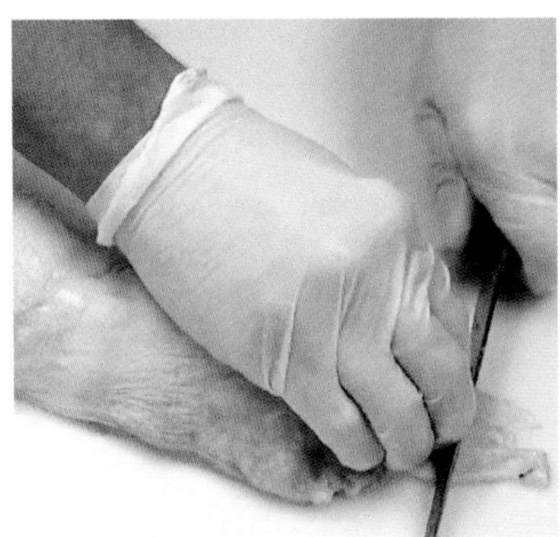

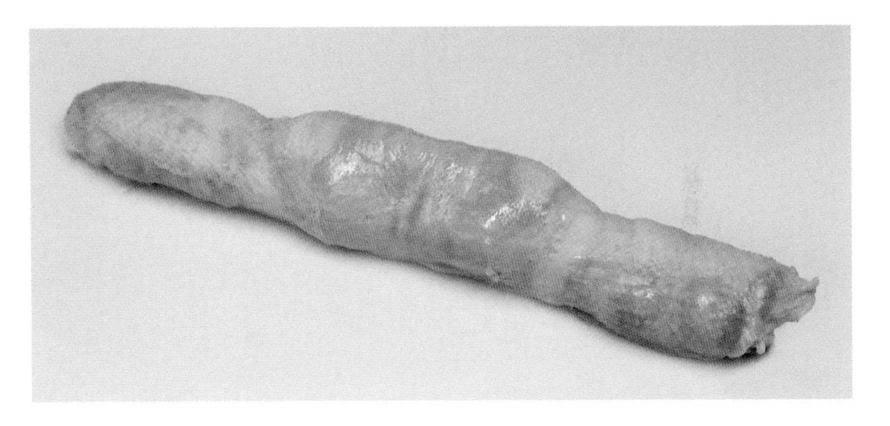

Whole deboned chickens are typically rolled into a tubular shape and trussed (tied with butcher's twine) to stay intact while cooking.

SECTION V
CUTTING AND TRIMMING SEAFOOD

- CUTTING SHRIMP
- CUTTING LOBSTERS
- CUTTING CEPHALOPODS
- CUTTING ROUNDFISH
- CUTTING FLATFISH
- SHUCKING BIVALVES

LEARNER RESOURCES
ATPeResources.com/QuickLinks™
Access Code: 583241

CUTTING SHRIMP

Paring Knife

Shrimp purchased in the shell with the head removed are typically prepared by peeling and deveining the shrimp. Deveined shrimp have had the intestinal tract removed. The intestinal tract, or sand vein, runs lengthwise down the back of the shrimp and is removed to enhance presentation. In larger shrimp, it often contains grit or has an unappealing flavor.

Peeled and deveined shrimp are frequently butterflied before cooking. Butterflying is a knife technique in which a food item is cut almost completely through the center, resulting in two halves of flesh that can be spread apart to lay flat. Butterflied items create a larger surface area for more even cooking, increased flavor development, and enhanced presentations.

PEELING AND DEVEINING SHRIMP

1. Hold a shrimp in the guiding hand with the head end facing away from the body and the underside facing the dominant hand.

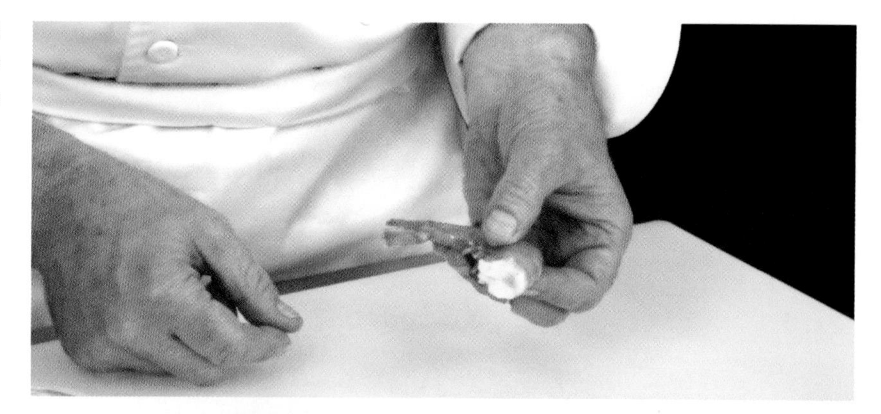

2. To peel the shrimp leaving the tail intact, start on the underside of the shrimp and grasp the edge of the shell between the thumb and index finger of the dominant hand. Pull the shell down and around the sides of the shrimp until a segment of the shell is removed. Repeat this step, leaving the shell segment nearest the tail intact.

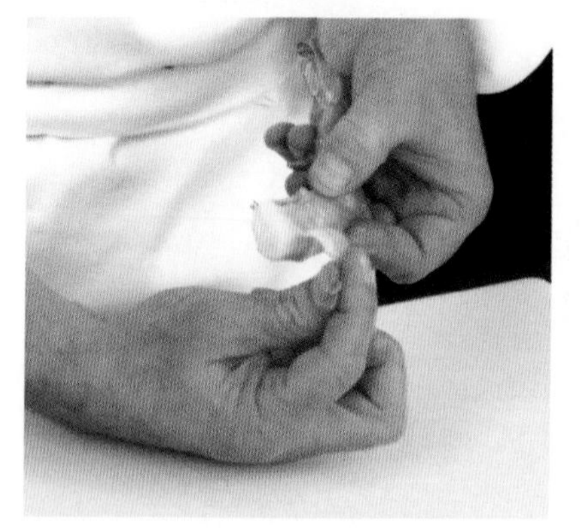

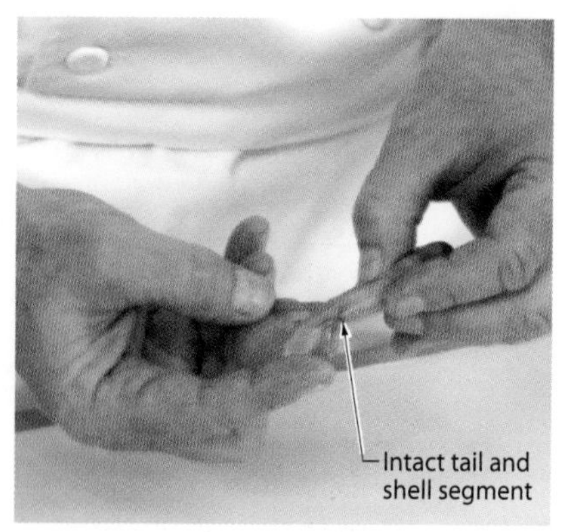

Note: If the tail is to be removed, pinch the shell segment that is closest to the tail between the thumb and index finger and pull away from the shrimp.

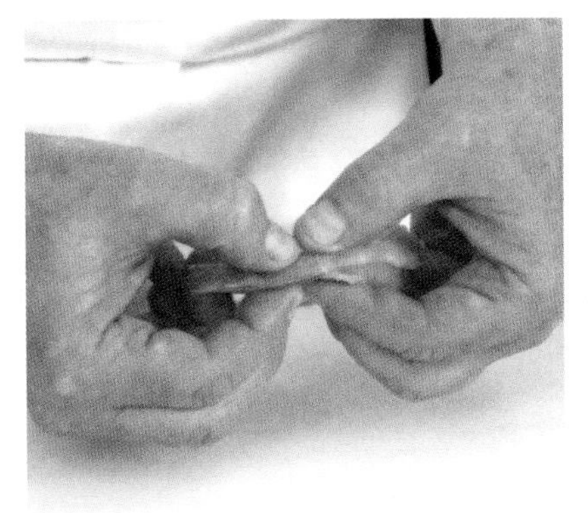

3. To start the deveining process, hold the shrimp in the guiding hand with the head end facing away from the body and the rounded back facing the knife hand side. Use a paring knife to make a shallow cut down the center of the shrimp's back from the head end to the tail that is just deep enough to reveal the intestinal tract.

4. To further reveal the intestinal tract, use the thumbs to carefully pry open the cut down the back of the shrimp. Insert the knife tip under the intestinal tract and pull it away from the shrimp.

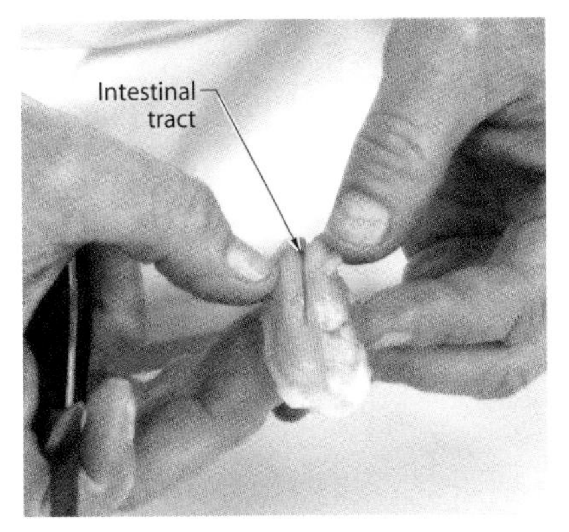

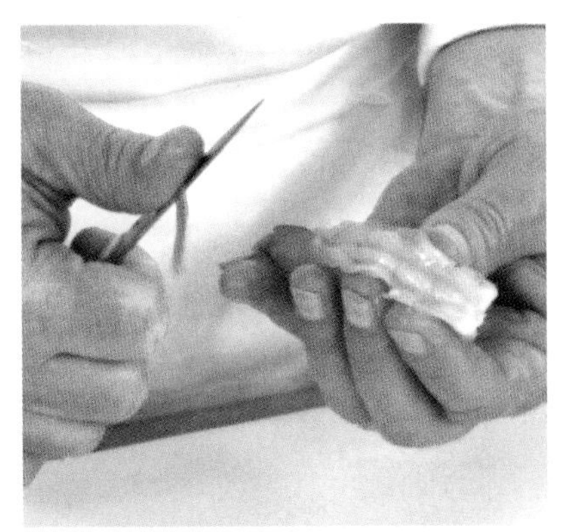

Finished peeled and deveined shrimp have had the shell and intestinal tract (sand vein) removed and contain a shallow lengthwise cut along the center of the back.

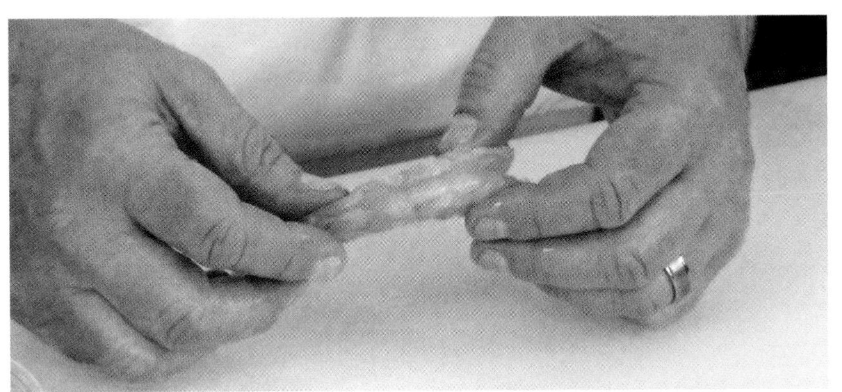

QUICK TIP

When doing large-scale production and efficiency is a must, place the shrimp on the cutting board to make the shallow cut down the center of the shrimp's back that reveals the intestinal tract.

BUTTERFLYING SHRIMP

1. Place a peeled and deveined shrimp on the cutting board with the head end facing away from the body and the rounded back facing the knife hand side.

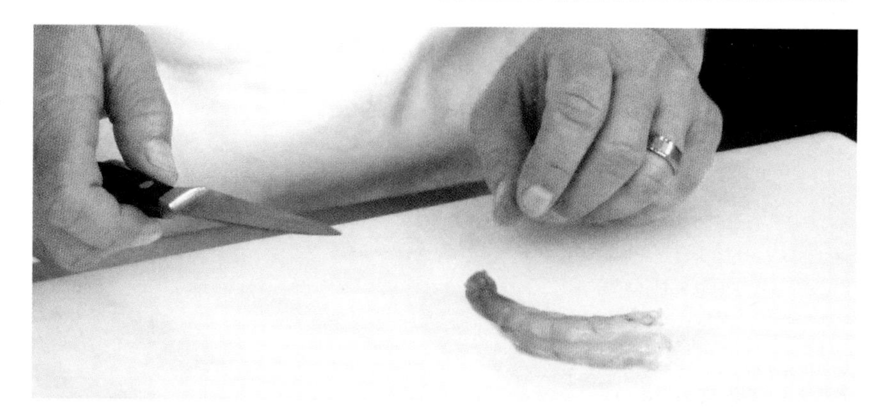

2. With the guiding hand held open and flat, place the fingertips on the shrimp to secure it in place. Starting at the head end and moving to the tail, use a paring knife to deepen the cut that was made to devein the shrimp. Repeat this cut as necessary to slice almost through to the underside of the shrimp.

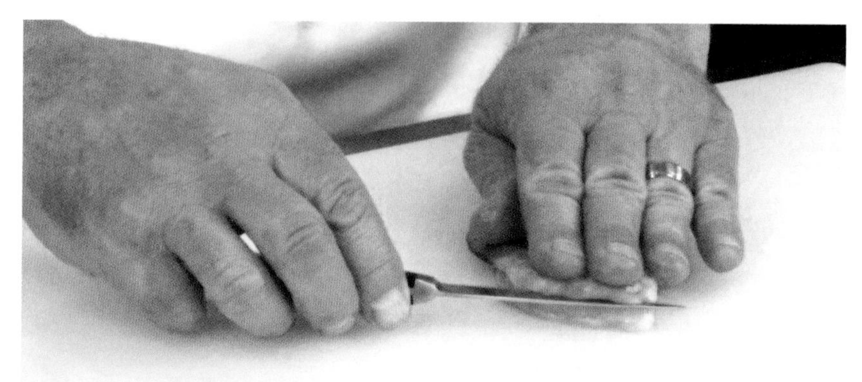

> **QUICK TIP**
>
> *Frozen shrimp is commonly packaged by count. For example, shrimp packed and labeled 21/25 count indicates that there is an average of 21–25 shrimp per 1 pound package. As a general rule, 1 pound of shell-on shrimp will yield between ½–¾ pound when peeled.*

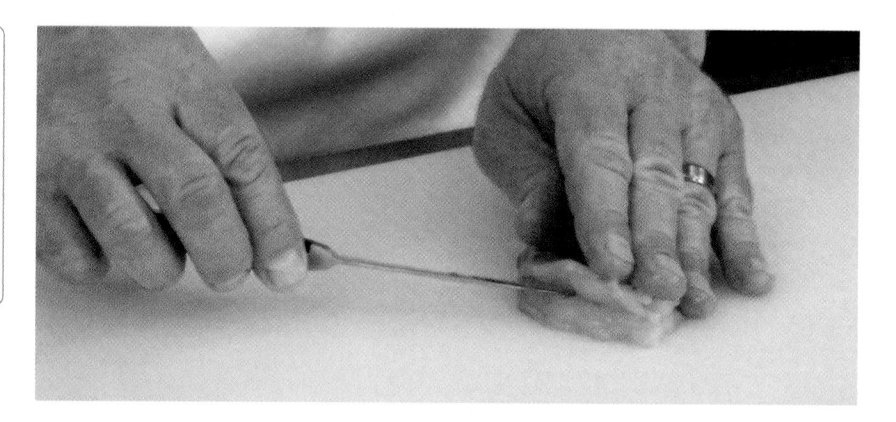

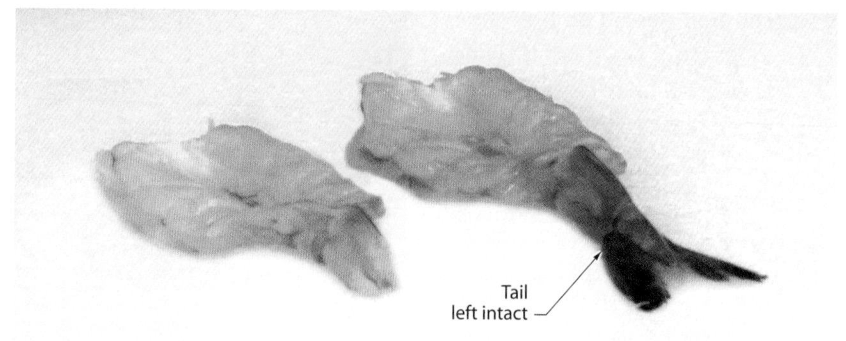

Finished butterflied shrimp are symmetrical halves of shrimp that lay flat on a surface with the inner flesh exposed.

VIDEO 35

CUTTING LOBSTERS

TECHNIQUE 35

Cutting lobster tails and splitting whole lobsters in half are two popular ways to prepare lobsters prior to cooking. Cut lobster tails create a refined and upscale presentation. With this technique, the shell of the lobster tail is cut lengthwise down its center and the lobster flesh is elevated by placing it on top of the shell.

While lobster tails are typically purchased frozen, whole lobsters are typically purchased live. It is important to keep lobsters alive until they are cooked because the flesh deteriorates and bacteria forms quickly after they die. When lobsters are split in half, the internal organs are removed, including the tomalley (liver and pancreas), stomach, intestinal vein, and often the coral (eggs). Split whole lobsters make an impressive plated presentation and are frequently served grilled or stuffed and baked.

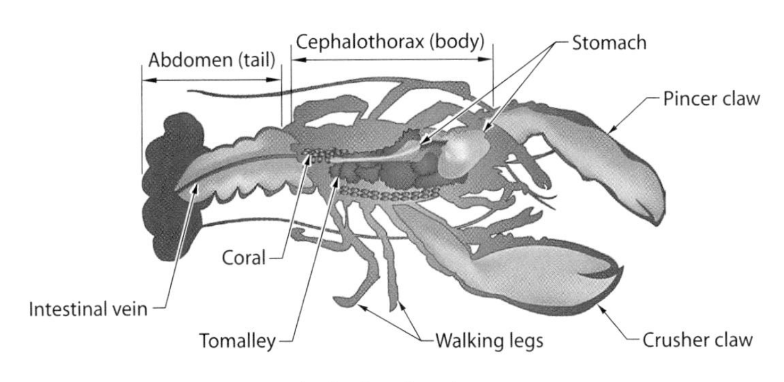

Lobster Anatomy

CUTTING LOBSTER TAIL SHELLS

1. Hold a lobster tail in the guiding hand with the top-side up and the tail fin facing away from the body. Use kitchen shears to cut down the center of the lobster shell, stopping just before reaching the tail fin.

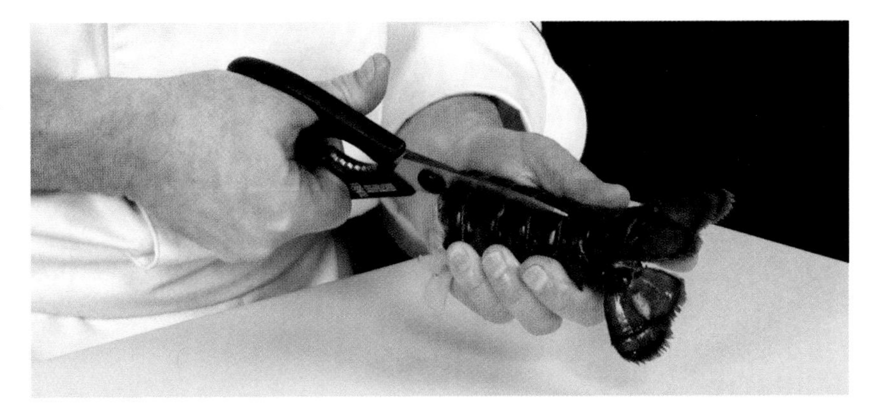

2. Use both hands to pry open the shell along the cut line to reveal the lobster flesh. Hold the lobster in the guiding hand and use the dominant hand to pull the flesh through the cut shell, keeping the flesh attached to the tail fin.

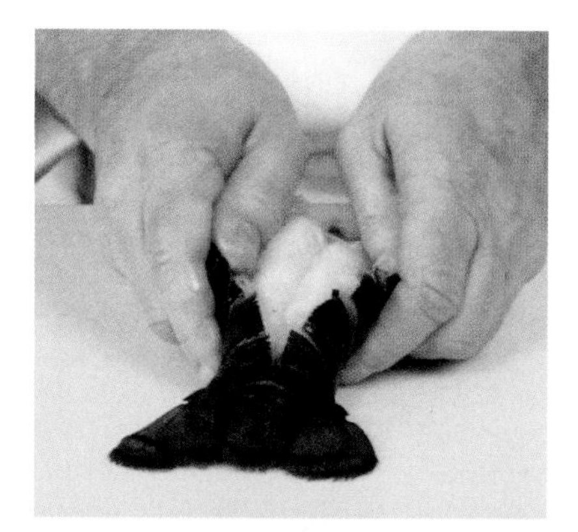

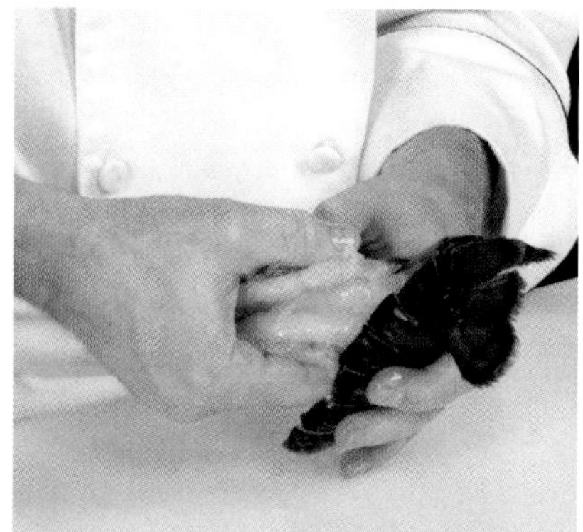

3. Set the lobster tail on the cutting board and position the lobster flesh so that it sits on top of the tail shell. Use the fingers to remove the stringlike intestinal vein.

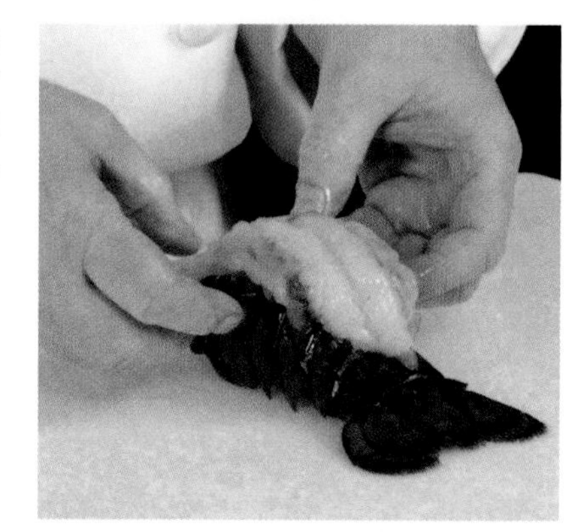

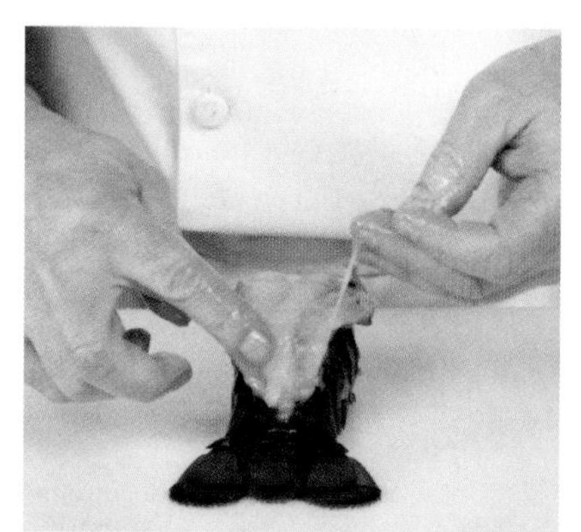

4. Gently flip the lobster flesh back toward the tail fin so that the underside of the flesh is facing up. Use a paring knife to make a shallow lengthwise cut along each side of the flesh. Place the flesh back on top of the tail shell and repeat the shallow lengthwise cuts on the top section of flesh.

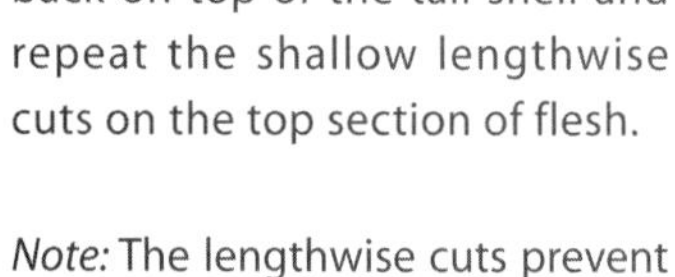

Note: The lengthwise cuts prevent the lobster from curling while cooking.

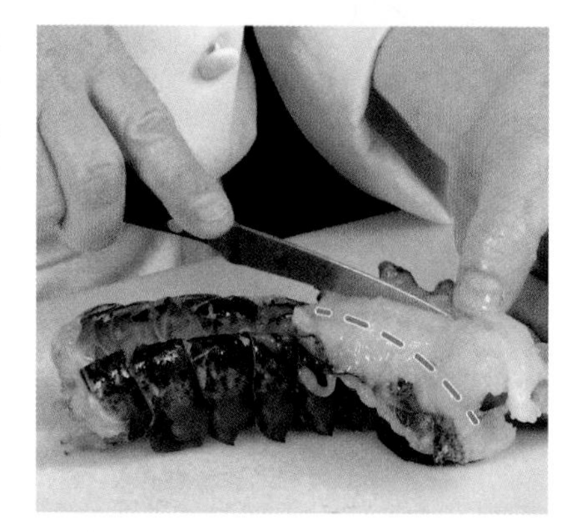

Finished cut lobster tails give the impression that the lobster flesh rests on top of an intact tail shell. This technique allows the flesh to cook evenly and elevates presentation.

SPLITTING WHOLE LOBSTERS IN HALF

* This technique was executed by a left-handed chef.

1. Place a live lobster top-side up on the cutting board with the head facing the knife hand side. Grasp the handle of a stiff boning knife or chef's knife with the knife hand. Hold the knife perpendicular to the cutting board with the edge of the blade facing the head end of the lobster and center the knife point between the eyes of the lobster at the base of the head.

2. Hold the tail with the guiding hand, then quickly drive the knife straight down through the head of the lobster. When the knife point makes contact with the cutting board, push the knife handle down toward the board so that the knife blade slices through the center of the lobster head and comes to rest on the cutting board.

3. Use kitchen shears to cut the rubber bands from the lobster claws.

4. Place the lobster top-side up with the tail facing the knife hand side. Grasp the handle of a chef's knife in the knife hand. Hold the knife perpendicular to the cutting board with the edge of the blade facing the tail of the lobster and insert the knife tip into the cut made at the base of the head in Step 2.

SPLITTING WHOLE LOBSTERS IN HALF (CONTINUED)

* This technique was executed by a left-handed chef.

QUICK TIP

If a lobster is to be boiled or grilled, insert a skewer into the underside of the tail fin and push the skewer through the center of the flesh until reaching the other end of the tail. This prevents the tail from curling as it cooks.

5. Cut the lobster in half by pushing the knife handle down toward the cutting board so that the knife blade makes a lengthwise cut through the center of the lobster. If additional force is needed to cut through the lobster, place the palm of the guiding hand on the spine of the knife and apply pressure while cutting.

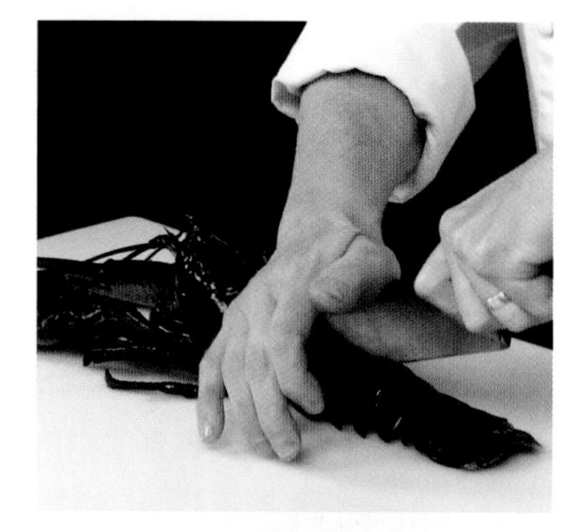

6. Use the hands to remove the green tomalley located in the body (cephalothorax), the stomach located near the head, and the stringlike intestinal vein that runs lengthwise through the lobster.

   Note: If coral is present, it can also be removed.

QUICK TIP

While some consider tomalley to be a delicacy, the U.S. Food and Drug Administration has warned consumers to avoid its consumption as it may harbor pollutants from the environment.

Finished split whole lobsters are halved lengthwise.

VIDEO 36

CUTTING CEPHALOPODS

TECHNIQUE 36

A cephalopod is a type of mollusk, such as squid, octopuses, and cuttlefish, that lacks an external shell, has arms around the head, has developed eyes, and often contains an ink sac. Some cephalopods have an internal shell as well as a birdlike beak that is used to crack the shells of their prey. When preparing cephalopods, the internal shell and beak are removed.

A squid has an internal shell commonly referred to as a pen. Squid also have an ink sac that is often extracted so that the ink can be used to color and flavor grain and pasta dishes. The flesh of squid is commonly cut into rings and sautéed, breaded and fried, or used to make stews. Squid is also cut into steaks and scored before it is grilled or pan fried. Scoring the squid steaks prevents them from curling during cooking. Squid, which is often referred to by its Italian name, calamari, develops a tough, rubbery texture if overcooked but is tender and flavorful when cooked properly.

Stiff Boning Knife

— or —

Chef's Knife

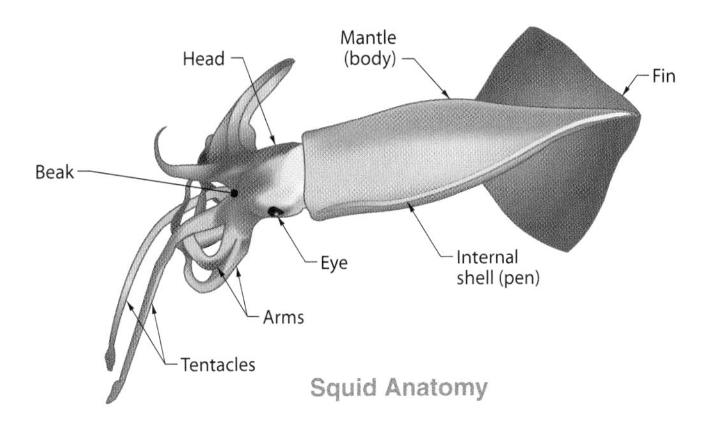

Squid Anatomy

PREPARING SQUID FOR CUTTING

* This technique was executed by a left-handed chef.

1. Place a squid on the cutting board with the head facing the dominant hand. Secure the mantle (body) with the guiding hand and use the dominant hand to gently pull on the head until both the head and the internal organs are removed from the mantle.

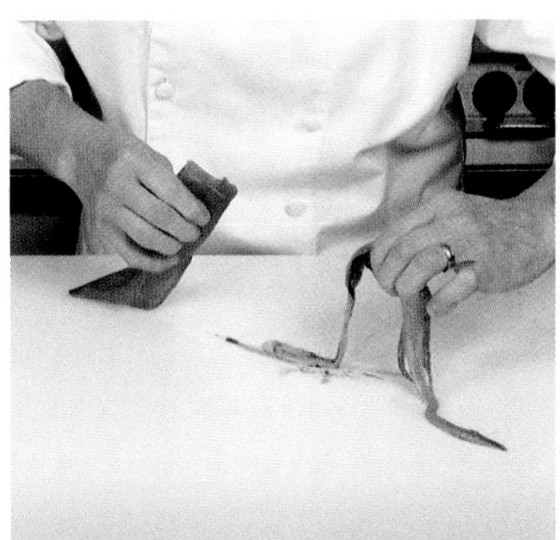

* This technique was executed by a left-handed chef.

2. Reach inside the mantle and remove the plasticlike internal shell. Discard the internal shell.

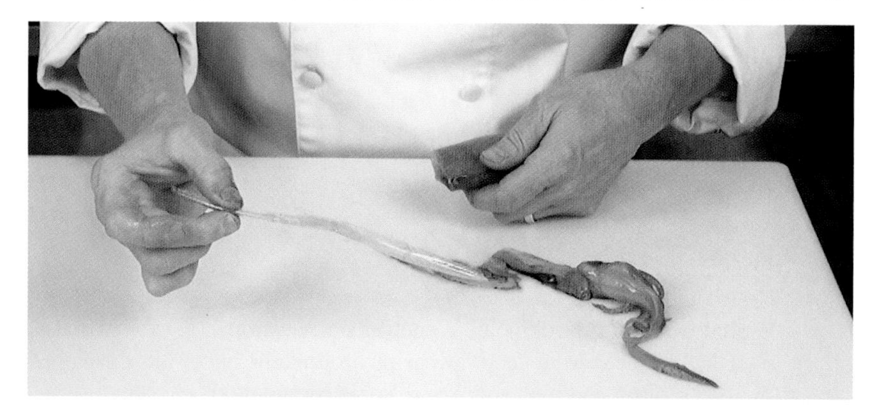

3. Hold the head of the squid with the underside of the arms and tentacles facing up. Use both hands to separate the arms and tentacles, revealing the beak located at the base of the head. Squeeze the area around the beak until it is fully exposed and can be pulled away and discarded.

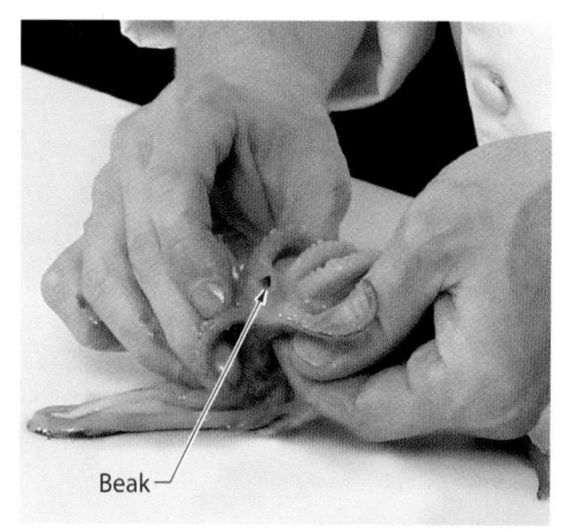

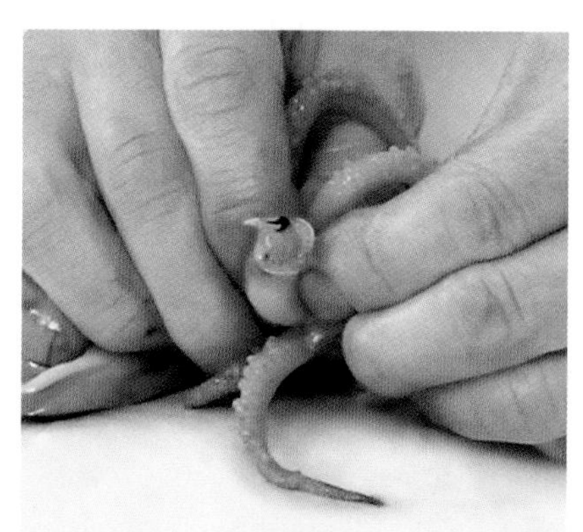

4. Place the head of the squid on the cutting board with the arms and tentacles facing the knife hand side. Hold the base of the head with the guiding hand and use a boning knife to cut just below the eyes, separating the head section from the arms and tentacles. Discard the head section attached to the internal organs. Rinse and reserve the arms and tentacles.

5. Place the mantle horizontally on the cutting board. Start at the cut side (head end) and use the fingers to peel off the dark skin that covers the mantle.

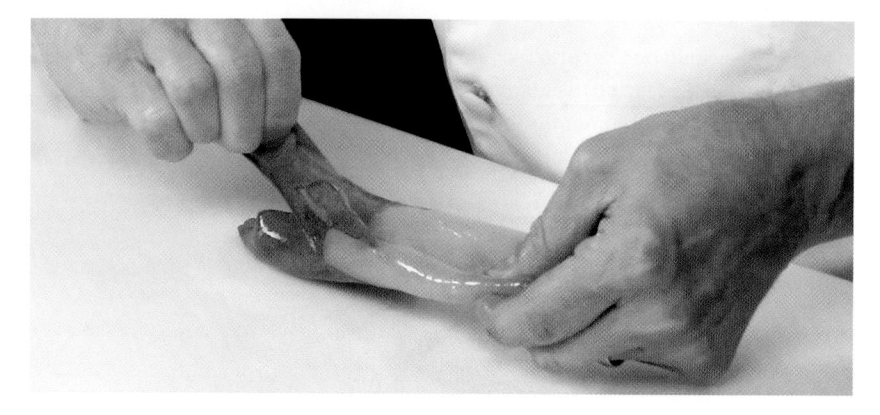

PREPARING SQUID FOR CUTTING (CONTINUED)

* This technique was executed by a left-handed chef.

6. Pull one of the fins toward the tapered end of the squid to remove it from the mantle. Repeat this step with the remaining fin. Rinse the inside and outside of the mantle before use.

   Note: The fins are edible but are typically not eaten because they are extremely tough.

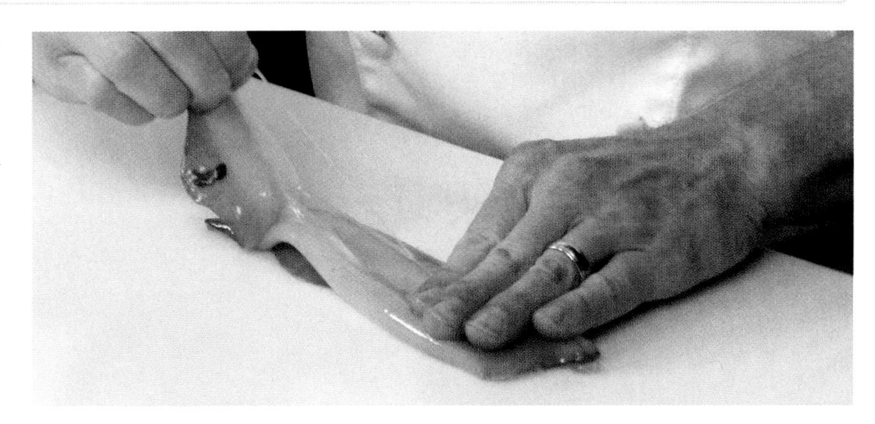

CUTTING SQUID RINGS

* This technique was executed by a left-handed chef.

1. Start with the mantle of a squid that has been cleaned and prepared for cutting. Place the mantle on the cutting board with the cut side (head end) facing the knife hand side.

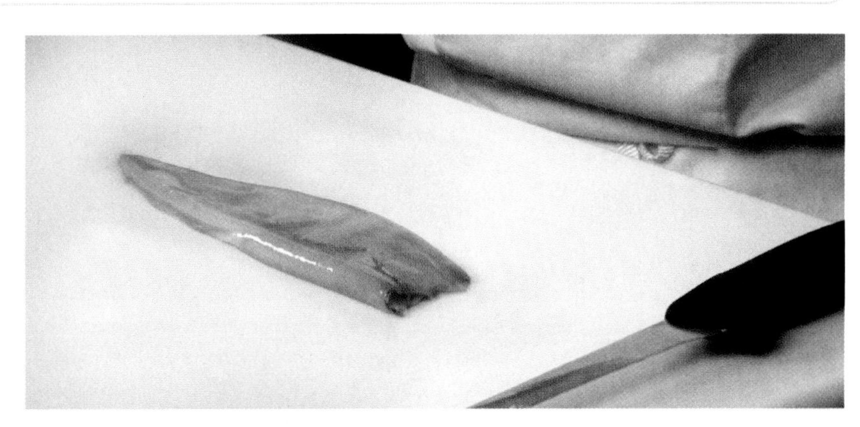

2. Hold the mantle in a claw grip with the guiding hand. Start at the cut side of the mantle and use a boning knife to make crosswise slices at equally spaced intervals until reaching the end of the mantle.

Finished squid rings have a hollow circular shape.

QUICK TIP

Squid rings are often fried or sautéed with the arms and tentacles and served as an appetizer.

CUTTING AND SCORING SQUID STEAKS

* This technique was executed by a left-handed chef.

1. Start with the mantle of a squid that has been cleaned and prepared for cutting. Place the mantle vertically (lengthwise) on the cutting board with the cut side (head end) farthest from the body.

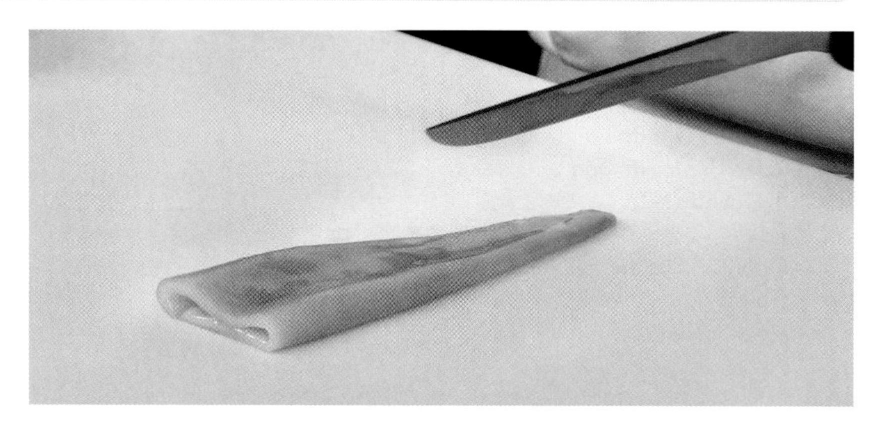

2. Start at the cut side of the mantle and use the guiding hand to slightly lift the top layer of flesh. To butterfly the mantle, use a boning knife and make a lengthwise cut down the center of the mantle that is deep enough to cut through just the top layer of flesh.

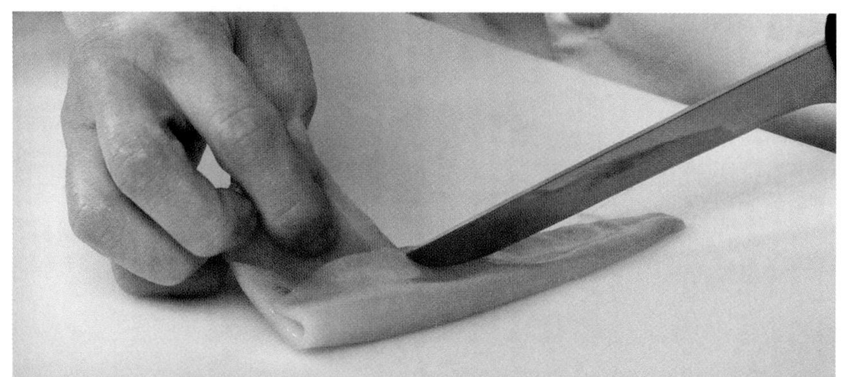

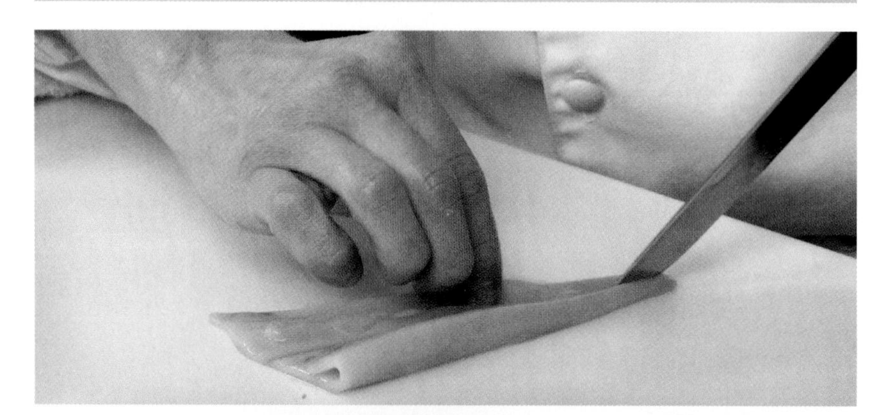

3. Open the butterflied mantle so that it spreads out and lies flat on the cutting board. Use the edge of the knife blade to gently scrape away any residual organs that were not removed when the squid was originally cleaned and prepared.

* This technique was executed by a left-handed chef.

4. Place the mantle with the tapered end facing the guiding hand side. Without cutting all the way through the flesh, score the mantle by holding the knife at a 45° angle to the flesh and making diagonal slices ½-1 inch apart starting at the wide end of the mantle and working toward the tapered end.

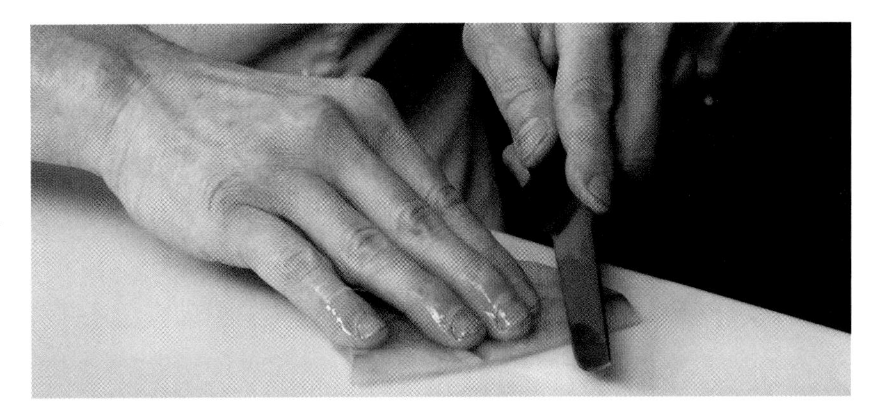

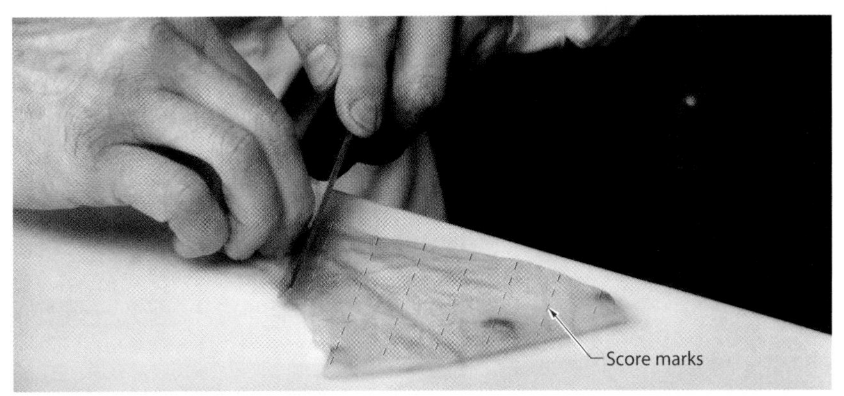

5. Rotate the mantle 90° so that the tapered end is closest to the body. Score the flesh again, this time starting the diagonal slices at the tapered end of the mantle and working toward the widest end to create a diamond (crosshatch) pattern on the mantle.

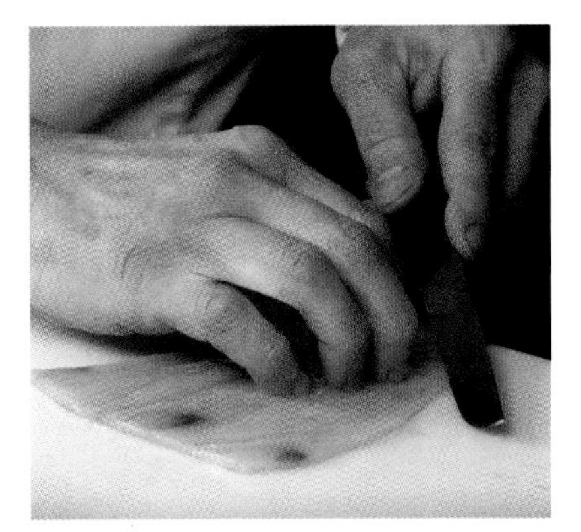

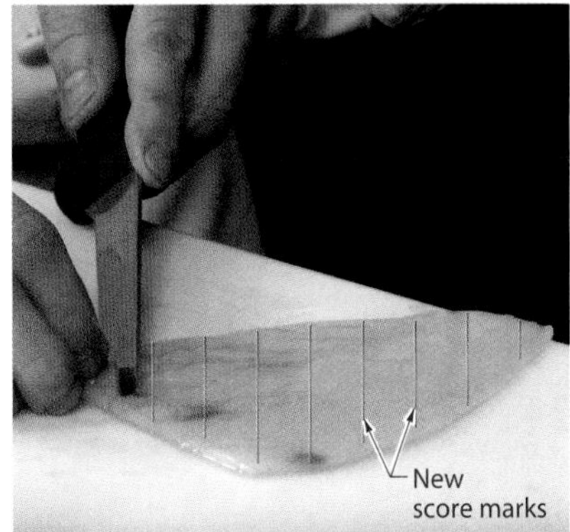

Finished scored squid steaks have flesh that has been butterflied to lay flat and crosshatch score marks for even cooking and enhanced presentation.

TECHNIQUE 37

CUTTING ROUNDFISH

VIDEO 37

A roundfish is any fish with a cylindrical body, an eye located on each side of the head, and a backbone that runs from head to tail in the center of the body. Examples of roundfish include bass, catfish, cod, tilapia, trout, tuna, salmon, and snapper.

Roundfish are regularly purchased whole or drawn and then filleted. A drawn fish is a fish that has had only the internal organs (viscera) removed. A fillet is a lengthwise piece of flesh cut away from the backbone. Each roundfish yields two fillets that can be prepared and served with or without the skin. If the skin remains, a fish scaler is used to remove the scales before filleting. Purchasing a whole or drawn fish for filleting often leads to better quality as this is the best way to verify freshness. Properly filleting a fish is also more cost-effective than purchasing fish already filleted, and the bones of some fish can be saved to make a flavorful stock.

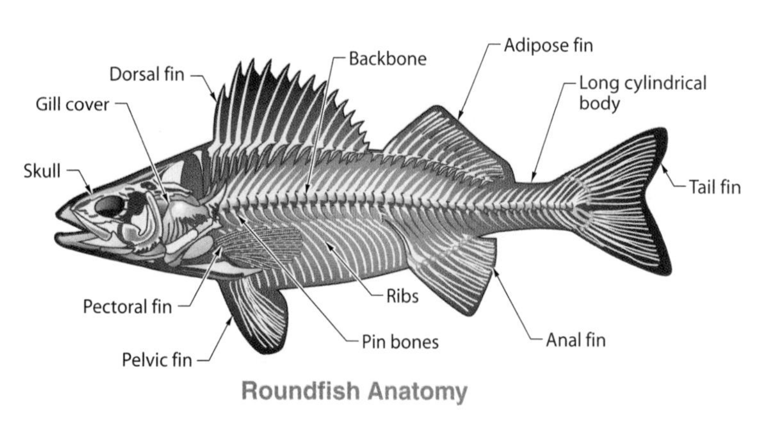

Roundfish Anatomy

TRIMMING AND FILLETING RED SNAPPERS

* This technique was executed by a left-handed chef.

1. Lay a drawn red snapper on the cutting board and use kitchen shears to remove all the fins except the tail fin.

   Note: A drawn fish has had the internal organs removed.

QUICK TIP

When purchasing fresh fish, consider the following factors:
- *Fish should have a slight smell of the sea, not a strong, fishy odor.*
- *The eyes should be clear and bulging, not cloudy and sunken.*
- *The gills should be bright red, not purplish-brown.*
- *The flesh should be intact and bounce back when gently pressed.*

2. Place the snapper across the cutting board with the tail facing the knife hand side and secure with the guiding hand. Use a boning knife or fillet knife to make a score mark across the base of the tail that is deep enough to reach the bone. Repeat this cut on the other side of the snapper.

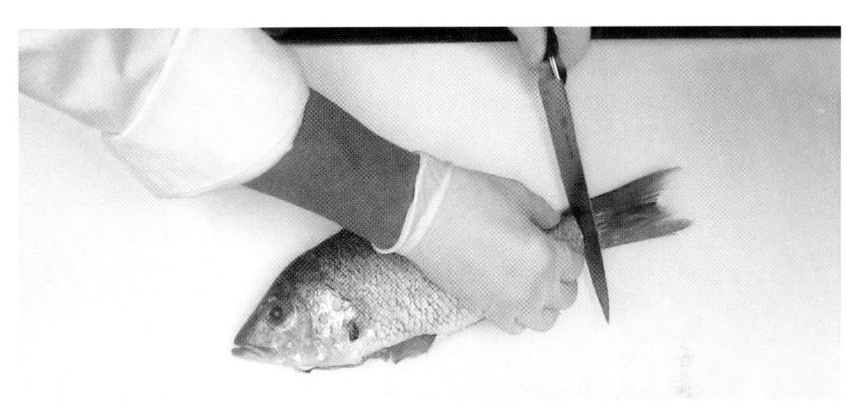

3. With the belly of the snapper facing away from the body, make a cut that follows the contour of the head, keeping the knife blade just behind the area where the fins have been removed. Continue cutting toward the top of the head but not through the backbone. Repeat this cut on the other side of the snapper.

4. To remove the head, use kitchen shears to cut through the backbone and any flesh that keeps the head attached to the body.

* This technique was executed by a left-handed chef.

5. Place the snapper with its back facing the knife hand side and the head end facing away from the body. Hold the knife parallel to the cutting board and place the guiding hand open and flat on the body of the snapper. Position the knife tip just above the removed dorsal fin and make a lengthwise cut ½ inch deep that extends from the head end to the tail end of the snapper.

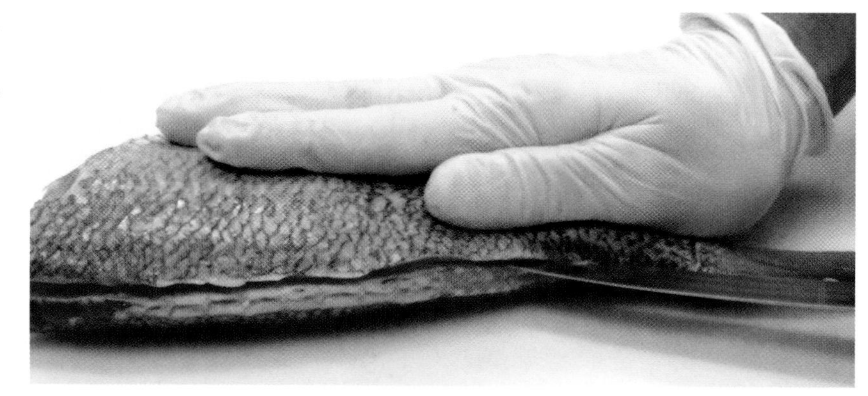

6. Reinsert the knife blade into the cut made in Step 5 and pull the knife back to slightly deepen the cut. Repeat this process until reaching the backbone.

7. Working from the midsection back toward the tail, cut along the backbone and through the bottom (belly area) of the snapper to separate this section of the fillet.

   Note: The fillet still remains attached to the backbone near the head end of the snapper.

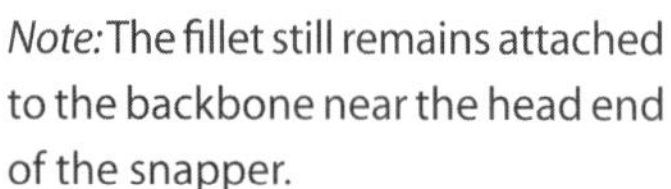

8. Lift the tail end of the fillet and place the knife blade under the flesh with the edge facing the head end and the blade parallel to the cutting board. Lay the fillet over the knife and grasp the tail with the guiding hand. Cut toward the head end to separate the fillet from the bones.

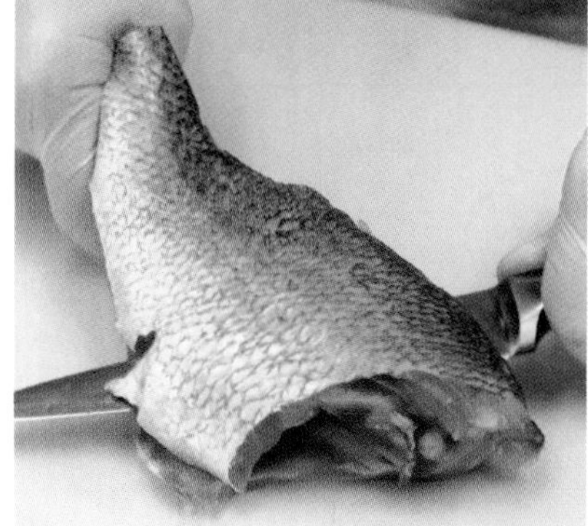

TRIMMING AND FILLETING RED SNAPPERS (CONTINUED)

* This technique was executed by a left-handed chef.

9. To fillet the other side, flip the snapper over and repeat Steps 5–8, this time making the lengthwise cuts from the tail end to the head end of the snapper.

   Note: For Step 8, rotate the snapper so that the head end is farthest from the body.

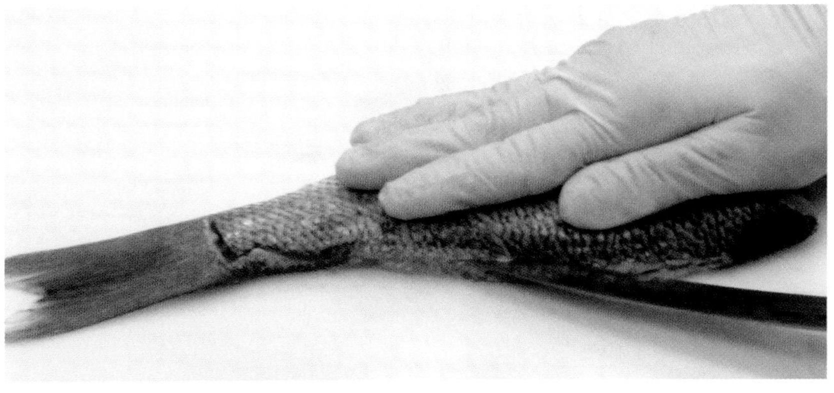

10. Remove the thin, belly portion and any remaining rib bones from each fillet by positioning the knife at the top of the rib bones and cutting down at a 45° angle toward the bottom of the fillet.

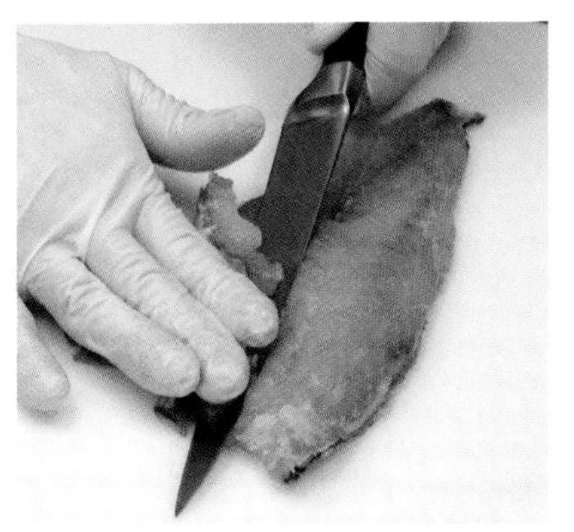

SKINNING RED SNAPPERS

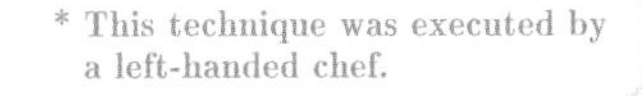

1. Place a red snapper fillet skin-side down across the cutting board and pinch the tail end between the index finger and thumb of the guiding hand. Angle the edge of the knife between the skin and flesh at the tail end of the fillet and cut 1–2 inches of flesh away from the skin.

SKINNING RED SNAPPERS (CONTINUED)

* This technique was executed by a left-handed chef.

2. Grasp the skin of the tail end tightly in the guiding hand. With the knife hand, use a sawing motion to cut in the direction of the head end while turning the knife blade parallel to the cutting board and flush against the skin of the fillet.

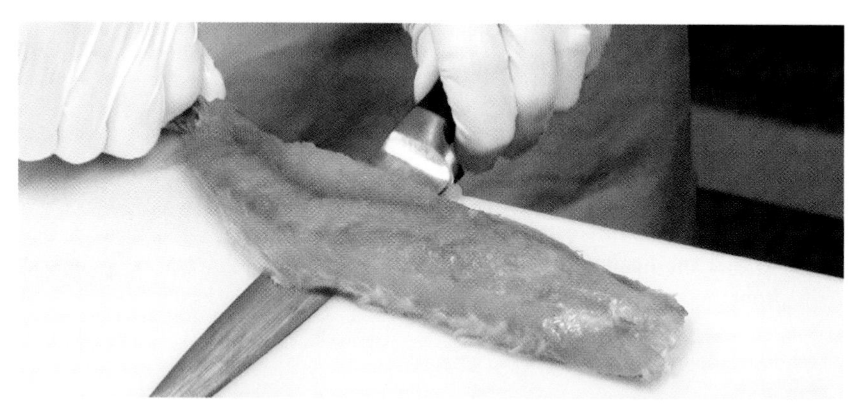

3. Maintain the parallel position of the knife blade and continue cutting until the flesh is separated from the skin. Repeat Steps 1–3 to skin the remaining fillet.

REMOVING THE PIN BONES AND BLOODLINE FROM RED SNAPPERS

* This technique was executed by a left-handed chef.

1. Run a finger down the surface of a red snapper fillet to locate the pin bones. Use tweezers or fish pliers to pull the pin bones from the fillet.

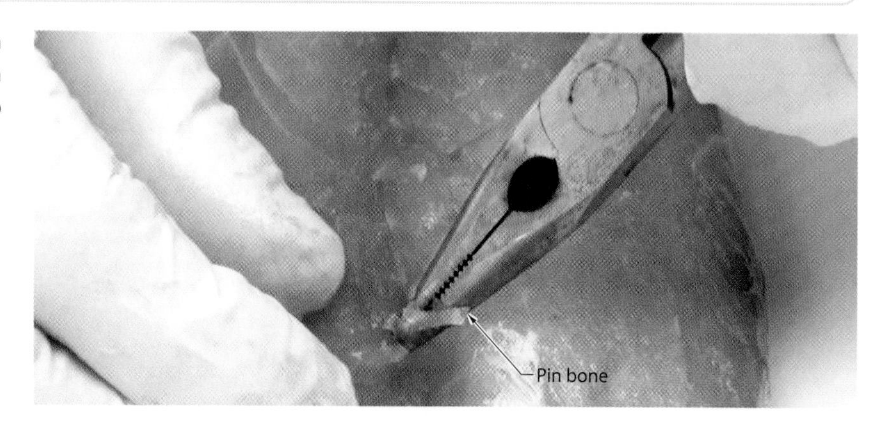

REMOVING THE PIN BONES AND BLOODLINE FROM RED SNAPPERS (CONTINUED)

* This technique was executed by a left-handed chef.

2. Position the fillet skinned-side up with one of the ends farthest from the body. Set the guiding hand open and flat on the fillet and place the edge of the knife blade alongside the bloodline. Pull the knife back, making a shallow cut just beneath the bloodline.

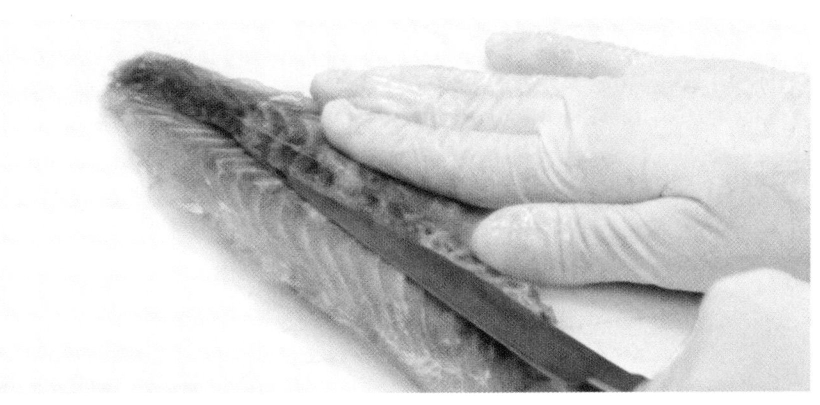

3. Rotate the fillet 180° and repeat Step 2 on the other side of the bloodline. Lift (or trim) the bloodline away from the fillet.

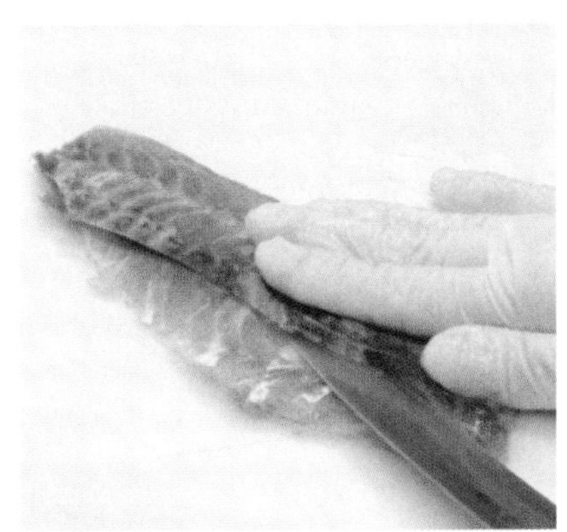

Filleting red snapper yields two fillets that are the boneless flesh of the fish. If desired, each fillet may be skinned and the bloodline may be removed.

TECHNIQUE 38

CUTTING FLATFISH

VIDEO 38

A flatfish is any thin, wide fish with both eyes located on one side of the head and a backbone that runs from head to tail through the midline of the body. The eyes of flatfish are located on the top side, which is typically dark greenish-brown, while the bottom side is pale. Flatfish are bottom-dwelling fish and swim with their eyes facing the surface of the water. Examples of flatfish include flounder, halibut, sole, and turbot.

Due to the location of the backbone, flatfish yield four fillets: two from the top side and two from the bottom side. After a flatfish is filleted, it may be skinned and prepared as desired.

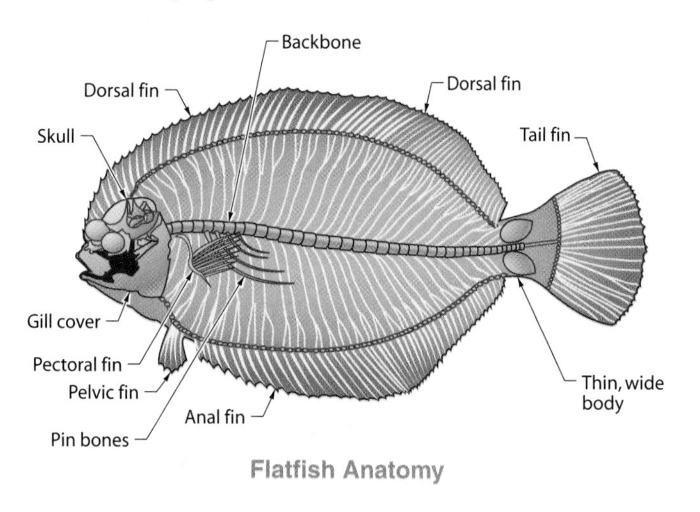

Flatfish Anatomy

TRIMMING AND FILLETING FLOUNDERS

* This technique was executed by a left-handed chef.

1. Lay a drawn flounder on the cutting board and use kitchen shears to remove all the fins except the tail fin.

   Note: A drawn fish has had the internal organs removed.

QUICK TIP *To descale a whole fish before filleting, grasp the fish by the tail and scrape each side of the fish from the tail to the head with a fish scaler or the spine of a knife.*

TRIMMING AND FILLETING FLOUNDERS (CONTINUED)

* This technique was executed by a left-handed chef.

2. Place the flounder bottom-side (light-side) up across the cutting board. Hold a flexible boning knife or fillet knife at a slight angle and make a shallow cut around the head. Repeat this cut on the top side of the flounder.

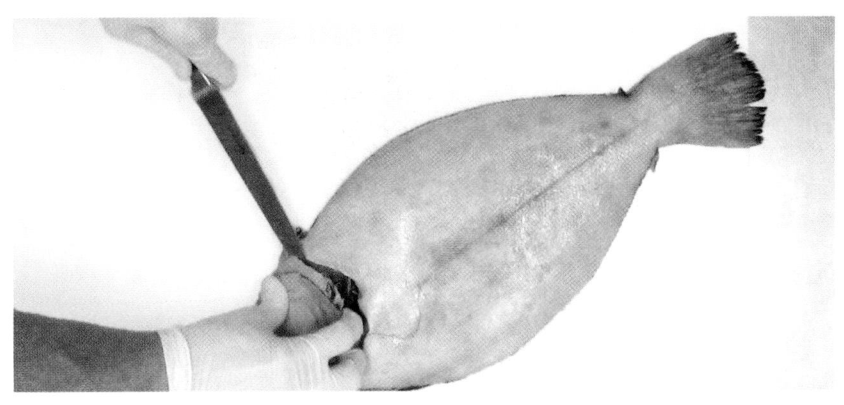

3. To remove the head, use kitchen shears to cut through the backbone and any flesh that keeps the head attached to the body.

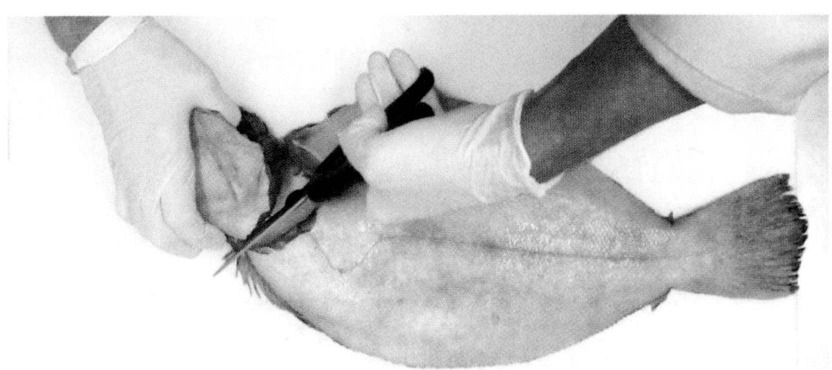

4. Place the flounder top-side up with the head end angled away from the body. Use the tip of the boning knife to make a lengthwise cut along the backbone that extends from the head end to the tail.

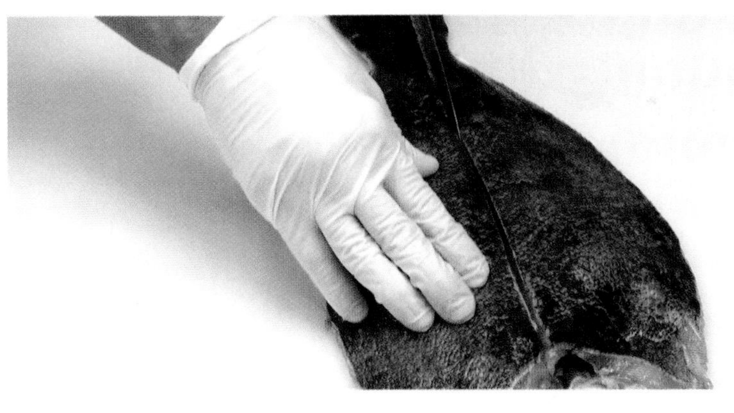

5. To remove the first fillet, angle the edge of the blade toward the guiding hand side and reinsert the knife blade into the cut along the backbone. Keep the side of the knife blade flush with the bones and gradually cut the flesh away using long strokes. Continue this step until the skin along the bottom of the fillet can be cut through.

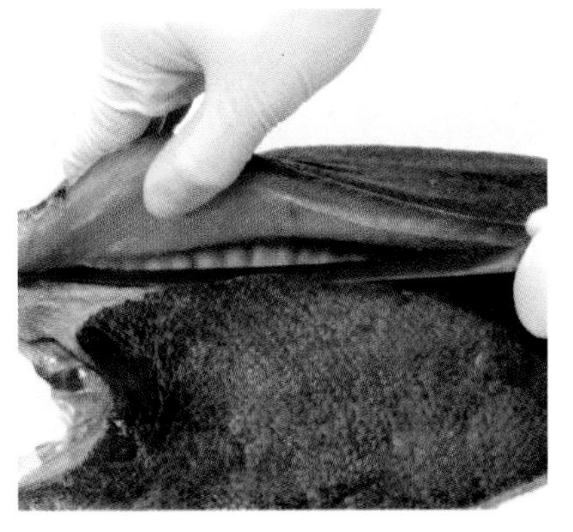

TRIMMING AND FILLETING FLOUNDERS (CONTINUED)

* This technique was executed by a left-handed chef.

6. Rotate the flounder so that the tail end is angled away from the body. To remove the second fillet, repeat Step 5, starting at the tail end.

   Note: A portion of the tail can remain with each fillet or removed if desired.

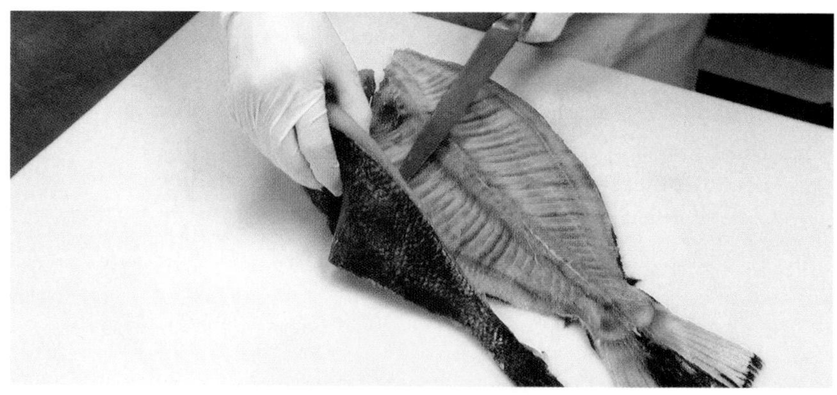

7. Place the flounder bottom-side up and repeat Steps 4–6 to remove the other two fillets.

   Note: There are pin bones located at the edge of the flounder that should be cut through.

8. Trim away the ribbed area of flesh along the rounded side of each fillet.

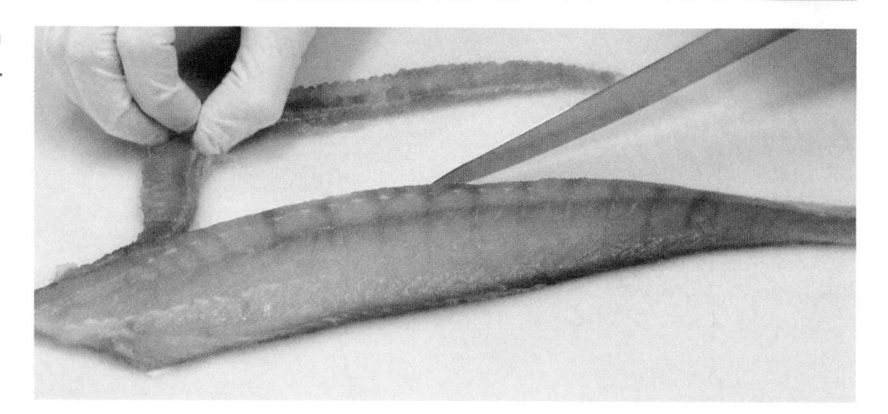

Filleting flounder yields four fillets that are the boneless flesh of the fish.

Note: To skin the flounder, refer to Technique 37: Cutting Roundfish and follow the procedure for Skinning Red Snappers.

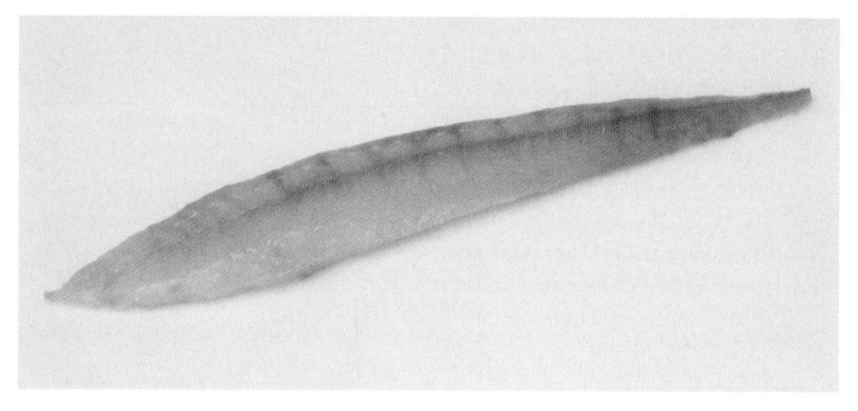

VIDEO 39

SHUCKING BIVALVES

TECHNIQUE 39

Shucking is the process of opening and removing the shell of a bivalve. A bivalve is a type of mollusk that has a top and a bottom shell connected by a central hinge such as an oyster, clam, mussel, or scallop. Adductor muscles keep the bivalve attached to the shell and allow the shell to open and close. Carefully scraping along the adductor muscles to release the bivalve from its shell allows the bivalve to stay intact and remain saturated in briny saltwater for maximum flavor. To protect the guiding hand when shucking bivalves, a wire mesh glove should be worn or a clean folded towel should be placed over the hand.

The tools and procedures for shucking oysters are slightly different than for shucking clams. More specifically, an oyster knife has a short, dull-edged blade and a tapered point. The tapered point is inserted into the hinge and used to pry the shell open. In contrast, a clam knife has a short, flat, sharp-edged blade and a rounded tip. The sharp blade opens the clam on the opposite side of the hinge by slicing between the top and bottom shells. Shucked oysters and clams can be served raw, steamed, baked, broiled, grilled, or in well-known dishes such as oysters rockefeller or clams casino.

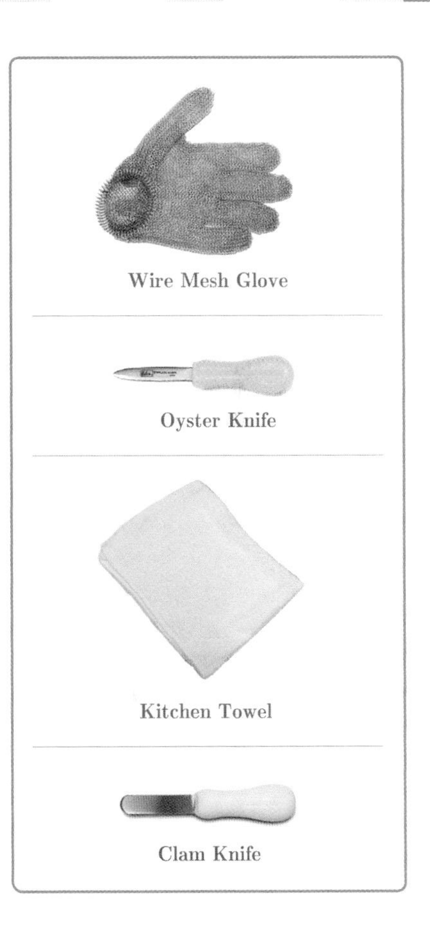
Wire Mesh Glove

Oyster Knife

Kitchen Towel

Clam Knife

SHUCKING OYSTERS

1. Place a wire-mesh glove on the guiding hand and then hold an oyster with the dark, flat underside of the shell facing up and the oyster hinge facing the knife hand side. Grasp the handle of an oyster knife and insert the knife point into the hinge.

   Note: A rubber glove can be placed over a wire-mesh glove to keep it clean and sanitary.

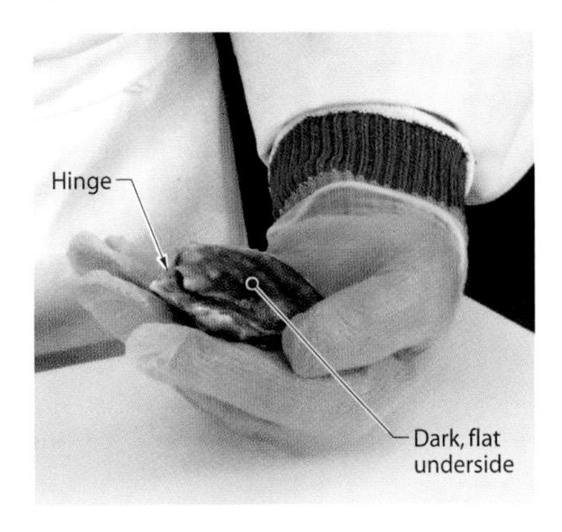

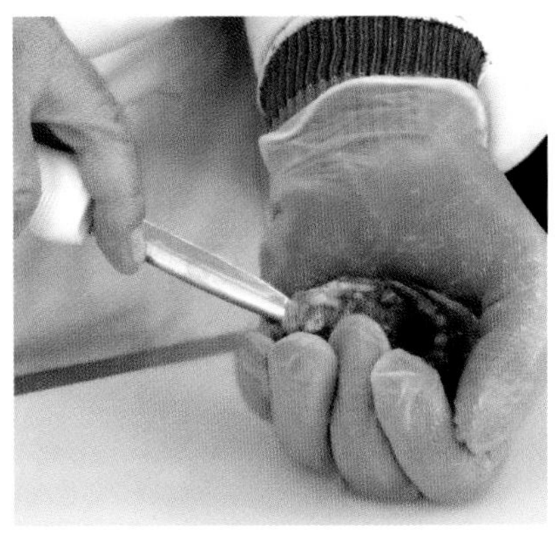

SHUCKING OYSTERS (CONTINUED)

2. Repeatedly twist the knife hand wrist to the right and left to rotate the knife blade until the hinge pries apart and the shells loosen from one another.

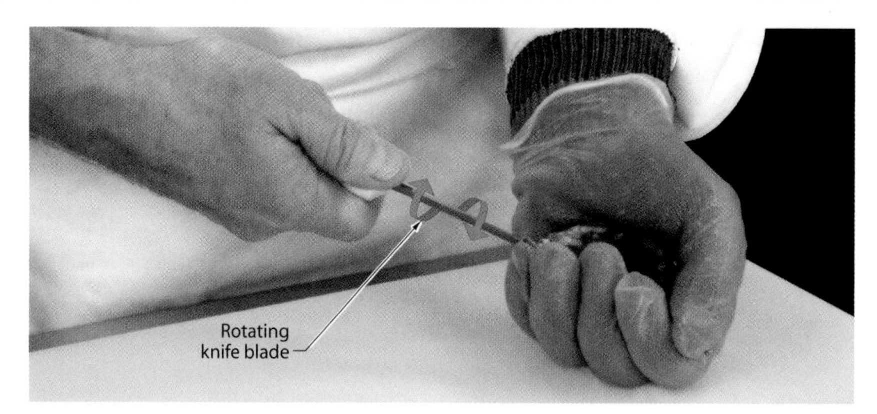

3. Keep the fingers wrapped around the knife handle and place the knife hand thumb opposite the hinge for leverage. Starting at the hinge side, insert the knife tip between the shells and slide the blade along the top shell, carefully scraping along the adductor muscle to release the oyster from the top shell.

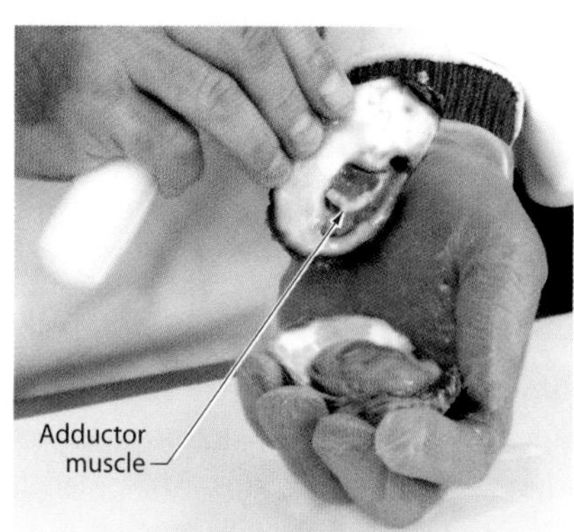

4. Rotate the shell, holding the oyster so that the hinge faces the palm of the guiding hand. Slide the knife tip along the bottom shell, carefully scraping along the adductor muscle to completely release the oyster from the shell. Keep as much liquid as possible inside the shell with the oyster.

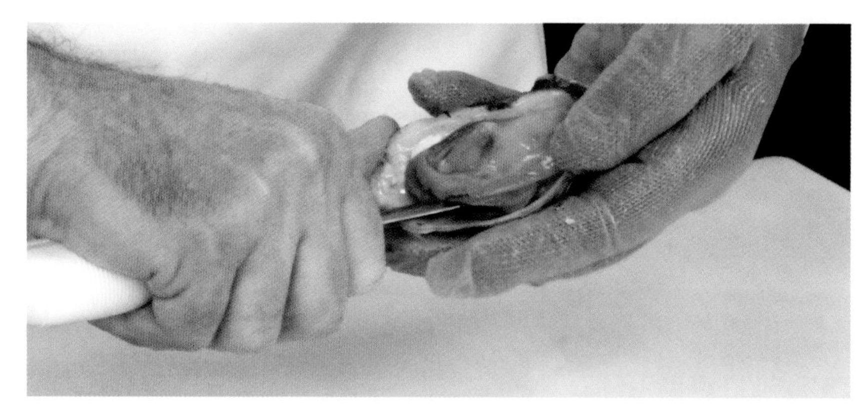

Finished shucked oysters are whole, intact oysters that are carefully detached from the shell to retain their briny moistness and plump shape.

SHUCKING CLAMS

1. Protect the guiding hand with a clean and dry folded towel. Hold a clam in the towel with the hinge gripped securely between the palm and the thumb of the guiding hand.

   Note: Instead of a towel, a wire mesh glove can be used to protect the guiding hand.

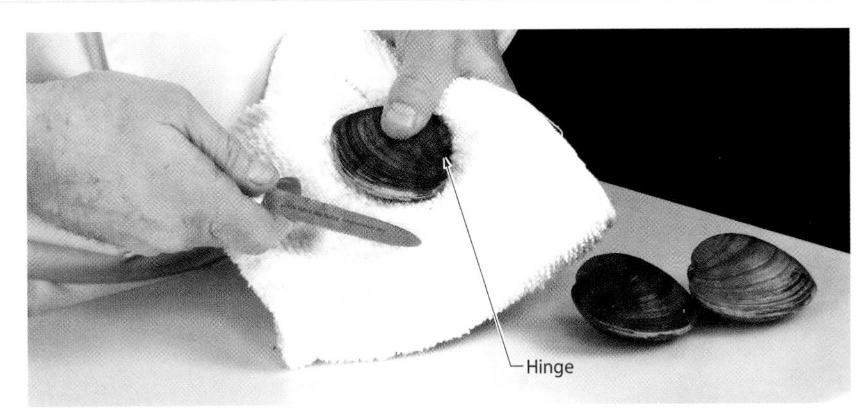

2. Grasp the handle of a clam knife and position the knife hand thumb near the hinge to be used for leverage. Work the edge of the knife into the seam between the top and bottom shell.

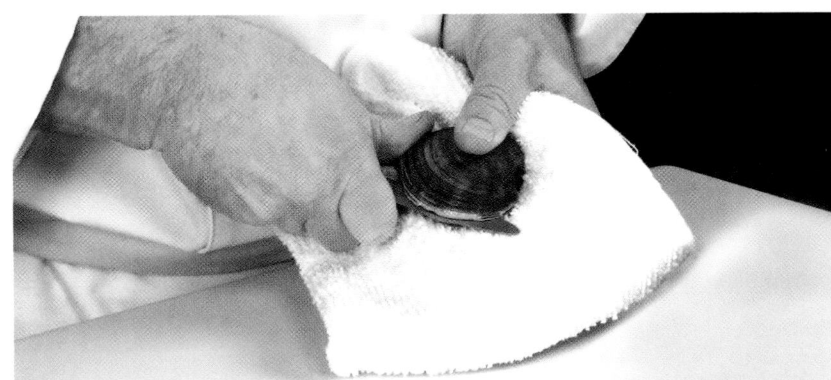

3. Wrap the knife hand thumb around the knife handle and twist the wrist so that the knife blade tilts upward and pries the shell open. Remove the knife and use both thumbs to pry the shells open further.

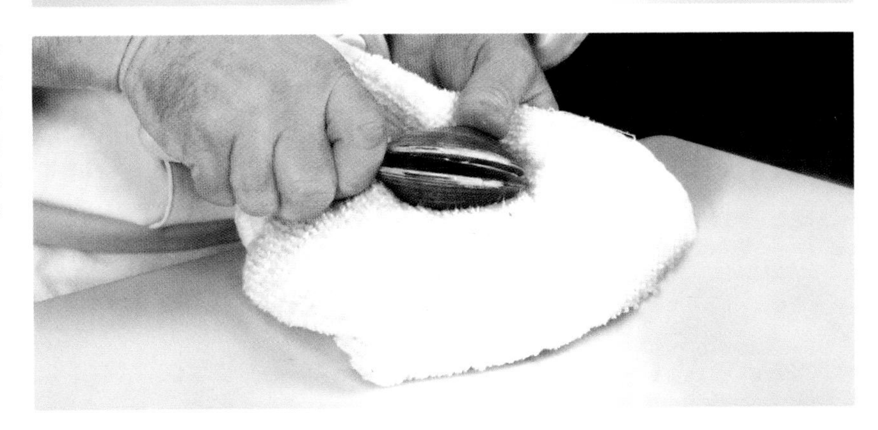

QUICK TIP

Discard bivalves with open shells or shells that do not close when tapped with a finger. These bivalves have died and are unsafe to eat.

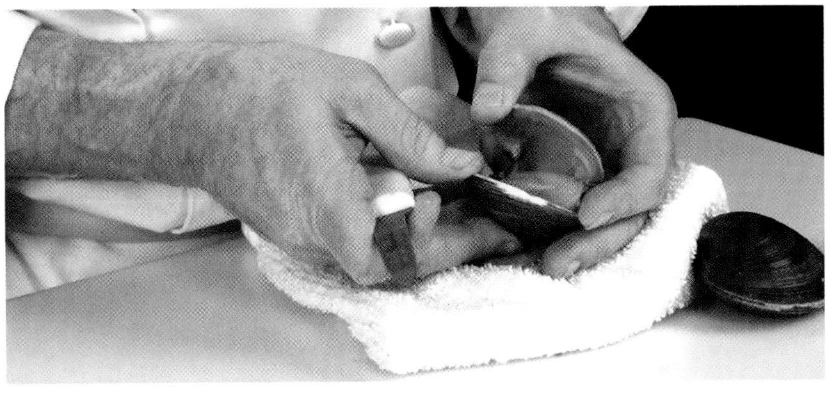

SHUCKING CLAMS (CONTINUED)

4. Use the knife tip to gently scrape the clam from one side of the shell, carefully sawing or scraping along the adductor muscles until the clam releases. Keep as much liquid as possible inside the shell with the clam.

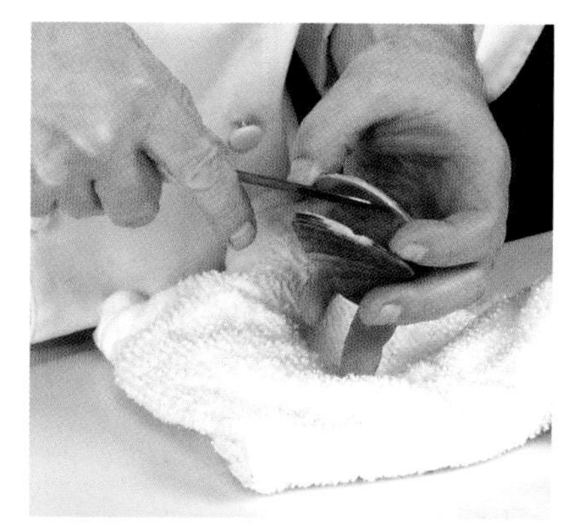

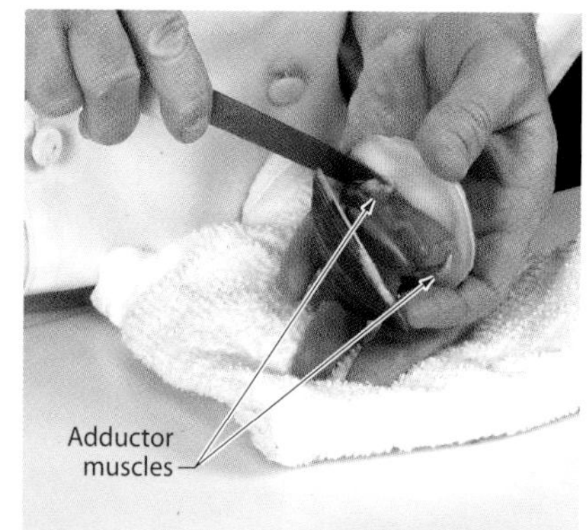

5. Repeat Step 4 to release the clam from the other side of the shell. Twist off the top shell.

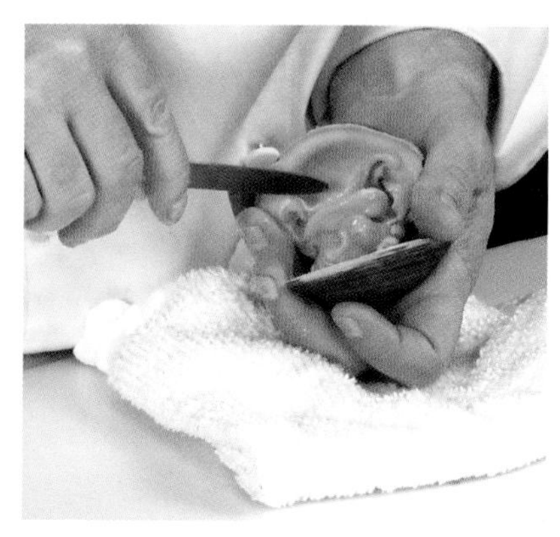

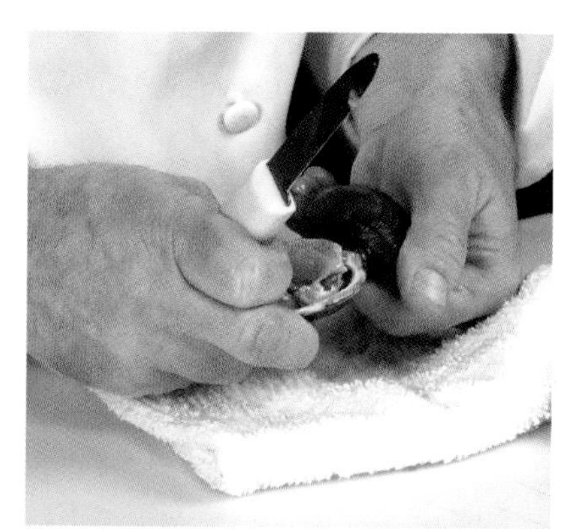

QUICK TIP

When shucking bivalves, place a bowl under the bivalve to capture any briny saltwater that may spill from the shell. Use this water to build flavor in dishes such as clam chowder.

Finished shucked clams are whole, intact clams that are carefully detached from the shell to retain their briny moistness and plump shape.

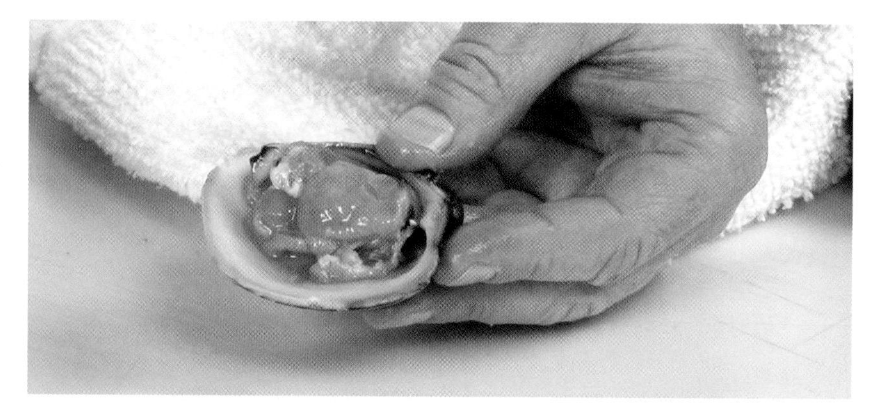

QUICK TIP

Because mussels have a thinner shell, they are not shucked like oysters or clams. To prepare mussels, the whiskerlike "beard" that extends outside their shells is pulled away and discarded. Then the mussels are typically opened by steam as they cook.

SECTION VI
FABRICATING MEATS

- TRIMMING TENDERLOINS
- FABRICATING BEEF TENDERLOINS
- FABRICATING WHOLE PORK LOINS
- FABRICATING LEGS OF LAMB
- FRENCHING RIB RACKS

LEARNER RESOURCES
ATPeResources.com/QuickLinks™
Access Code: 583241

TECHNIQUE

40 TRIMMING TENDERLOINS

VIDEO 40

The tenderloin is a tapered strip of muscle in animals such as cows and pigs that extends into both the short loin and sirloin and runs lengthwise with the backbone below the ribs. The procedure for trimming tenderloins is commonly used on beef and pork tenderloins but can also be applied to other tenderloins such as lamb, bison, and venison.

The tenderloin is comprised of three muscles: the main muscle (psosas major), the "chain" or "side muscle" (psosas minor), and the "wing" (ilacus). The chain runs along one side of the main muscle from the thin (short loin) end to the thick (sirloin) end. The chain has much fat, gristle, and connective tissue and is typically removed from the main muscle. The wing is wider, shorter, and more tender than the chain and often remains attached to trimmed tenderloins.

To promote tenderness, excess fat, gristle, tissue, and silverskin are trimmed from tenderloins. Silverskin is tough, rubbery, silver-white connective tissue that does not break down when heated. Once trimmed, tenderloins may be cooked whole or fabricated into portioned cuts.

Stiff Boning Knife

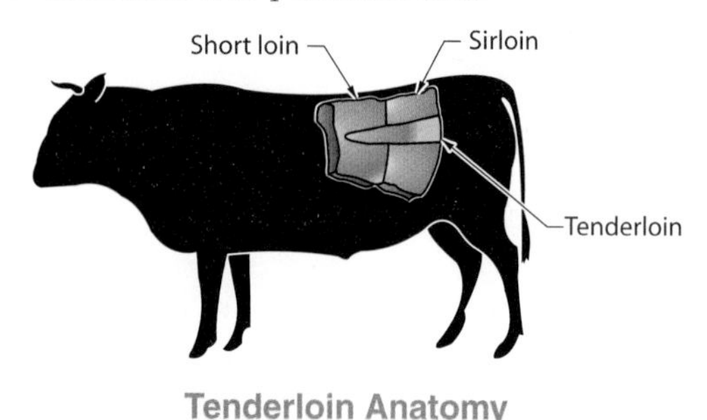

Tenderloin Anatomy

TRIMMING BEEF TENDERLOINS

* This technique was executed by a left-handed chef.

1. Angle a beef tenderloin top-side up across the cutting board with the thin (short loin) end closest to the knife hand side. Pull the outer layer of fat from the top of the tenderloin down toward the chain to reveal the seam between the chain and the main muscle.

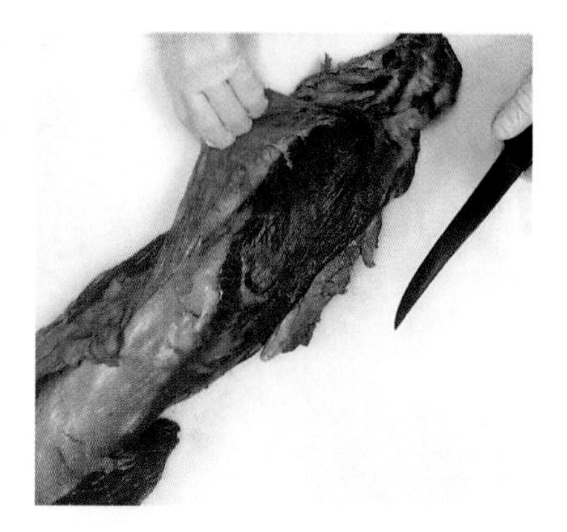

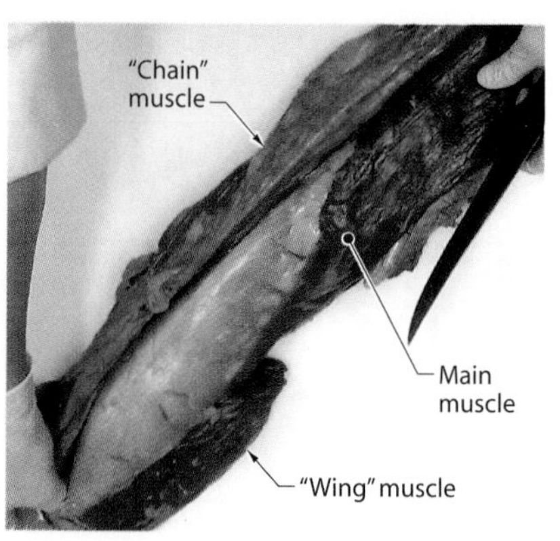

* This technique was executed by a left-handed chef.

2. To remove the entire chain, start from the thin end of the tenderloin and use a stiff boning knife to cut along the seam between the chain and the main muscle.

   Note: Near the thick end of the tenderloin, the chain begins to narrow and has less visible fat. If preferred, cut the chain at this point. Then remove the rest of the chain (side muscle) to be used as desired.

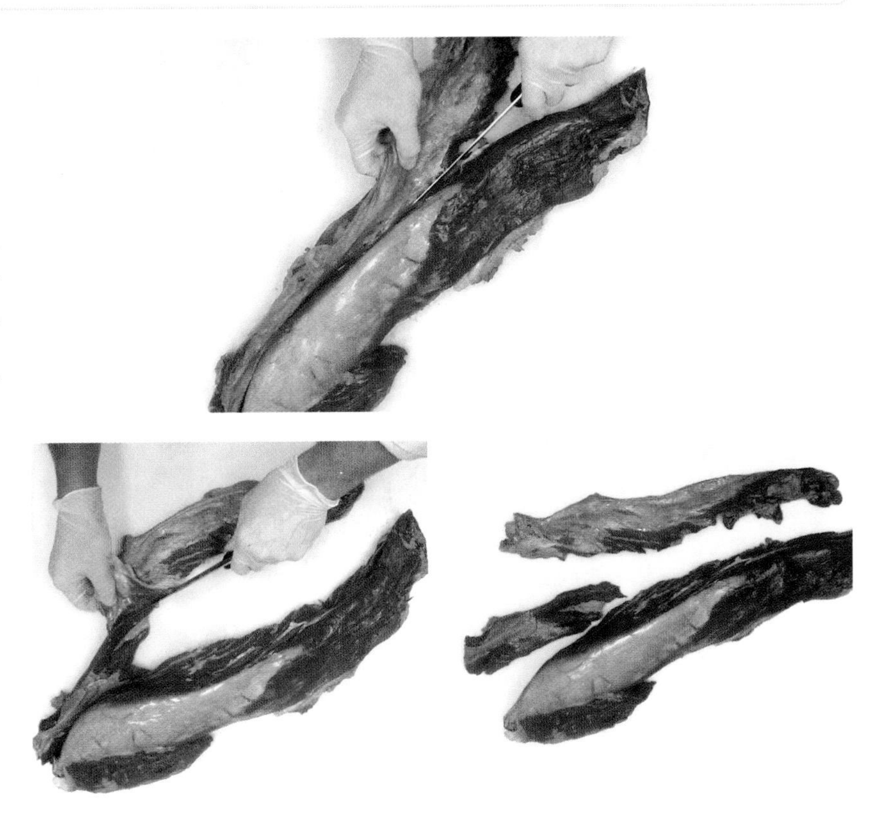

QUICK TIP

After trimming fat, gristle, tissue, and silverskin from the chain, the meat can be pounded and grilled, cubed for stews and kabobs, sliced for stir fries, or ground for a variety of uses.

3. Working on the top side of the tenderloin, use the fingers to peel off the slightly transparent, thin membrane that covers the silverskin.

* This technique was executed by a left-handed chef.

4. Place the tenderloin horizontally across the cutting board with the thin end toward the knife hand side. Hold the knife parallel to the board with the edge facing the thin end. Insert the knife tip under an edge of silverskin, angle the knife blade slightly upward, and cut through the silverskin.

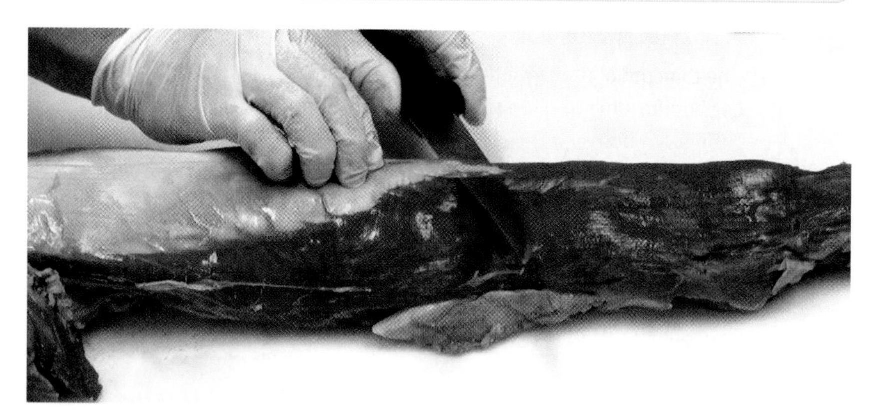

5. Hold the knife parallel to the cutting board with the edge now facing the thick end of the tenderloin. Place the blade under the cut strip of silverskin and use the guiding hand to pull the silverskin over the knife blade. Hold the cut strip of the silverskin against the tenderloin and cut toward the thick end to release a strip of silverskin.

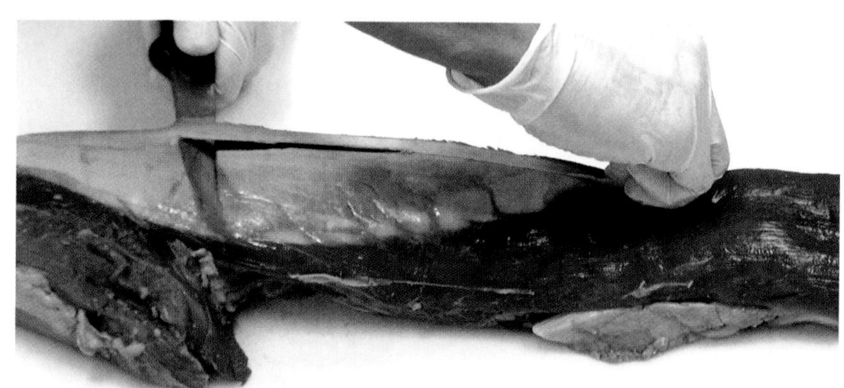

6. Run an index finger under the wide strip of remaining silverskin to loosen it. Repeat Steps 4–5 to remove the silverskin.

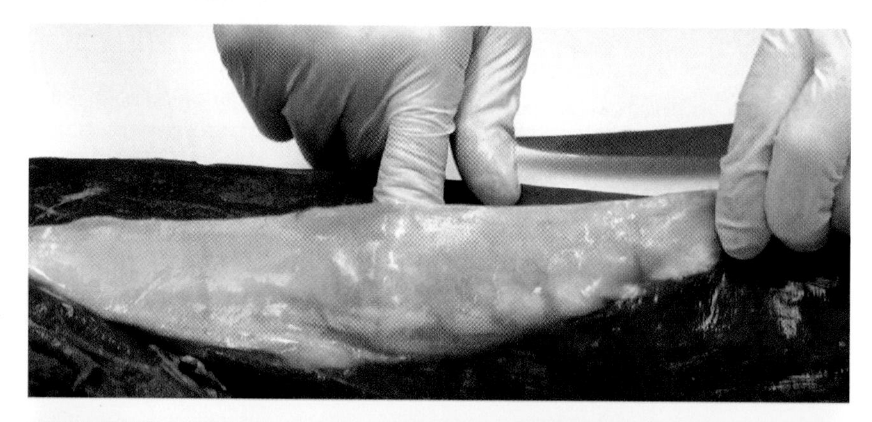

7. Inspect the top and sides of the tenderloin and trim any excess silverskin, fat, gristle, and tissue. Turn the tenderloin over (top-side down) and repeat this step at the thick end of the tenderloin.

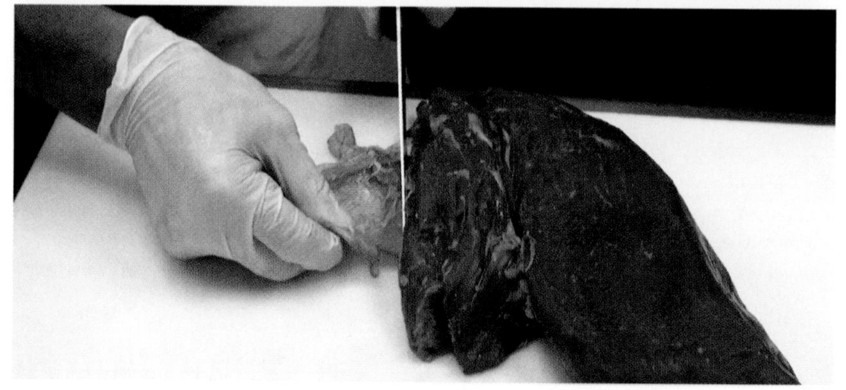

TRIMMING BEEF TENDERLOINS (CONTINUED)

* This technique was executed by a left-handed chef.

8. Angle the tenderloin so that the thick end is near a corner of the cutting board farthest from the body. Pull the knife back along the grooved area where the ribs were once located to remove a layer of fat, gristle, tissue, and any remaining silverskin.

   Note: Use caution not to cut too deep so that the flesh stays as intact as possible.

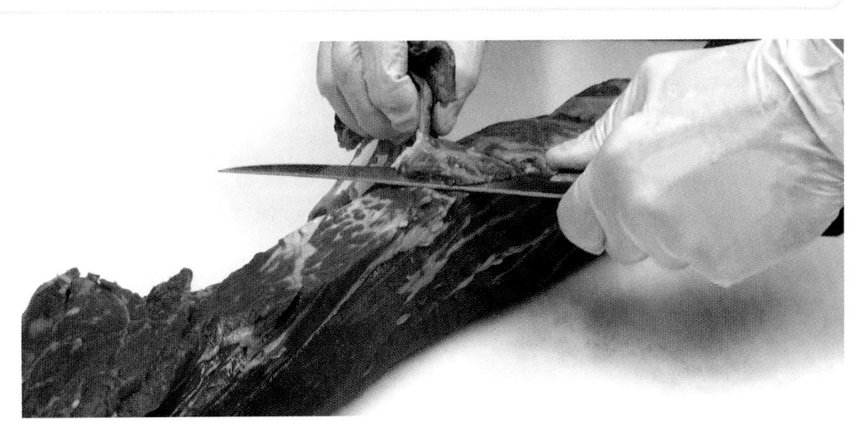

Finished trimmed tenderloins are cleaned of excess fat, gristle, tissue, and silverskin and commonly have had the chain (side muscle) removed.

Note: Any pieces of flesh removed during the trimming procedure can be reserved, trimmed, and used as desired.

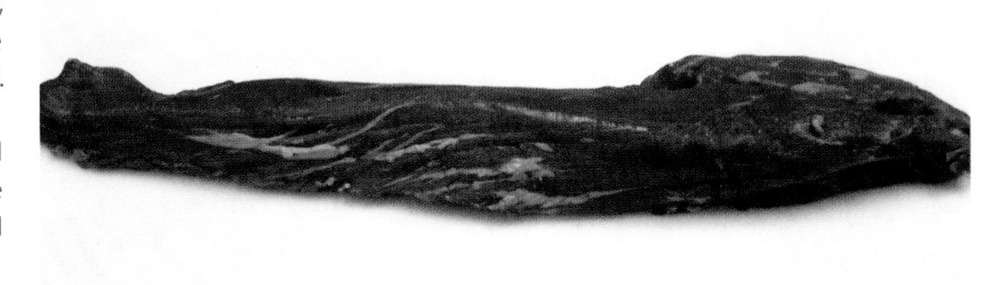

TECHNIQUE 41

FABRICATING BEEF TENDERLOINS

VIDEO 41

Beef tenderloins are commonly trimmed and then fabricated into portioned cuts. Cuts for roasting, such as chateaubriand, may be cut from the center of the tenderloin or from the thicker (sirloin) end.

A beef tenderloin filet is generally 1½–2 inches thick and is cut from either end of the tenderloin or from the center. A medallion is a pounded disc-shaped cut of meat weighing approximately 2–4 ounces. It is often fabricated from the ends of the tenderloin. Filets and medallions are ideal for quick-cooking methods such as broiling, grilling, and sautéing.

Scimitar

— or —

Chef's Knife

— or —

Messermeister

Stiff Boning Knife

Paderno World Cuisine

Meat Pounder

CUTTING BEEF TENDERLOINS INTO PORTIONS

* This technique was executed by a left-handed chef.

1. Place a trimmed beef tenderloin top-side up on the cutting board with the thick (sirloin) end facing the knife hand side. Secure the tenderloin with the guiding hand and use a scimitar to slice approximately 2 inches off the end. Reserve for a medallion.

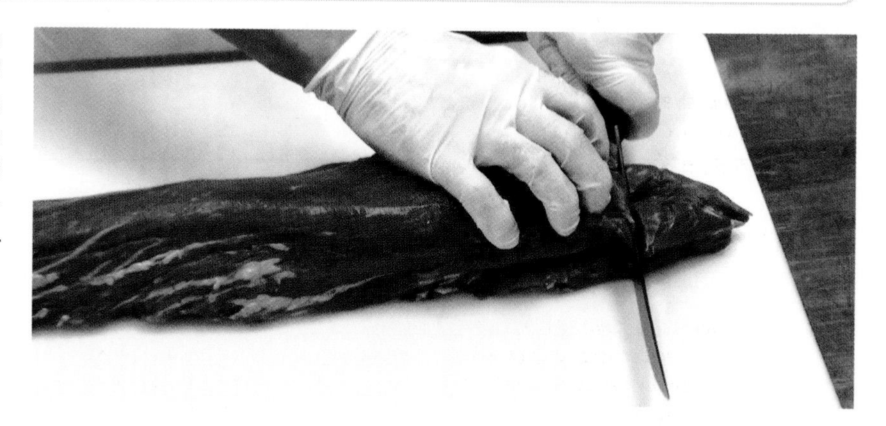

CUTTING BEEF TENDERLOINS INTO PORTIONS (CONTINUED)

* This technique was executed by a left-handed chef.

2. To cut a chateaubriand, secure the tenderloin with the guiding hand and make a crosswise slice through the tenderloin approximately 6 inches from the sirloin end.

   Note: The chateaubriand can also be cut into filets.

3. To cut filets, secure the tenderloin with the guiding hand and make crosswise slices from the sirloin end to the loin end that are approximately 1½–2 inches thick. As the tenderloin starts to thin, cut the filets thicker to keep them consistent in terms of weight. Reserve the end piece for a medallion.

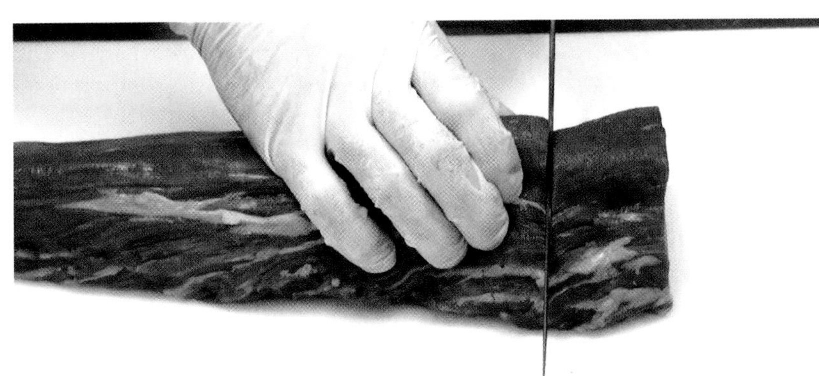

4. To make medallions, wrap one of the reserved end pieces in a clean kitchen towel and use the towel to form the meat into a disc shape. Use a meat pounder or the smooth side of a mallet (tenderizer) to pound the meat until it is approximately 1 inch thick.

   Note: Medallions can also be produced from irregular or thinner cuts made during fabrication.

Finished chateaubriand, filets, and medallions cut from trimmed tenderloins weigh approximately 1–3 pounds, 5–16 ounces, and 2–4 ounces, respectively.

QUICK TIP

Beef tenderloin cuts from the sirloin end are ideal for roasting in the oven. When cooked properly, roasting produces a well-browned exterior and moist interior.

TECHNIQUE 42

FABRICATING WHOLE PORK LOINS

A pork loin is a primal cut that extends along the greater part of the backbone from about the second rib through the rib and loin area of a hog. A primal cut is a large cut from a whole or a partial carcass. A pork loin is commonly divided into four sections: the blade end (near the shoulder), rib end, center loin, and sirloin end. Fatback is the layer of fat that runs along the back of the hog and is generally trimmed during fabrication.

Fabricating whole pork loins is cost effective and results in cuts such as the tenderloin, back ribs, roasts, and chops. To remain cost effective, it is important to remove as much flesh as possible by cutting along the natural curvature of the bones.

Once fabricated, the cuts can be prepared in a variety of ways. For example, ribs are commonly barbequed over indirect heat, roasts are often tied and roasted in a shallow pan, and pockets can be cut into chops for stuffing.

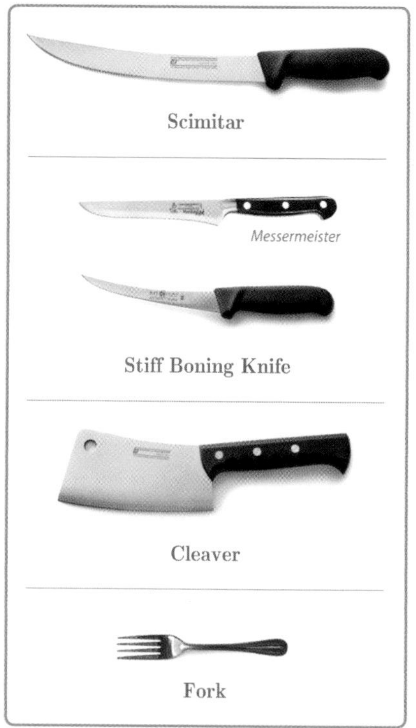

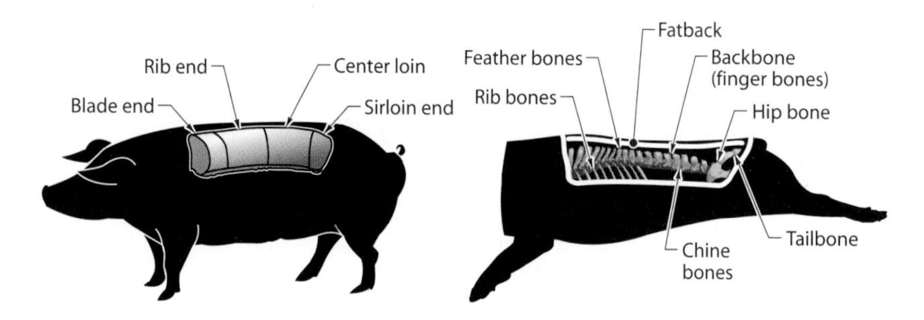

Pork Loin Anatomy

CUTTING TENDERLOINS FROM PORK LOINS

* This technique was executed by a left-handed chef.

1. Place a pork loin fat-side down across the cutting board with the sirloin end facing the knife hand side. Starting at the tapered end of the tenderloin, keep the tip of a scimitar pressed firmly against the chine bones and make smooth, even strokes along the tenderloin.

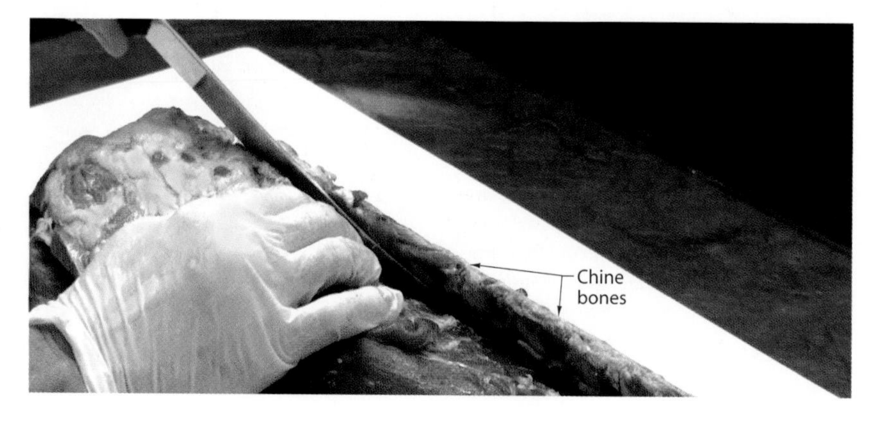

CUTTING TENDERLOINS FROM PORK LOINS (CONTINUED)

* This technique was executed by a left-handed chef.

2. Cut the tenderloin away from the finger bones while using the guiding hand to gently pull the tenderloin away from the loin.

   Note: The finger bones are the small bones that extend under the tenderloin.

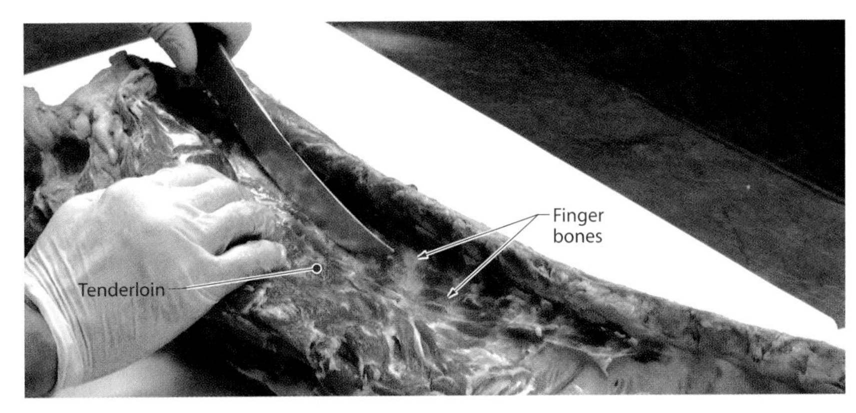

3. Cut the connective tissue between the tenderloin and the loin while lifting the tenderloin with the guiding hand.

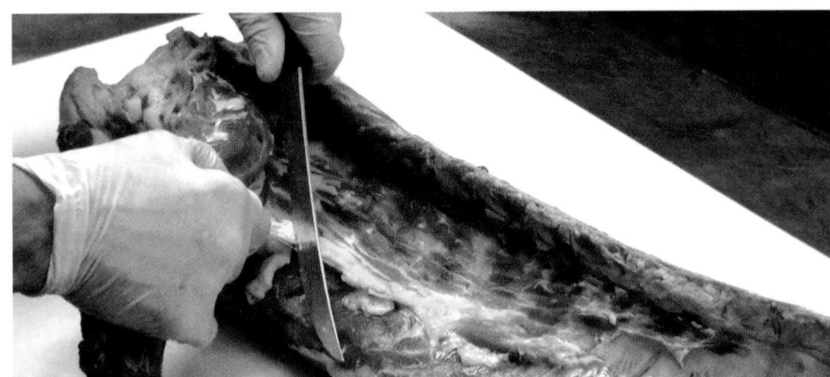

4. Cut along the hip bone to separate the tenderloin from the loin.

Finished pork tenderloins are boneless, tapered strips of flesh that have been cut away from the chine bones, finger bones, and hip bone of a pork loin.

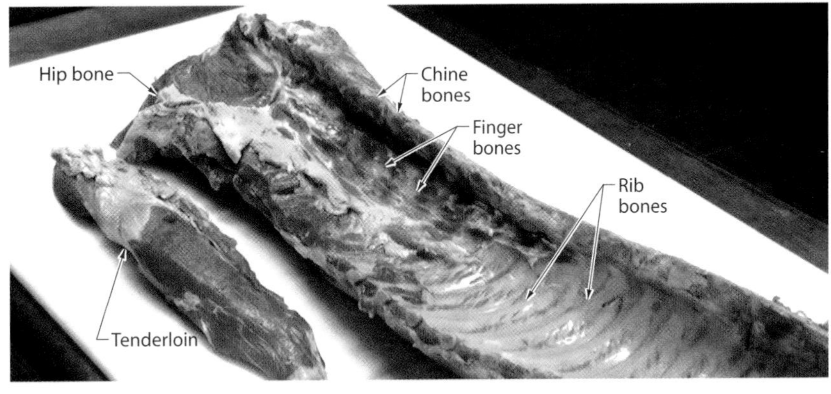

DEBONING PORK LOINS

* This technique was executed by a left-handed chef.

1. After removing the tenderloin, place the backbone of a pork loin across the cutting board with the blade (shoulder) end facing the knife hand side. Start at the rib end and insert the blade of a stiff boning knife along the top side of the ribs. Keep the blade against the ribs and pull the knife toward the blade end using smooth, even strokes to expose the bones.

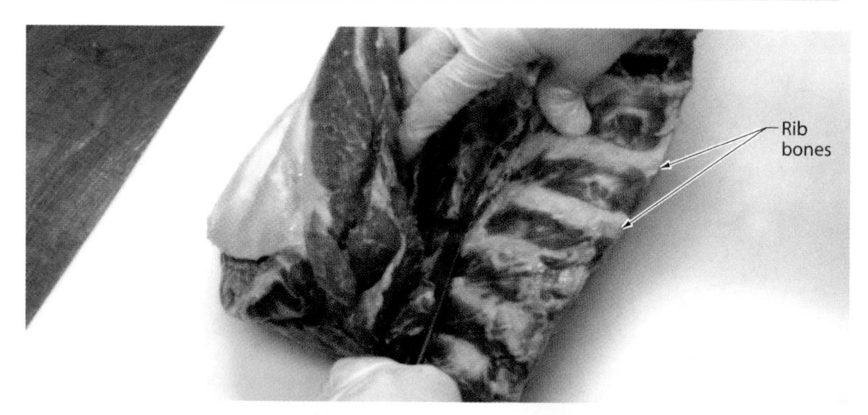

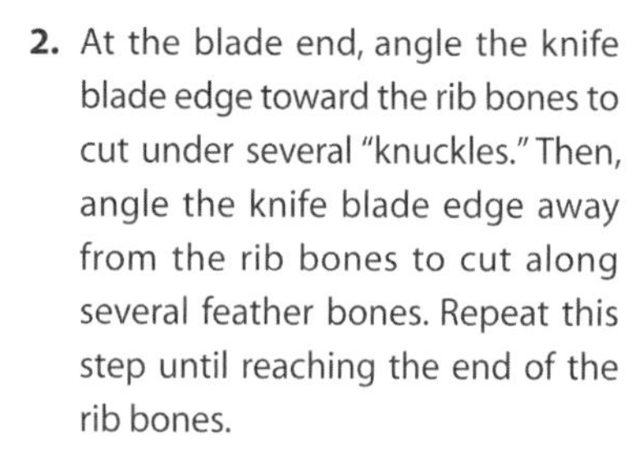

2. At the blade end, angle the knife blade edge toward the rib bones to cut under several "knuckles." Then, angle the knife blade edge away from the rib bones to cut along several feather bones. Repeat this step until reaching the end of the rib bones.

   Note: Knuckles are the rounded areas at the end of each rib bone just above the feather bones.

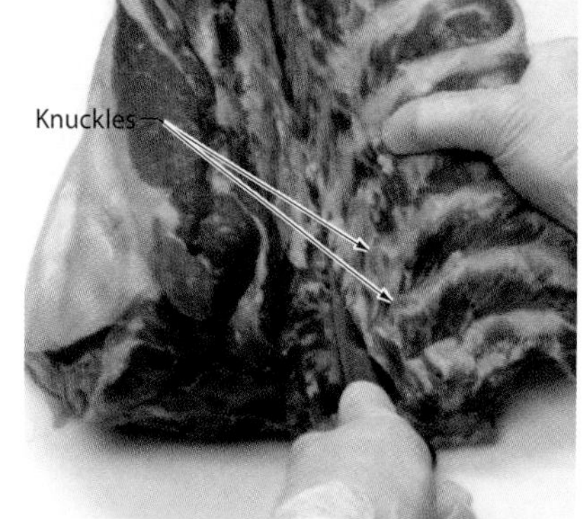

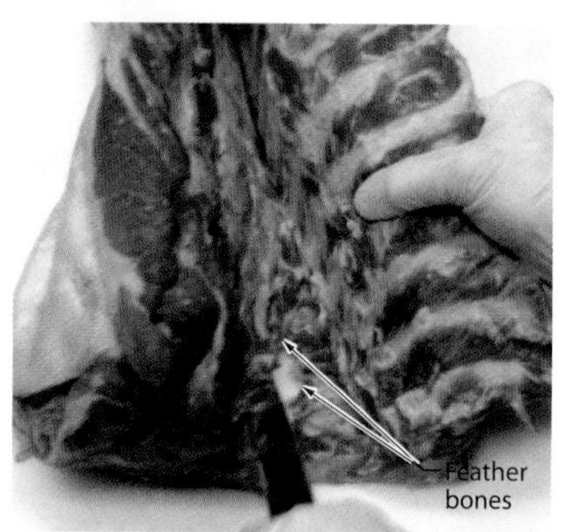

3. Follow the same procedure outlined in Step 2 to cut along the contour of the finger and chine bones.

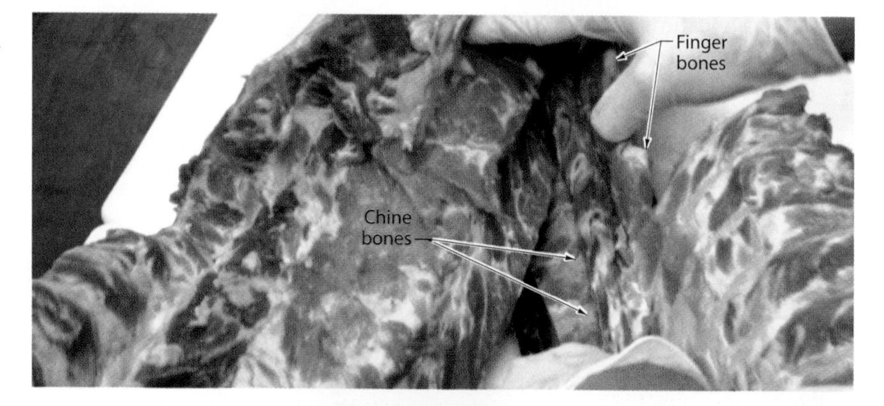

4. Following the contour of the bone, cut the flesh from along the hip bone.

DEBONING PORK LOINS (CONTINUED)

* This technique was executed by a left-handed chef.

5. To finish deboning the loin, cut along the edge of the tailbone.

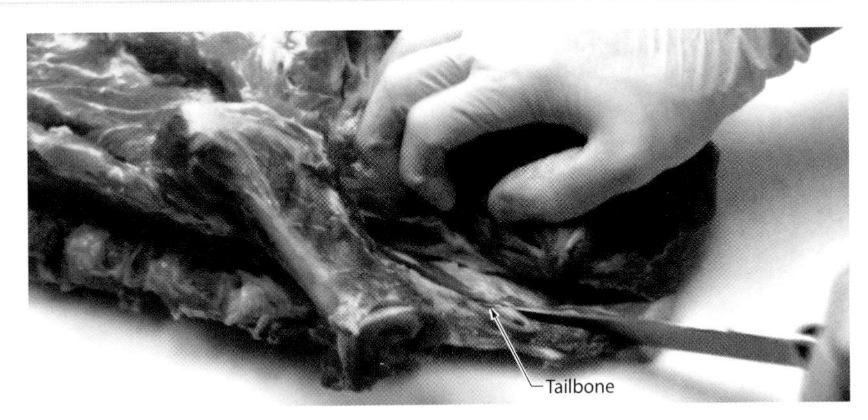

Finished deboned pork loins are void of the tenderloin, rib bones, feather and finger bones, chine bones, hip bone, and tailbone.

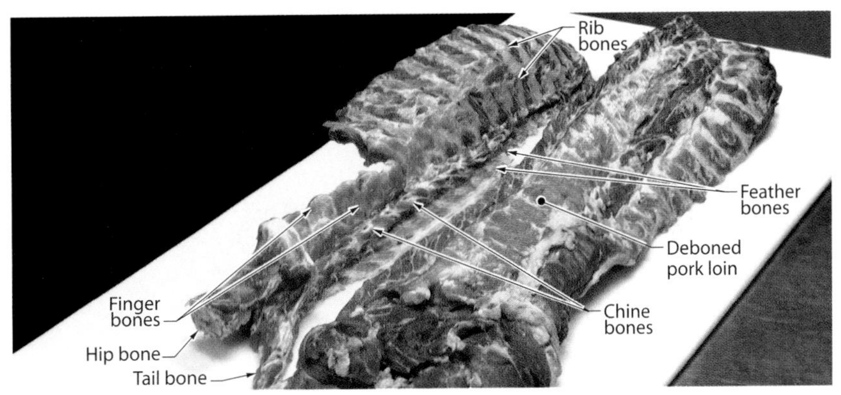

FABRICATING PORK RIBS

* This technique was executed by a left-handed chef.

Note: Place a clean, folded kitchen towel under the ribs to stabilize them during fabrication.

1. Start with the rib and chine bones from a deboned pork loin. With the ribs facing the knife hand side, grip the chine and finger bones with the guiding hand and stand the ribs upright. Grasp a cleaver in the knife hand and strike through each joint between the rib and chine bones until the entire rib section separates from the chine bones.

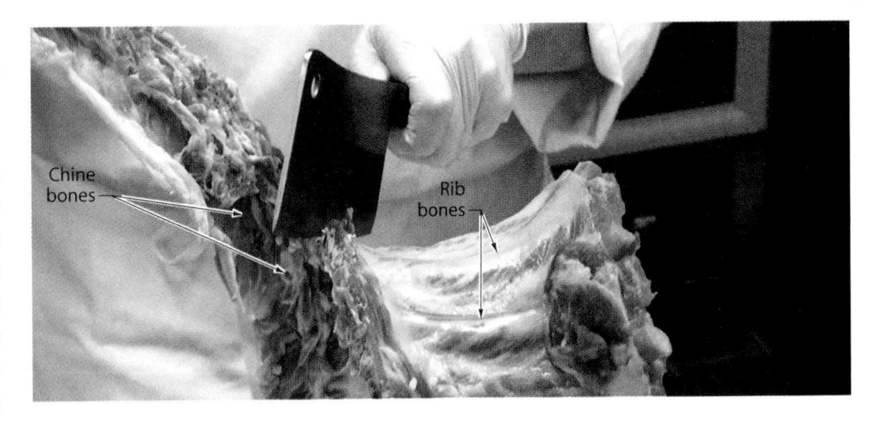

QUICK TIP

If a great deal of force is required to remove the ribs from the chine bones, the cleaver is not striking through the rib joint and should be repositioned to align with the joint.

* This technique was executed by a left-handed chef.

2. Place the ribs top-side down across the cutting board. Start at one end of the rack and insert the tines of a fork under the rib membrane and over the bone. Lift the fork so that the membrane starts to separate from the bones.

3. Secure the ribs with one hand. Grasp the edge of the membrane with the other hand and pull it off the rack.

   Note: If the membrane is slippery, use a clean, dry paper towel to get a better grip.

Fabricated pork back ribs are removed from the loin and include 11–13 ribs that are approximately 3–6 inches long.

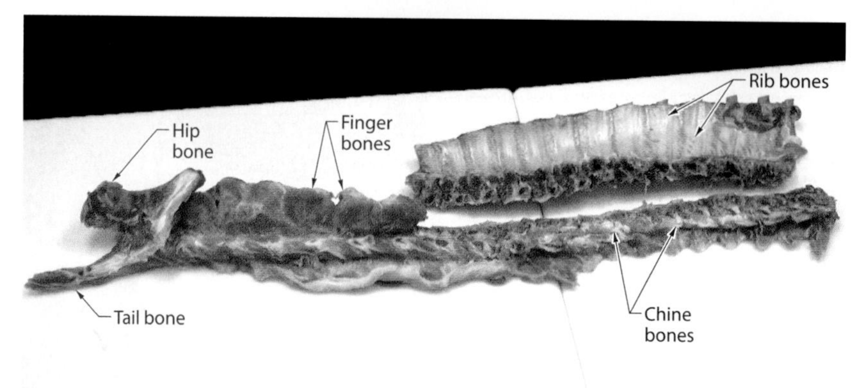

QUICK TIP

Pork is smoked and barbequed more often than any other meat. The smoking process flavors the pork as it cooks over burning wood. Pork is typically basted in barbeque sauce during the grilling or smoking process. Different types of barbeque sauce are served in various regions of the United States.

CUTTING PORK LOIN ROASTS AND CHOPS

* This technique was executed by a left-handed chef.

1. Place a deboned pork loin fat-side down across the cutting board. To cut a sirloin roast, use a stiff boning knife to cut through the natural indentation where the hip bone was located.

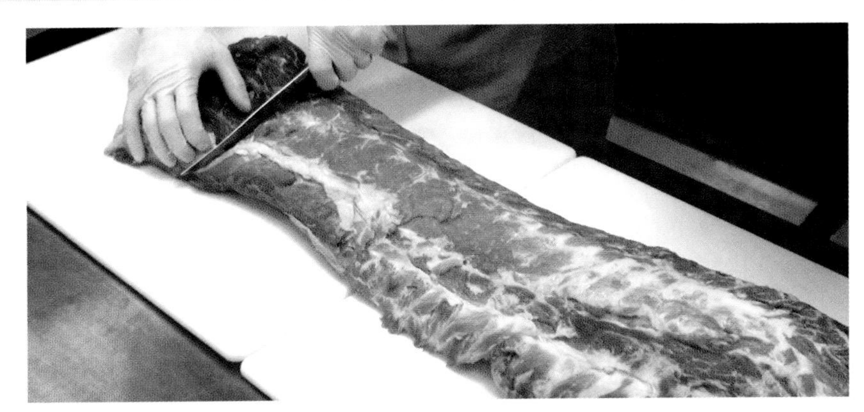

2. To cut a blade roast, use the markings where the ribs were located as a guide and cut through the loin approximately four ribs back from the blade end.

3. Turn the loin fat-side up with the tapered (shorter) side near the edge of the cutting board closest to the body. Make a lengthwise cut along the tapered side that removes a 2–3 inch wide strip of loin and creates an even edge.

   Note: The removed strip of loin can be ground or cut into smaller pieces to use as desired.

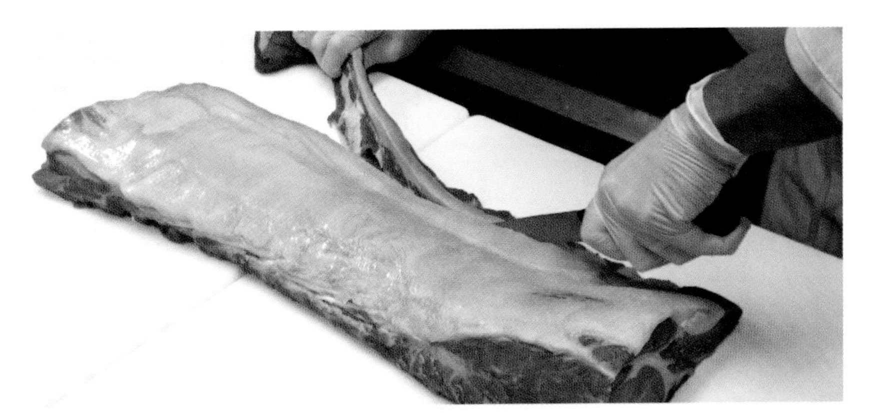

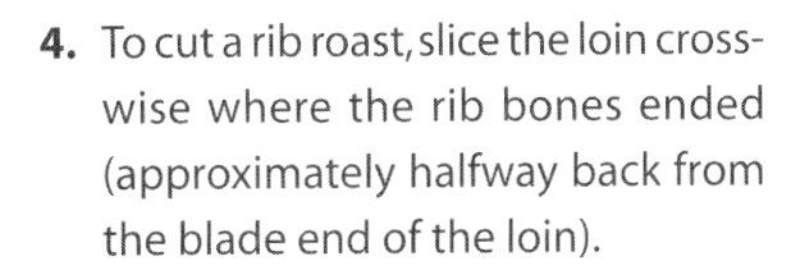

4. To cut a rib roast, slice the loin crosswise where the rib bones ended (approximately halfway back from the blade end of the loin).

   Note: The remaining cut is the center loin, which is commonly fabricated into lean, boneless chops.

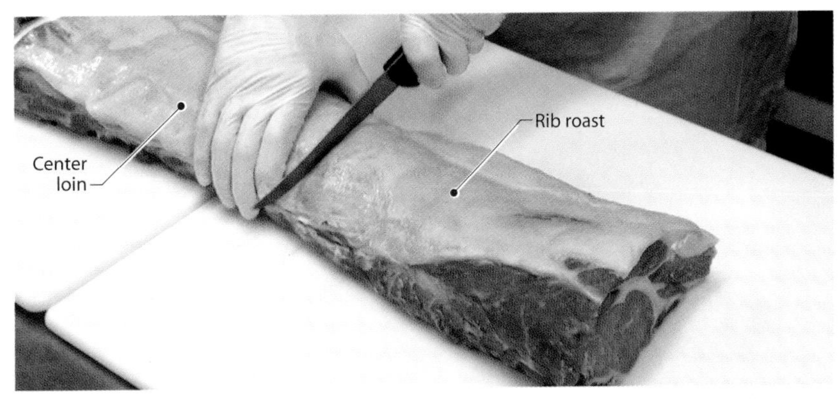

* This technique was executed by a left-handed chef.

5. To cut chops, place the center loin fat-side up across the cutting board. Slice crosswise through the loin at equally spaced intervals to produce consistently sized chops.

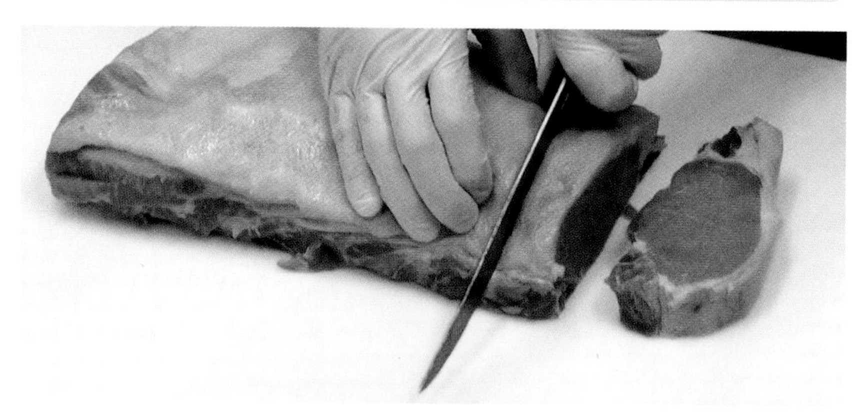

6. If necessary, trim excess fat from the chops.

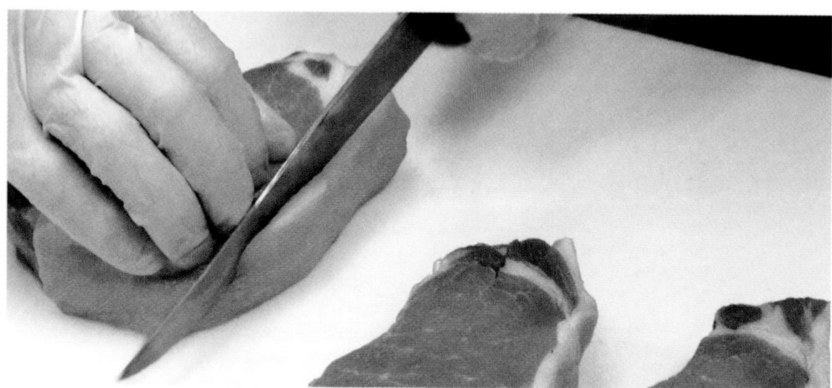

QUICK TIP

In addition to center-cut chops, pork chops can also be fabricated from the blade, rib, and sirloin sections of the loin. Each chop has a unique appearance with varying degrees of marbling (lines of fat that run throughout the meat).

Finished pork loin roasts include sirloin roasts, blade roasts, and rib roasts. The center loin is often cut into chops.

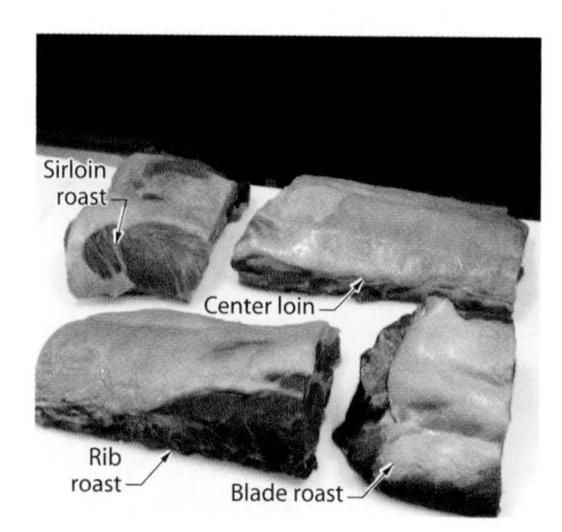

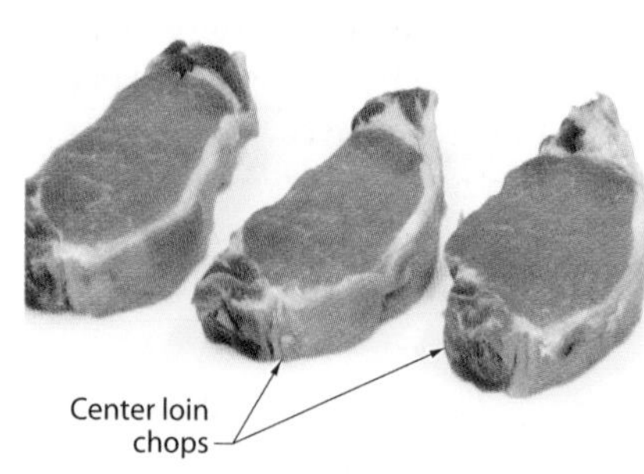

CUTTING PORK CHOP POCKETS

* This technique was executed by a left-handed chef.

1. Place a deboned pork chop on the cutting board with the fatback facing the knife hand side. Place the guiding hand open and flat on top of the chop. With the edge of the blade facing the body, hold a boning knife parallel to the cutting board with the knife point centered along the side of the chop.

2. Hold the knife blade at a 45° angle to the chop and push the knife tip toward the opposite side of the chop without puncturing through the opposite side.

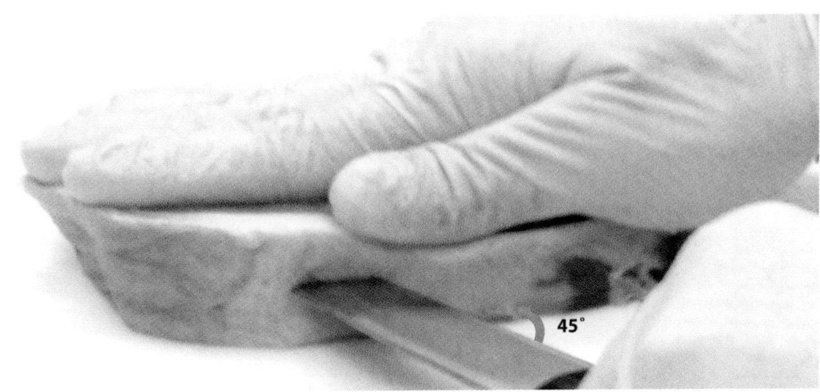

3. Keep the incision from Step 2 approximately 1½–2½ inches and use a slight sawing motion to carefully cut through the middle of the chop without puncturing through the opposite side.

   Note: The chop can be butterflied at this step by cutting through the entire length of the fatback side. Stop the cut just before slicing through the opposite side to allow the chop to be opened flat like a book.

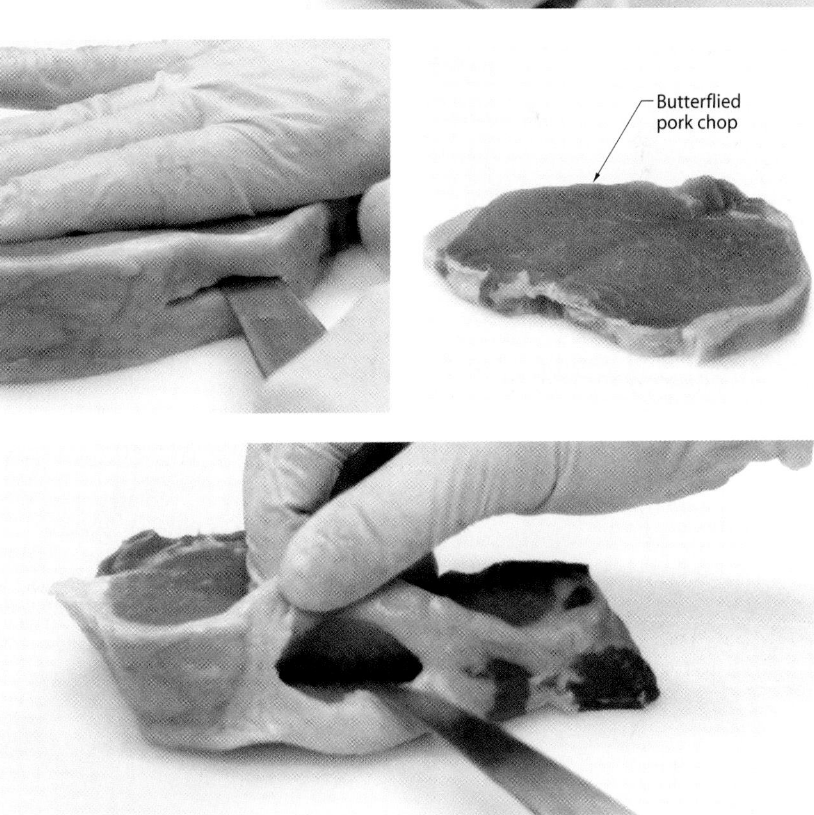

Finished pork chop pockets have an unseen incision through the middle of the chop, creating a compartment that can be seasoned and stuffed.

TECHNIQUE

43 FABRICATING LEGS OF LAMB

VIDEO 43

A leg of lamb is a primal cut of lamb that contains the last portion of the backbone, hip bone, aitchbone, femur (leg bone), hindshank, and tailbone. A leg of lamb includes part of the sirloin, the top round, bottom round, and knuckle meat.

The first step in fabricating a leg of lamb is to trim away the fell and excess fat. Fell is a thin, tough membrane that lies directly under the hide and over the fat layer. After removing the fell and fat, the leg is often prepared partially deboned using a butterflying technique. With this technique, only the hindshank (shank bone) remains and the thigh lies open and flat against the cutting surface. A partially deboned leg of lamb is commonly stuffed, rolled, and tied before it is cooked.

Stiff Boning Knife

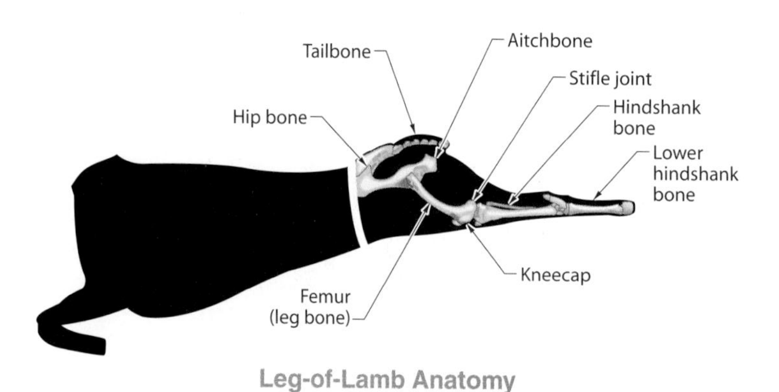

Leg-of-Lamb Anatomy

PARTIALLY DEBONING LEGS OF LAMB

* This technique was executed by a left-handed chef.

1. Place a leg of lamb fat-side up on the cutting board. Using a stiff boning knife, trim away the fell, excess fat, and silverskin with the side of the knife blade. Rotate the leg as necessary during this step.

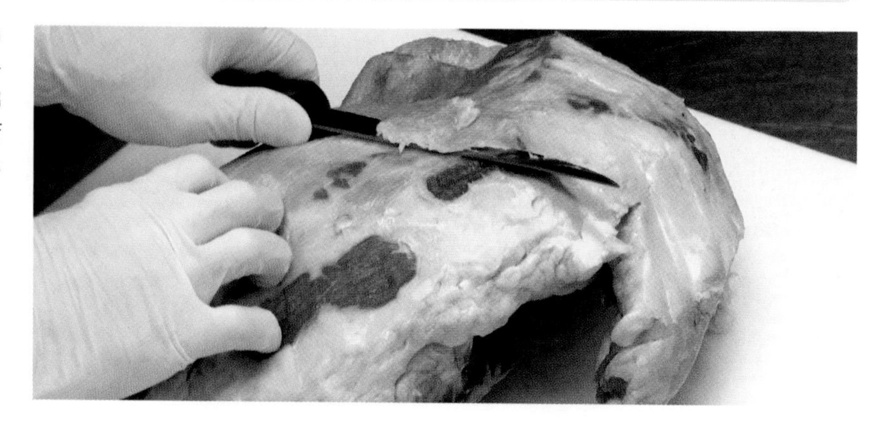

PARTIALLY DEBONING LEGS OF LAMB (CONTINUED)

* This technique was executed by a left-handed chef.

Note: During Step 1, there may be times when it is helpful to do the following:

- Use the knife point to break through the fell, fat, or silverskin layer before trimming it away.
- Use the guiding hand to pull the strips of fell, fat, or silverskin taut as they are trimmed away.

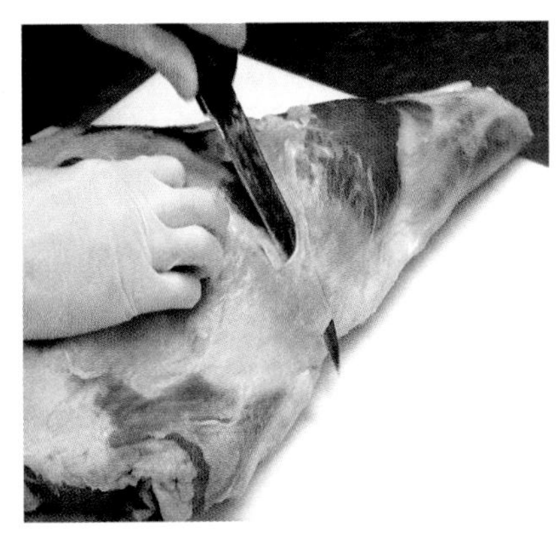

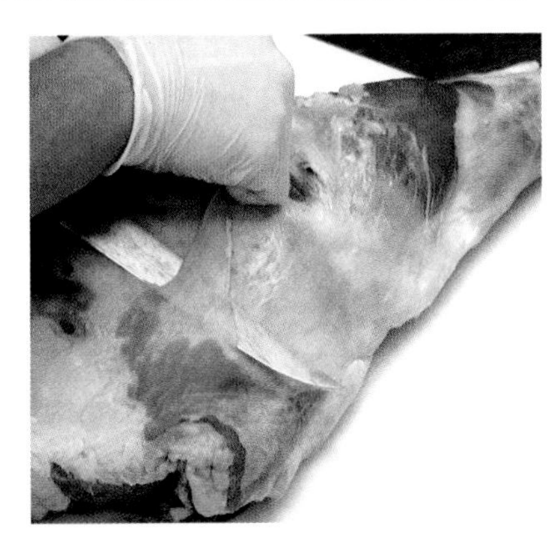

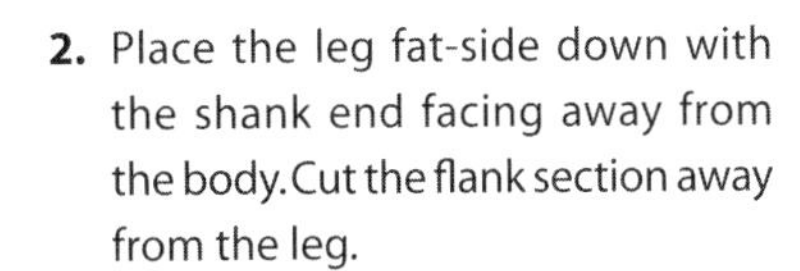

2. Place the leg fat-side down with the shank end facing away from the body. Cut the flank section away from the leg.

   Note: The flank is a portion of muscle near the aitchbone that is loosely connected to the underside of the leg.

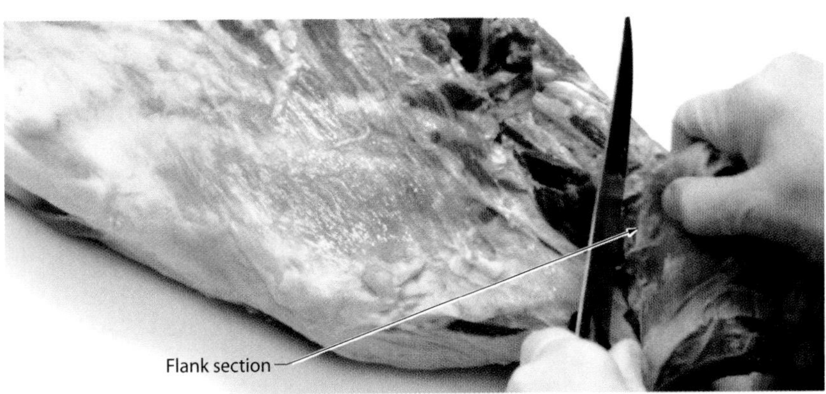

3. Insert the knife tip along the outside edge of the aitchbone. Cut along the top of the aitchbone until reaching the ball joint between the aitchbone and femur (leg bone).

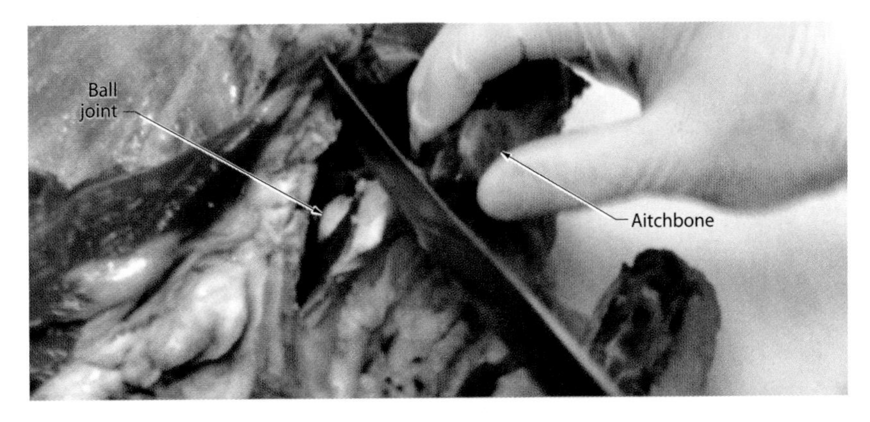

4. Insert the knife tip just below the ball joint and cut back lengthwise to expose the rest of the aitchbone and hip bone. Cut along the left side of these bones to remove the flesh.

* This technique was executed by a left-handed chef.

5. Cut through the ball joint and socket. Then, continue cutting under the aitchbone until it separates from the rest of the leg.

6. Rotate the leg so that the shank end faces the body. Start at the femur ball joint and cut back through the flesh to expose the femur from end to end.

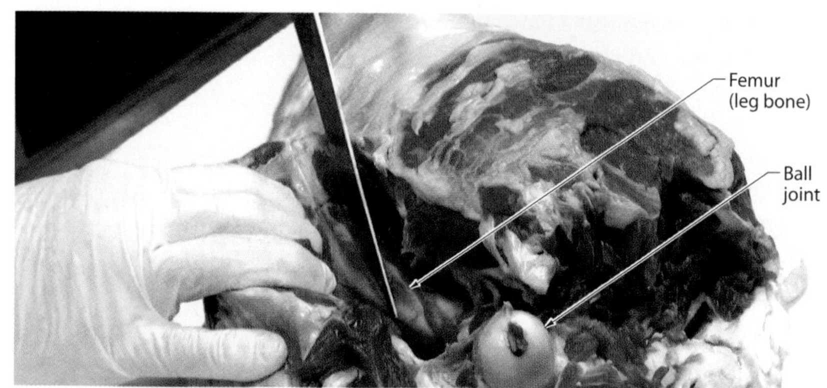

7. Keeping the side of the knife blade flush with the bone, cut the flesh from each side of the femur.

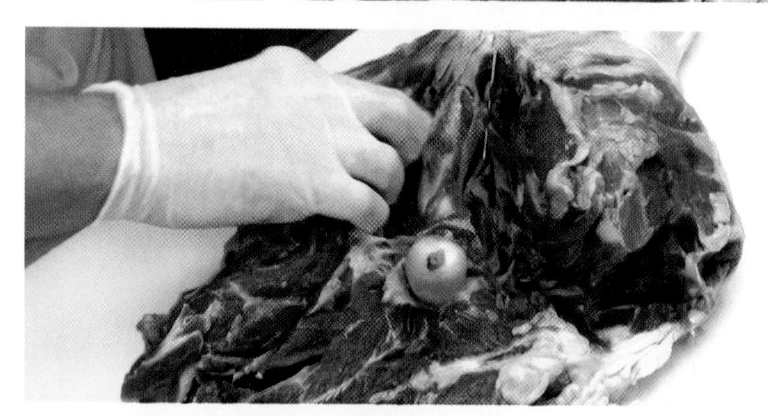

8. Rotate the leg so that the shank end faces the guiding hand side. Cut the flesh from around the ball joint and under the femur until it remains attached to the leg only by the stifle joint.

   Note: The stifle joint connects the femur to the shank.

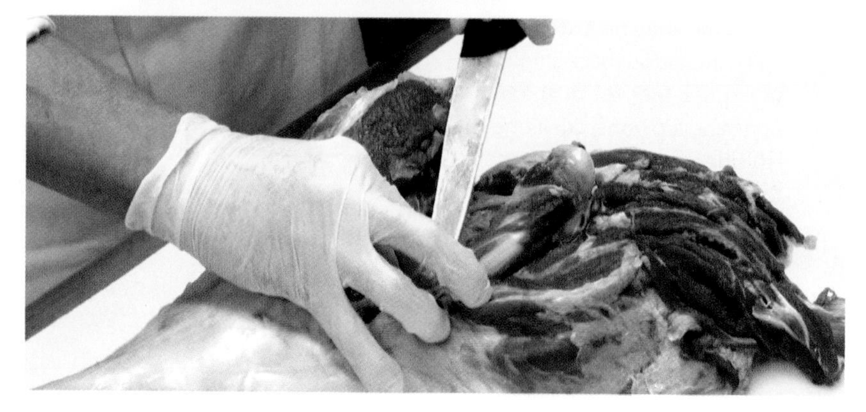

9. Reposition the leg so that the shank end is angled toward the body. To remove the femur, cut around and through the stifle joint. Use the guiding hand to lift and rotate the femur as needed during this step.

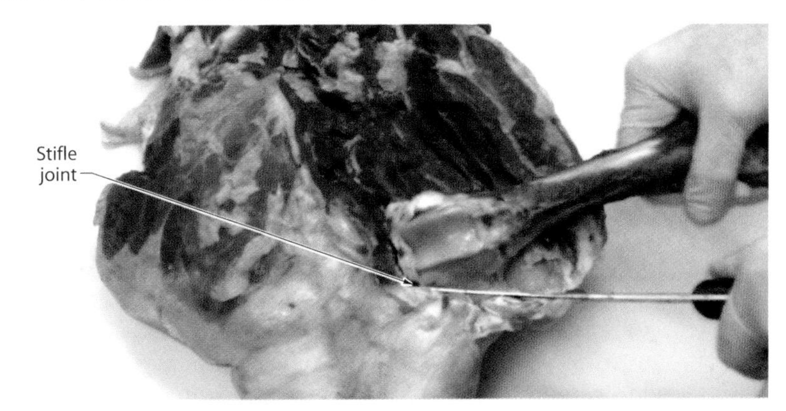

10. Cut the kneecap away from the leg.

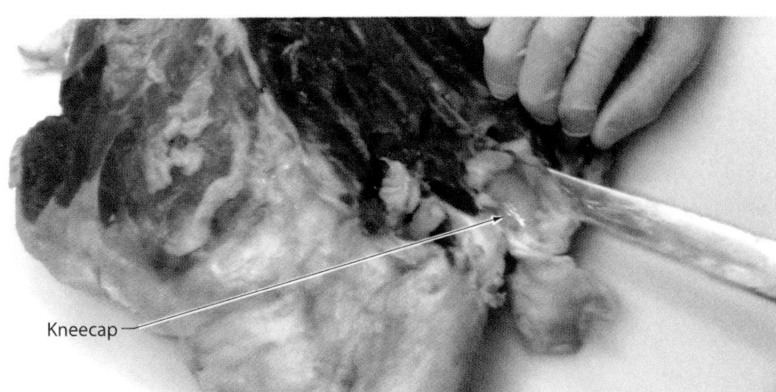

11. Locate the fat pocket under the top-round section of the lamb. Carefully cut around the fat pocket to remove it from the leg. Inspect the leg and remove any excess fat or connective tissue.

    Note: The removed fat pocket contains a gland that may leave an unappealing aroma and flavor when the leg is cooked.

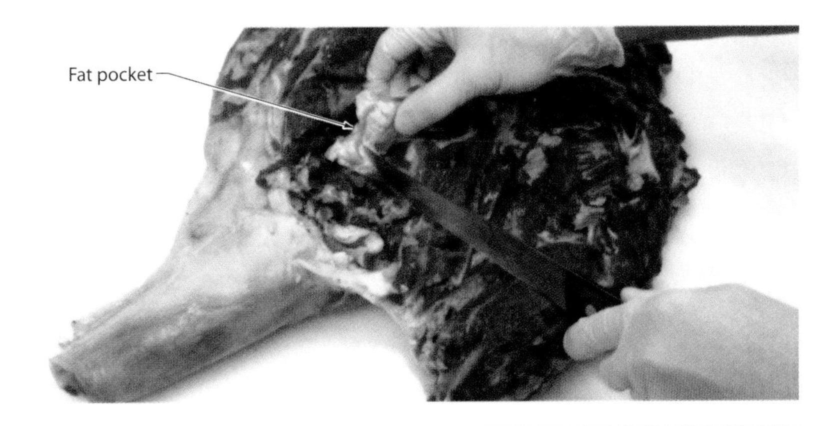

Finished partially deboned legs of lamb are trimmed of fell, fat, and connective tissue and lack all bones except the shank bone.

FRENCHING RIB RACKS

Frenching is the process of removing the meat, fat, and membranes from the end of a bone. The frenching technique is commonly applied to rib racks of lamb, beef, veal, and pork. A rib rack typically comprises nine rib bones located between the shoulder and loin.

When frenching rib racks, the fat cap is removed first. Fat cap is the fat that surrounds a muscle. Rib racks are commonly frenched by cutting the flesh from in between the bones. Then the bones are scraped with the edge of the knife until the connective membranes are loose enough to be pulled away. What remains at the end of the clean bones is referred to as the eye of the rack or the edible portion of meat.

Once frenched, a rib rack is typically cut into single or double chops and grilled or broiled. A frenched rib rack can also be served as a crown roast. A crown roast generally comprises one or two frenched racks formed into a circle to resemble a crown.

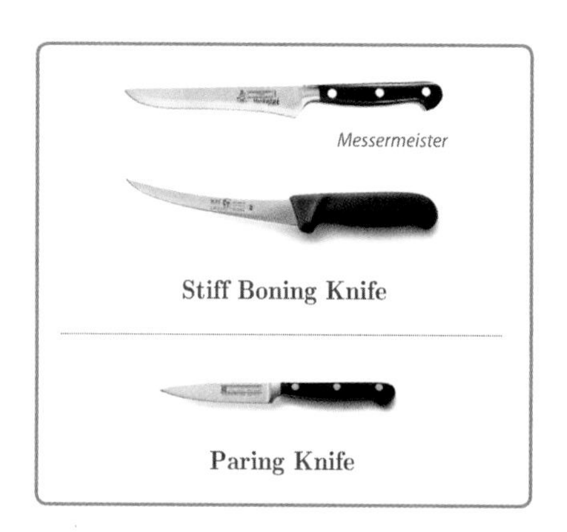

FRENCHING LAMB RACKS

* This technique was executed by a left-handed chef.

1. Stand a lamb rack upright with the eye of the rack against the cutting board and the fat cap positioned toward the guiding hand side. Keep the blade of a stiff boning knife pressed firmly against the rib bones and make small cuts between the rib bones and the fat cap.

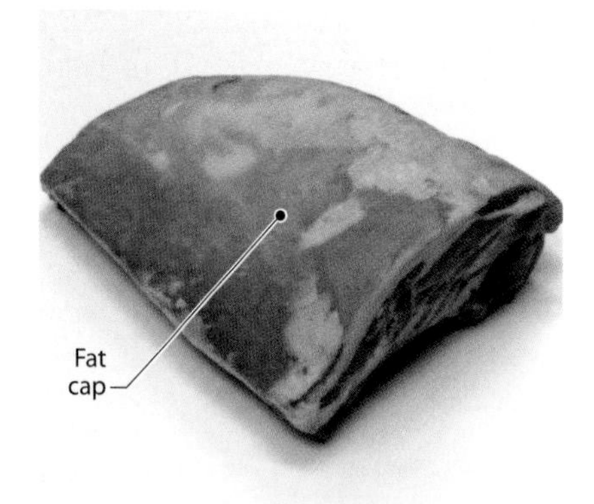

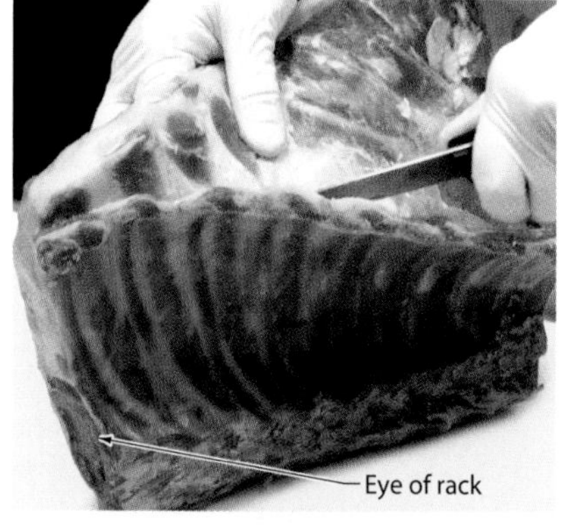

QUICK TIP

All lamb used in foodservice operations must be purchased from USDA inspected plants. At the time of slaughter, the lamb carcass or the inspection tag is stamped with the round USDA inspection stamp, indicating the lamb was slaughtered at an inspected plant. This stamp does not indicate anything about the quality of the meat.

FRENCHING LAMB RACKS (CONTINUED)

* This technique was executed by a left-handed chef.

2. Continue with Step 1 while periodically pulling the fat cap away from the bones until the fat cap can be cut away from the rack.

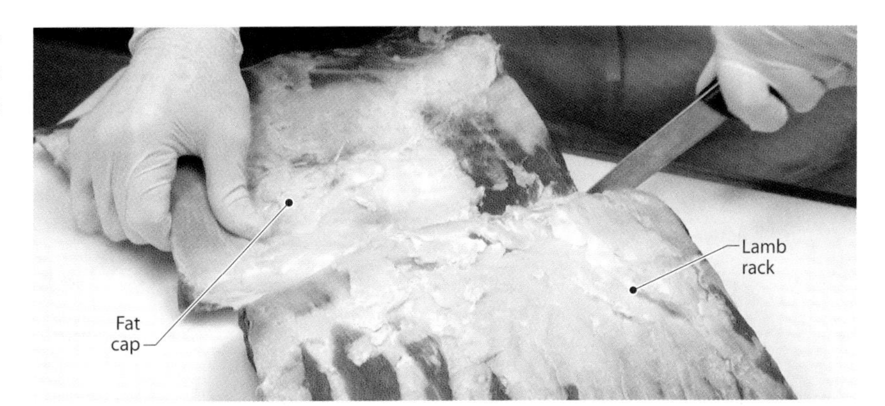

3. Position the rack fat-side down on the cutting board with the ribs facing the guiding hand side. Cut along the bottom edge of the chops to remove the feather bones and tendon.

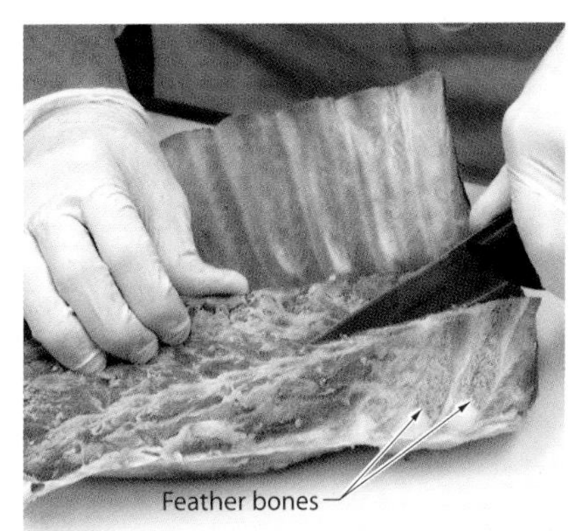

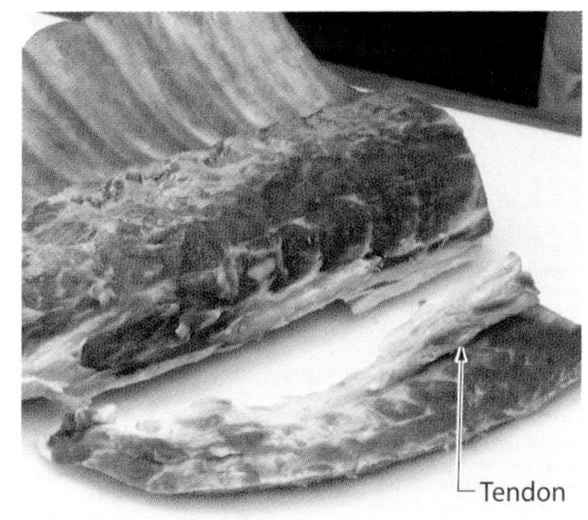

4. Hold the eye of the rack slightly above the cutting board with the guiding hand. At one end of the rack, make a cut along the outside rib bone 1–2 inches from the eye. Repeat this cut on the other side of the rack.

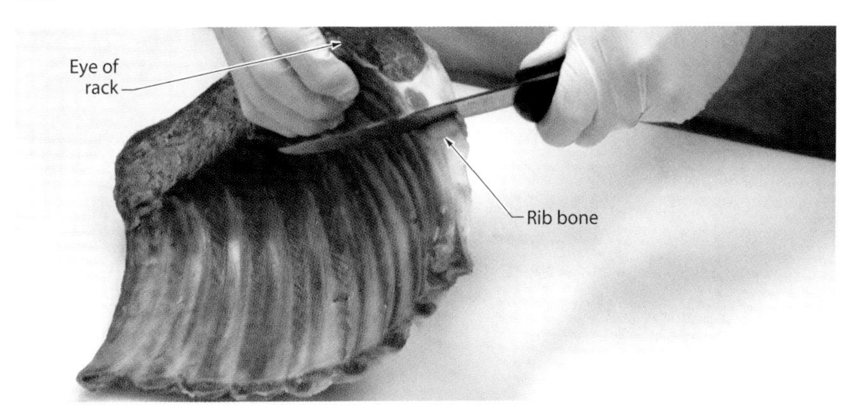

5. Place the rack fat-side up with the ribs facing the knife hand side. Start at the far end of the rack and insert the knife tip into the cut made in Step 4. Cut across the rack and down to the bone by pulling the knife toward the body until reaching the second cut mark from Step 4.

   Note: The cuts made in Steps 4–5 are used as guides for upcoming cuts.

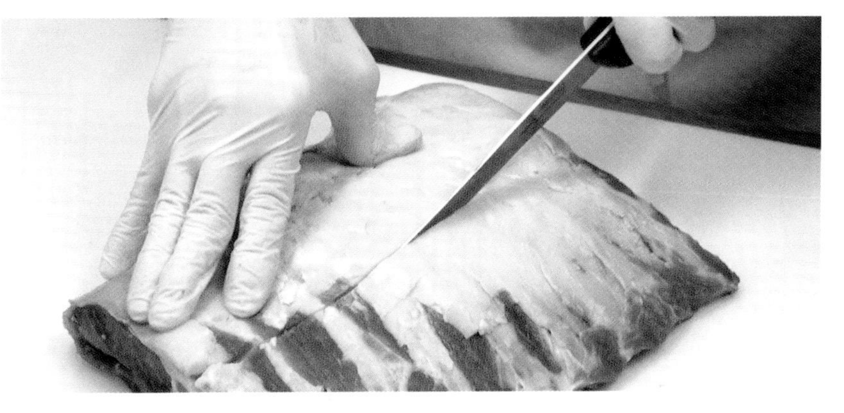

* This technique was executed by a left-handed chef.

6. Start at the far end of the rack and insert the knife blade into the cut mark made in Step 5. Angle the blade to follow the contour of the rib bones and cut down along the top of the ribs to remove excess fat. Repeat this step across the rack.

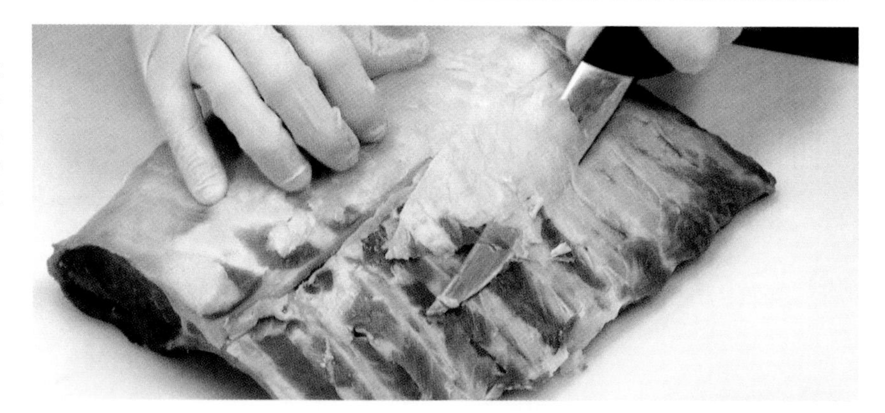

7. Turn the rack so that the bones are facing away from the body and pull the side of the knife blade across the eye of the rack to remove a thin layer of fat.

8. Stand the rack upright with the eye of the rack against the cutting board and the fat side facing the body. Insert the knife blade into the cut mark along the outside of the first rib bone. Cut up the side of the bone to remove the flesh.

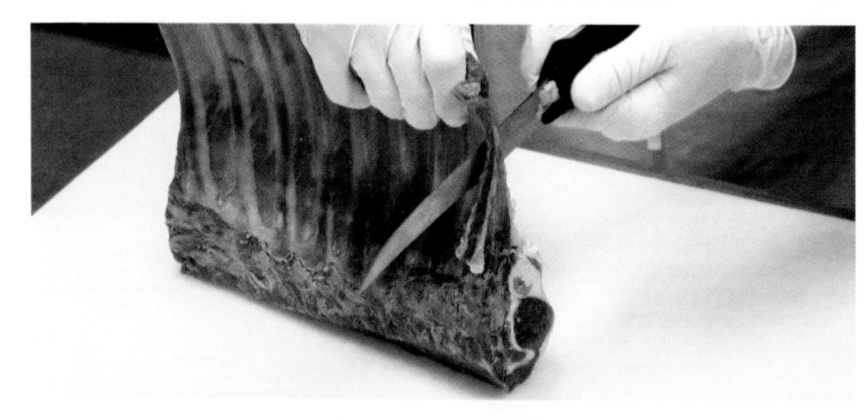

9. To remove the flesh from in between the bones, cut down the inside of the first rib bone to the cut mark and up the side of the adjacent bone. Repeat this process until the flesh is removed from in between all bones.

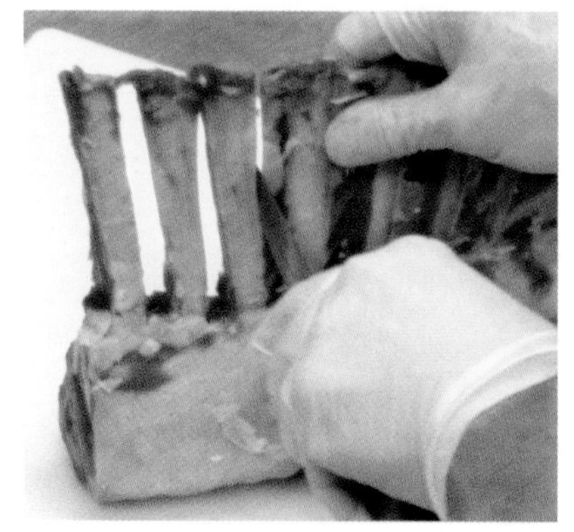

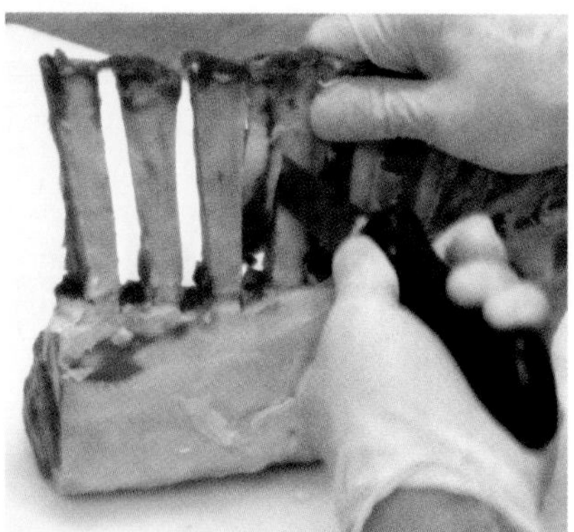

* This technique was executed by a left-handed chef.

10. Place the rack fat-side up on the cutting board with the bones facing away from the body. Starting at the cut mark, use the edge of a paring knife to scrape halfway along the top, sides, and underside of each bone to break through the membrane layer. Use the guiding hand to rotate the rack as needed.

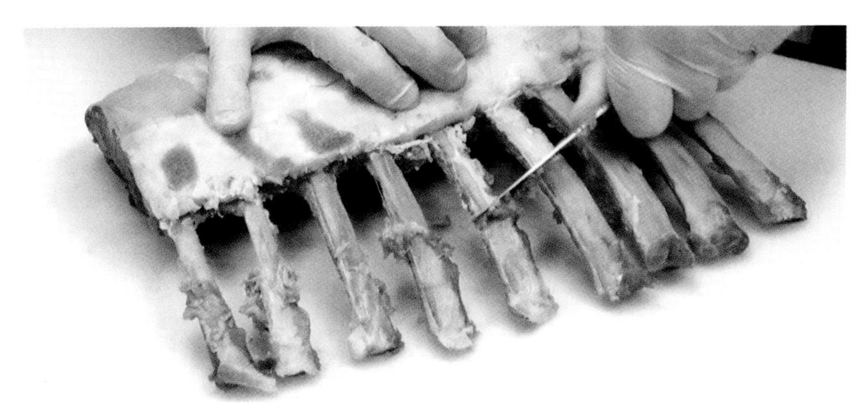

11. Place the rack fat-side up on the cutting board with the bones facing the body. Use Method A or Method B to remove the membrane and bits of flesh along each bone.

Method A:

Use a clean kitchen towel to pull the membrane and any remaining flesh from each bone.

Method B:

Wrap kitchen twine tightly around the base of a bone at the cut mark. Firmly pull the twine toward the end of the bone to remove the membrane and any remaining flesh.

Note: If necessary, scrape the bones with the edge of the paring knife or use the towel to remove any traces of flesh remaining on the bones.

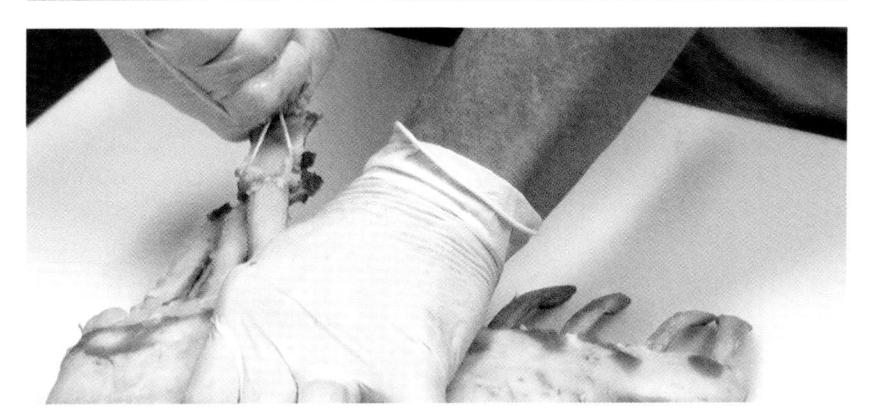

Finished frenched lamb racks have a thin layer of fat covering the eye of the rack and exposed rib bones that are free of fat, flesh, and any connective membranes.

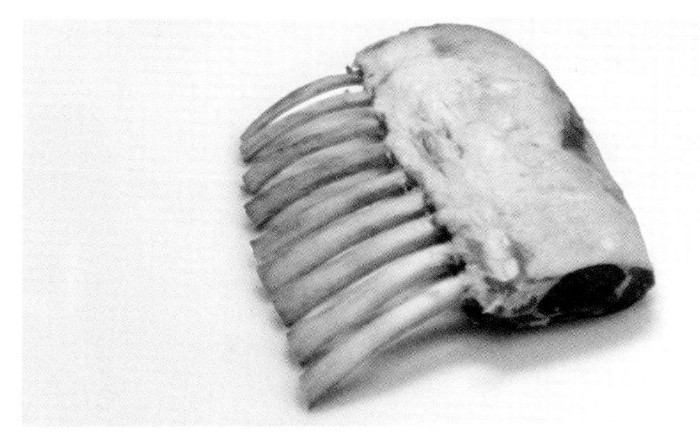

SECTION VII

CARVING AND CUTTING COOKED PROTEINS

- CARVING WHOLE POULTRY
- CUTTING WHOLE COOKED LOBSTERS
- CARVING MEATS ACROSS THE GRAIN ON THE BIAS
- CARVING PRIME RIB ROASTS
- CARVING HAM
- CARVING LEGS OF LAMB

LEARNER RESOURCES

ATPeResources.com/QuickLinks™
Access Code: 583241

TECHNIQUE 45

CARVING WHOLE POULTRY

VIDEO 45

All poultry, including chickens, turkeys, and ducks, are similar in terms of anatomy. Therefore, the procedure used to carve a whole turkey can also be used to carve other types of poultry.

Poultry is carved by following the natural shape of the legs, wings, and breasts. The legs and wings are removed by separating the ball joint from the socket and cutting through the joint that connects them to the carcass. A leg joint is cut through to separate the drumstick and thigh, and wing joints are cut through to separate the drumette, wing tip, and wing paddle. The breast can remain on the carcass and be cut into slices or removed from the carcass and sliced on a cutting board.

Messermeister

Note: Cooked and rested poultry and meats are typically carved while still hot and secured with a chef's fork. For demonstration purposes, the poultry shown in this technique has been carved cold.

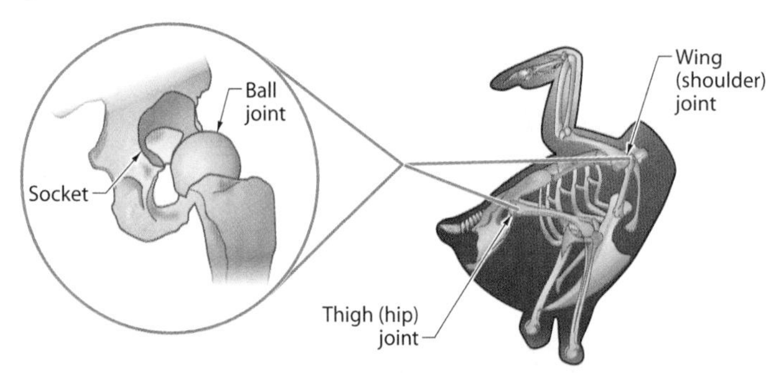

Ball and Socket Joints

CARVING TURKEY LEGS

* This technique was executed by a left-handed chef.

1. Place a cooked and rested turkey breast-side up on the cutting board with the wings facing away from the body.

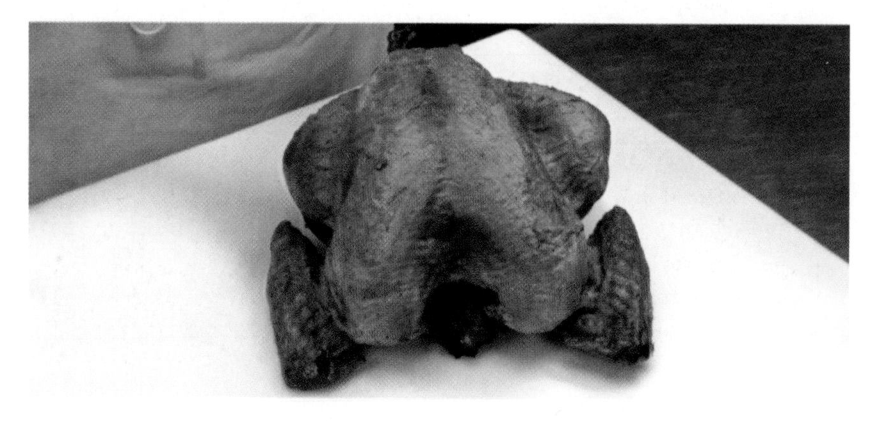

* This technique was executed by a left-handed chef.

2. Hold a leg with the guiding hand and use a stiff boning knife to cut through the skin that connects the leg to the carcass.

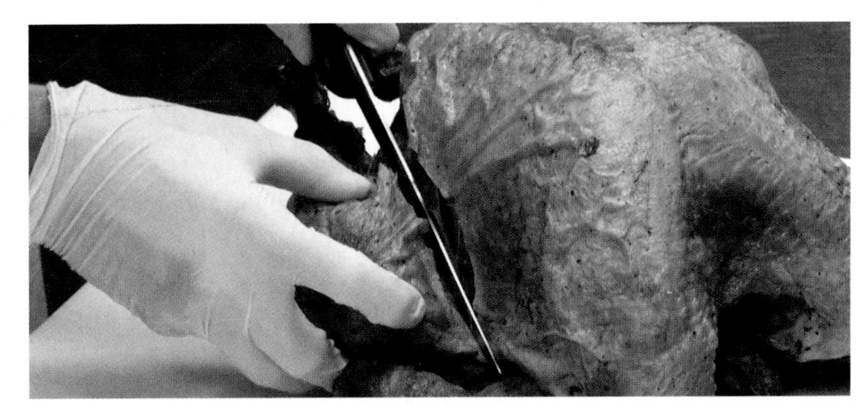

3. Pull the leg away from the carcass to dislodge the thigh (hip) ball joint from the socket.

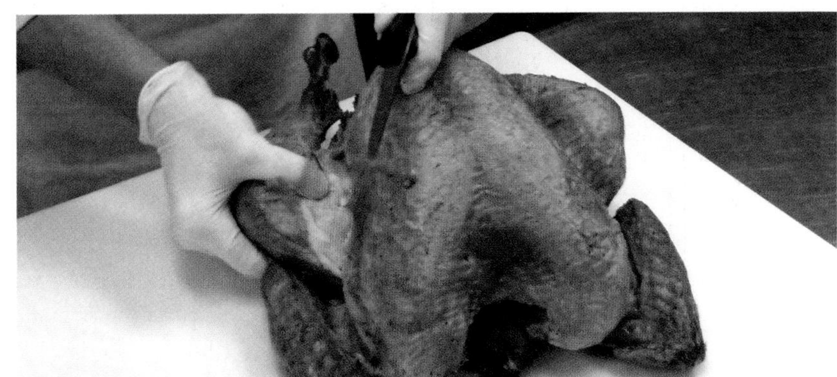

4. Cut through the thigh joint and down through the skin to separate the leg from the carcass.

5. To separate the thigh from the drumstick, hold the end of a drumstick bone with the guiding hand and stand the leg upright on the cutting board. Position the knife blade at the indentation between the drumstick and thigh. Remove the thigh by cutting straight down through the joint connecting the drumstick to the thigh.

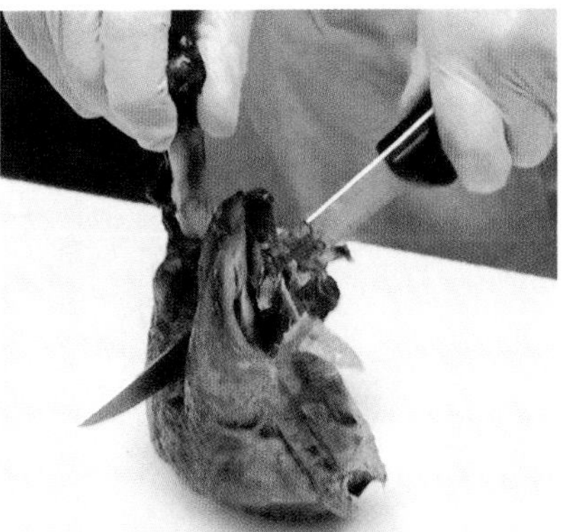

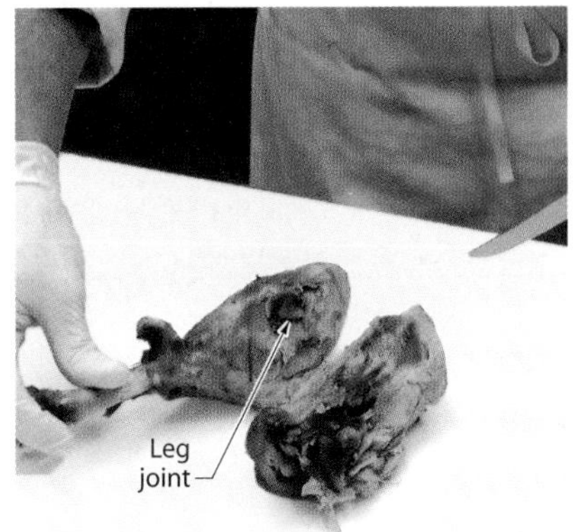

* This technique was executed by a left-handed chef.

6. To carve the drumstick into slices, hold the drumstick upright with the guiding hand. Make vertical slices down the side of the drumstick until the knife blade is flush with the bone.

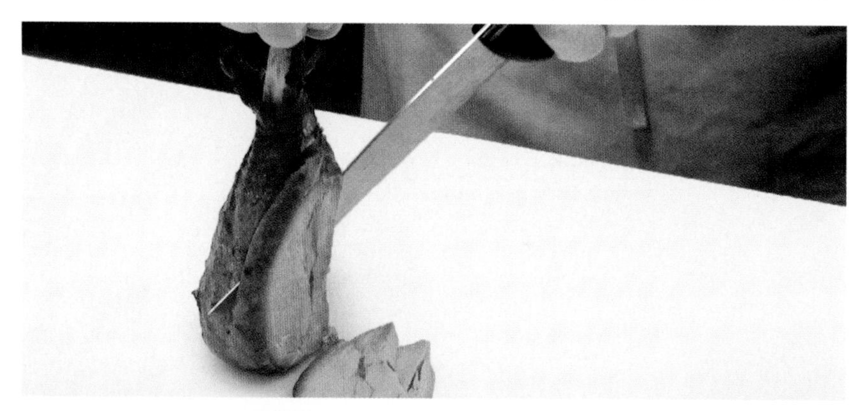

7. Rotate the drumstick toward the body and continue making vertical slices down the side of the drumstick until the knife blade is once again flush with the bone. Repeat this step until all the flesh is removed from the bone.

8. To carve the thigh into slices, place the thigh skin-side up on the cutting board. Secure the thigh bone with the guiding hand and cut slices of equal thickness until the knife blade nears the bone.

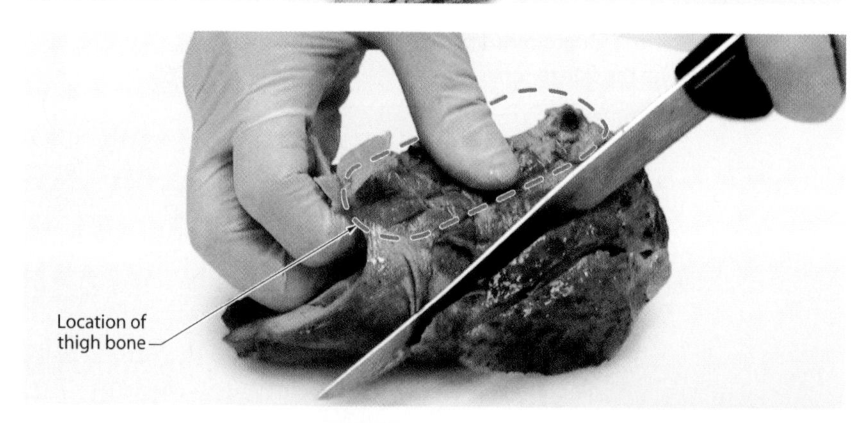

9. Hold the end of the thigh bone with the guiding hand and stand the thigh upright. Cut down the sides of the thigh bone to remove any remaining flesh. Repeat Steps 1–9 to carve the other leg.

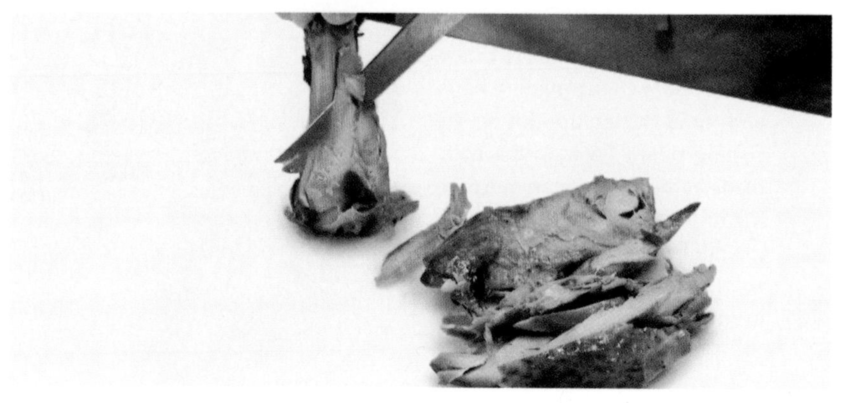

REMOVING TURKEY WINGS

* This technique was executed by a left-handed chef.

1. Place the turkey breast-side up with the wings near the side of the cutting board closest to the body. Secure the bird with the guiding hand and position a stiff boning knife where a wing attaches to the carcass.

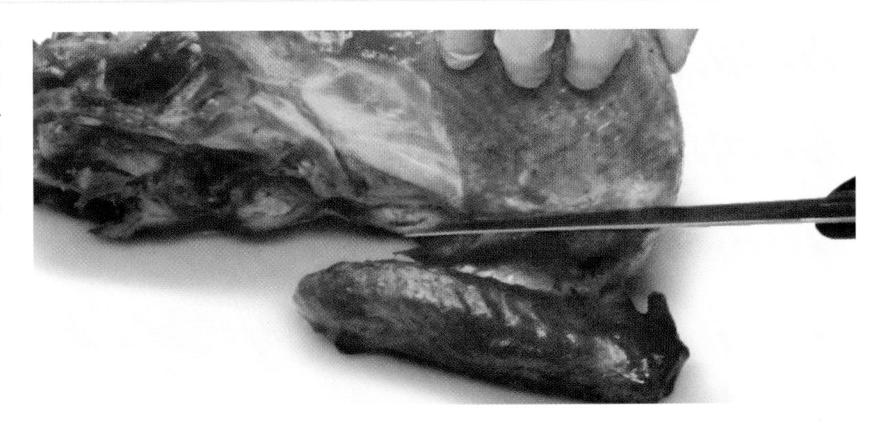

QUICK TIP

Let a cooked turkey rest 20–30 minutes before carving. This allows the juices that were forced to the center of the turkey during the cooking process to redistribute throughout the turkey, keeping it moist and flavorful.

2. Cut through the skin and flesh until the knife reaches the wing (shoulder) joint.

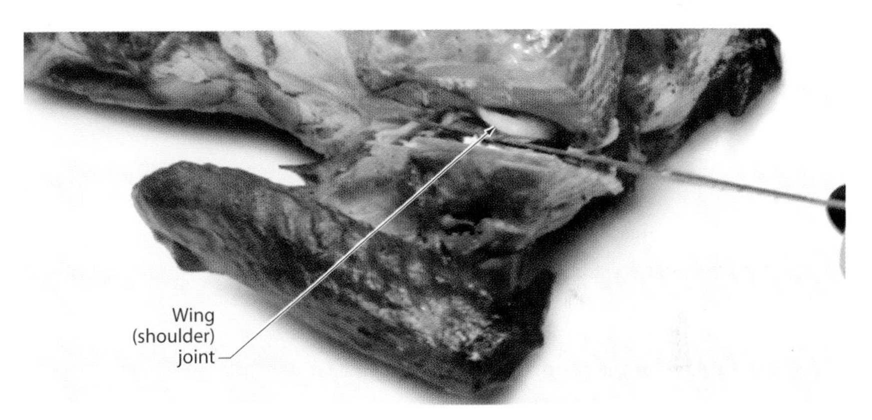

3. Pull the wing away from the bird to dislodge the ball joint from the socket.

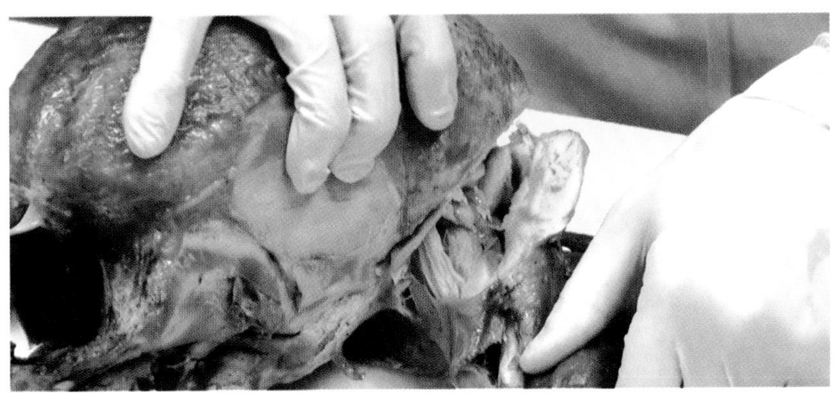

* This technique was executed by a left-handed chef.

4. Cut through the wing ball joint and down through the skin to separate the wing from the carcass.

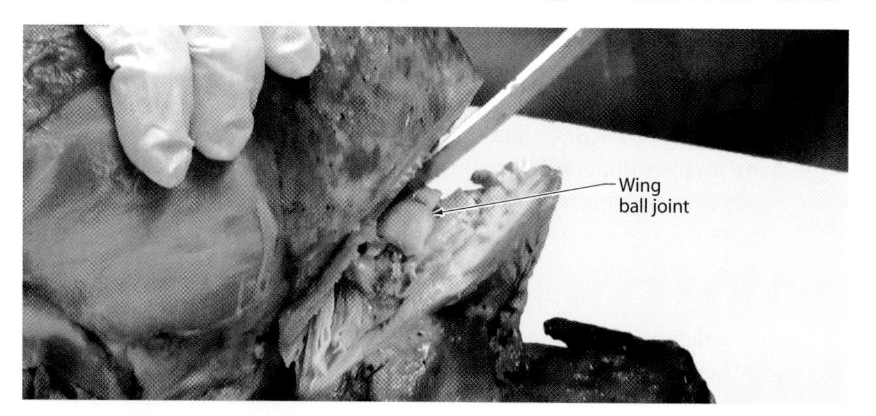

5. If desired, cut through the joints where the wing naturally bends to separate the wing into three separate portions: the drumette, the wing paddle, and the wing tip. Repeat Steps 1–5 to remove the other wing.

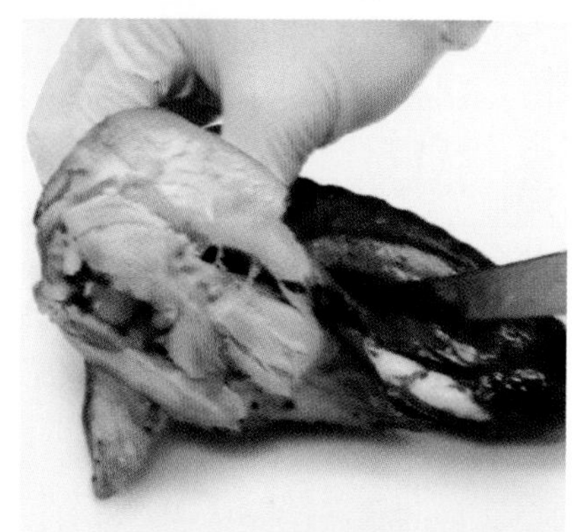

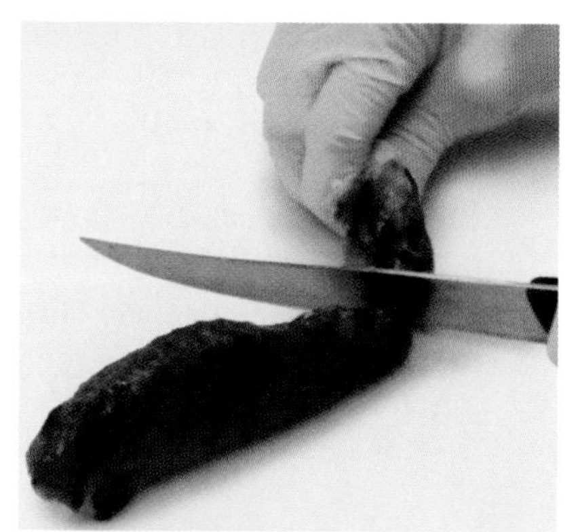

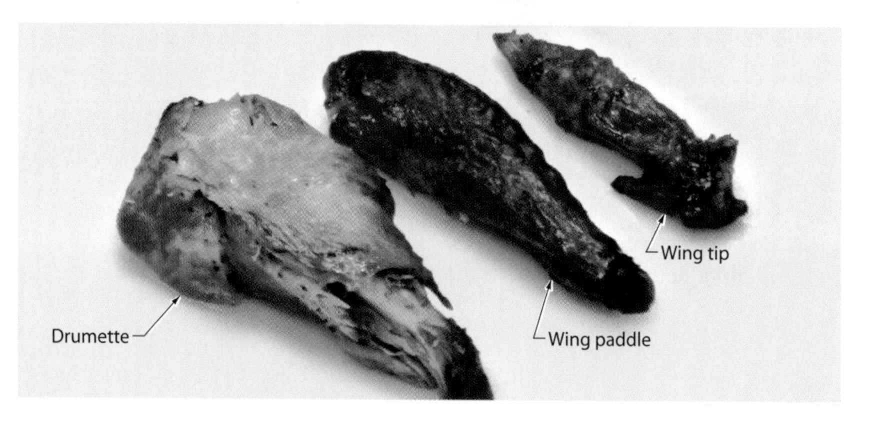

QUICK TIP

Poultry wings are commonly referred to by section, with the drumette known as the first section, the wing paddle as the second section, and the wing tip as the third section. There are also many names for the wing paddle, including the flat, midjoint wing, and winglet.

CARVING TURKEY BREASTS

* This technique was executed by a left-handed chef.

1. Place the turkey breast-side up with the tail end near the side of the cutting board closest to the body and secure with the guiding hand. Hold a slicer parallel to the cutting board and make a horizontal cut through the flesh along the bottom of the breast just above the wing joint.

   Note: All slices will stop at this horizontal base cut.

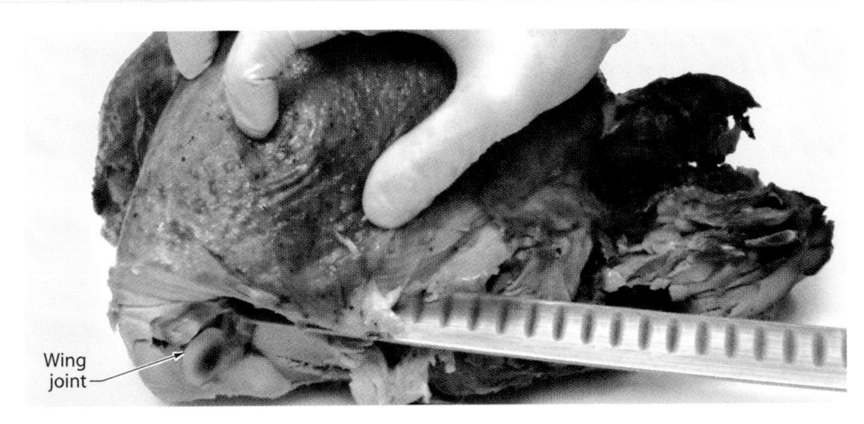

2. Position the slicer on the breast slightly above the area where the wing was attached. Keep the blade perpendicular to the base cut from Step 1 and cut downward to remove a thin slice of flesh from the breast.

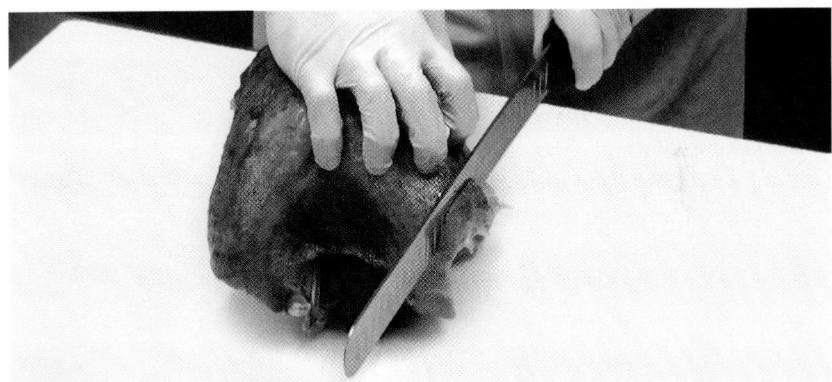

3. Reposition the slicer at a slightly higher point on the breast and cut downward to make another slice. Repeat this step until the entire breast is cut into slices of equal thickness. Repeat Steps 1–3 to carve the other breast.

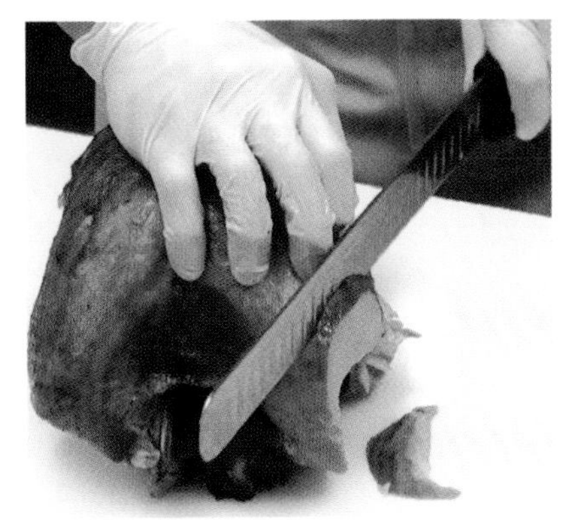

QUICK TIP

If stuffing is to be served with turkey, it is safer to bake the stuffing outside the turkey rather than placing it in the cavity of the raw bird to bake. This reduces the risk of contaminating the stuffing with microorganisms from the raw turkey.

REMOVING AND CARVING TURKEY BREASTS

* This technique was executed by a left-handed chef.

1. Place the turkey breast-side up with the neck end on the side of the cutting board closest to the body and secure with the guiding hand. Use a stiff boning knife to cut along one side of the breast bone.

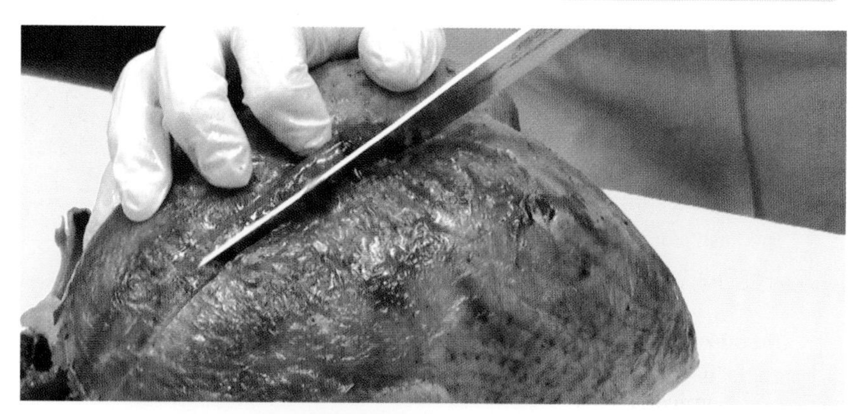

2. Continue cutting down the side of the breast bone and along the keel bone and rib cage while pulling the flesh away from the carcass with the guiding hand.

   Note: Because the keel bone divides the two breast muscles, it is used to help guide the cut.

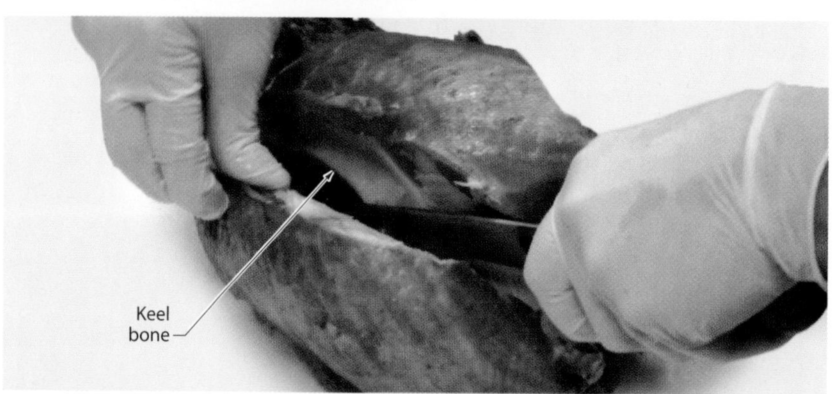

3. Cut along the bottom of the breast near the rib cage to separate the breast from the carcass.

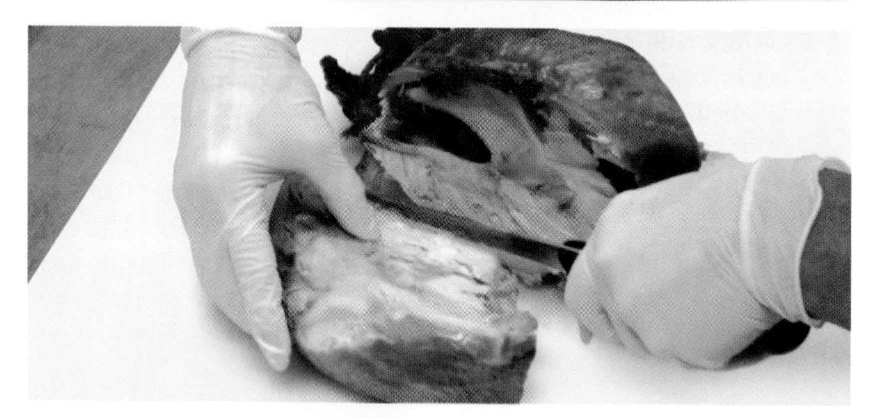

4. Place the breast skin-side up with the tapered end toward the knife hand side and secure with a claw grip. Use a slicer to cut the breast crosswise into slices of equal thickness. Repeat Steps 1–4 to remove and carve the other breast.

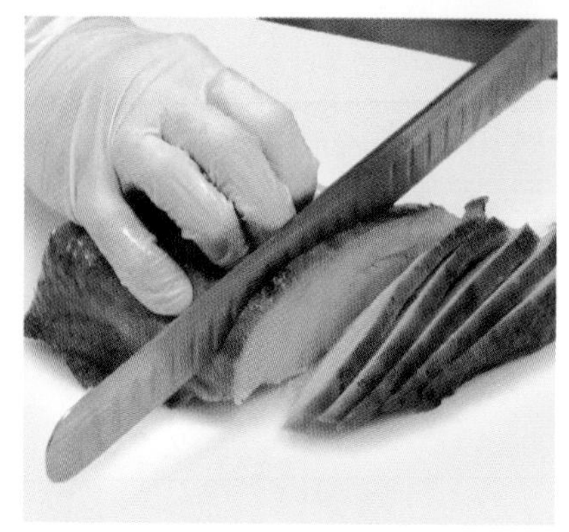

CUTTING WHOLE COOKED LOBSTERS

46 TECHNIQUE

Removing cooked lobster meat from the shell is done efficiently with kitchen shears, a chef's knife, and one's hands. Lobster meat is typically extracted from the tail, knuckles, and claws but can be removed from the walking legs and body (cephalothorax) if desired. Once removed, lobster meat is ready for use in a variety of preparations. For example, lobster meat is often added to spring rolls and salads and is the featured ingredient in lobster bisque and lobster roll sandwiches. The shells of lobsters can be reserved and used to make a flavorful stock for soups and sauces.

Kitchen Shears

Chef's Knife

REMOVING COOKED LOBSTERS FROM THE SHELL

* This technique was executed by a left-handed chef.

1. Hold the top side of a cooked lobster with the guiding hand. Use the dominant hand to break off the claws where they attach to the body (cephalothorax). Set the claws aside.

2. Hold the lobster tail with one hand and the body with the other. Twist and bend the tail until it separates from the body.

QUICK TIP

To remove lobster meat from the walking legs, first bend each leg until it separates from the body. Then break each leg apart at the joint. Finally, squeeze the leg sections to extrude the meat.

* This technique was executed by a left-handed chef.

3. Hold the top side of the lobster tail with the guiding hand. Use kitchen shears to cut along each side of the thin, transparent shell on the underside of the tail. Set the scissors down and pull the shell away from the underside of the lobster.

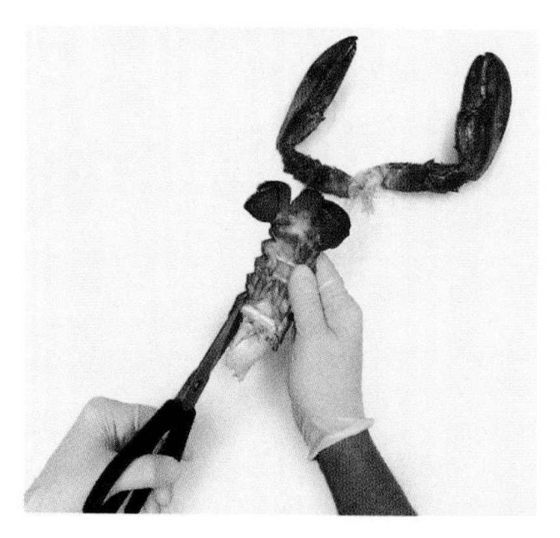

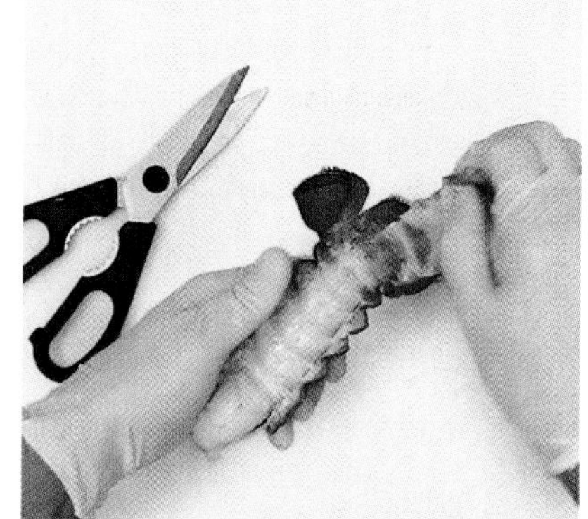

4. Carefully pull the lobster meat away from the tail shell.

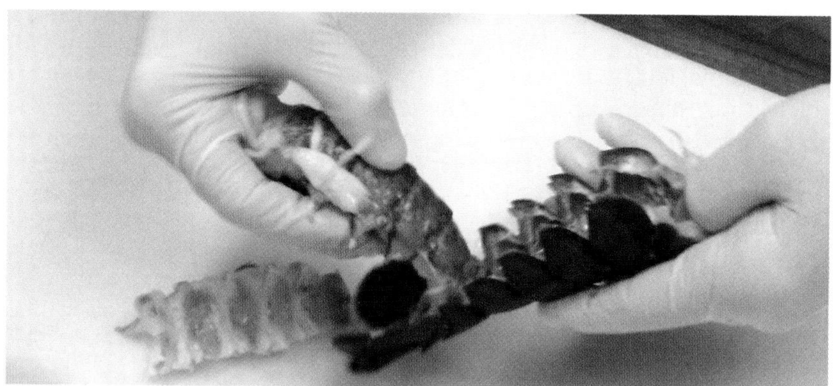

5. To break the knuckles away from the claws, use the hands to bend and twist the joints along each leg.

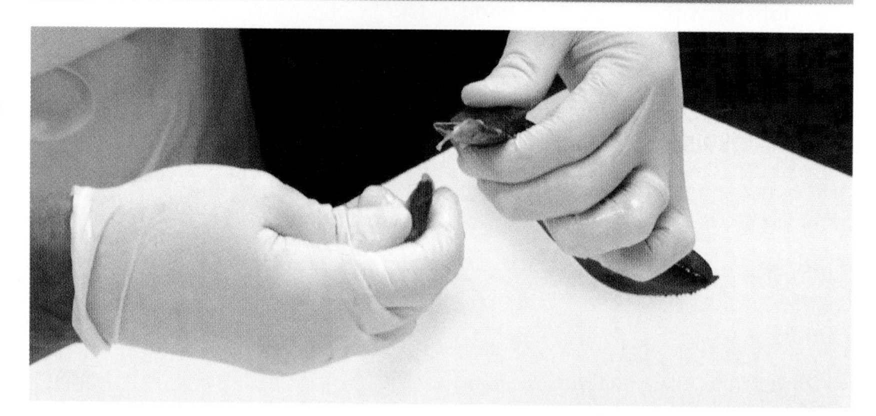

6. Cut open each knuckle with kitchen shears. Set the scissors down, pry each knuckle shell open, and pull the meat from the knuckle shells.

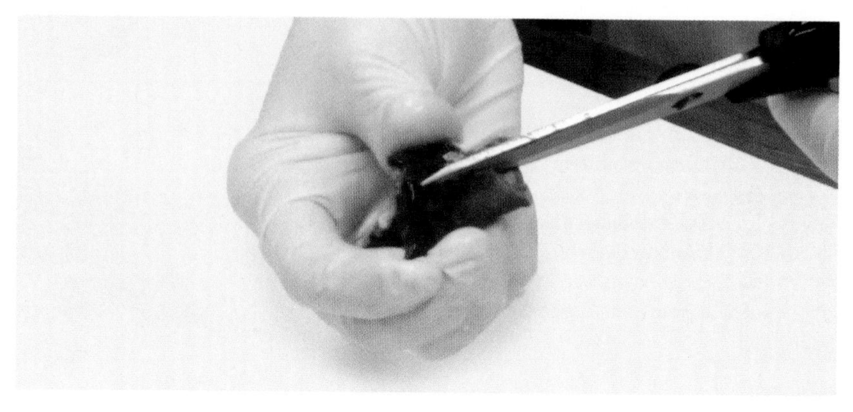

REMOVING COOKED LOBSTERS FROM THE SHELL (CONTINUED)

* This technique was executed by a left-handed chef.

7. Hold a claw between both hands. Gently bend the smaller side of the claw down and up several times to break the shell at the joint. Carefully pull the smaller shell away to reveal a section of claw meat.

   Note: To keep the meat intact, use caution when bending and removing the small claw shell.

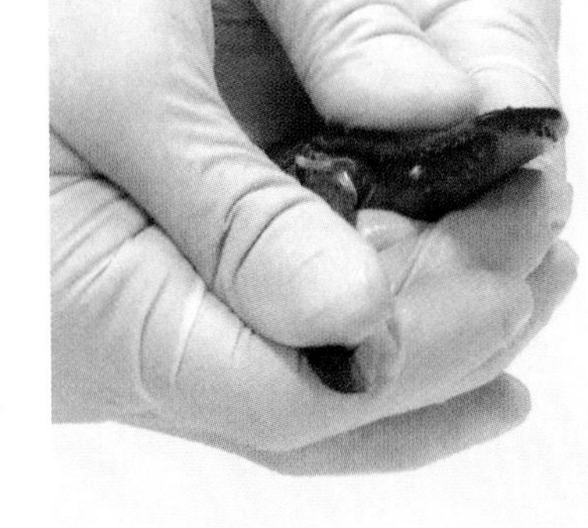

8. Place the claw on the cutting board and hold the pincers with the guiding hand. Strike the widest part of the claw with the spine of a chef's knife to crack the shell.

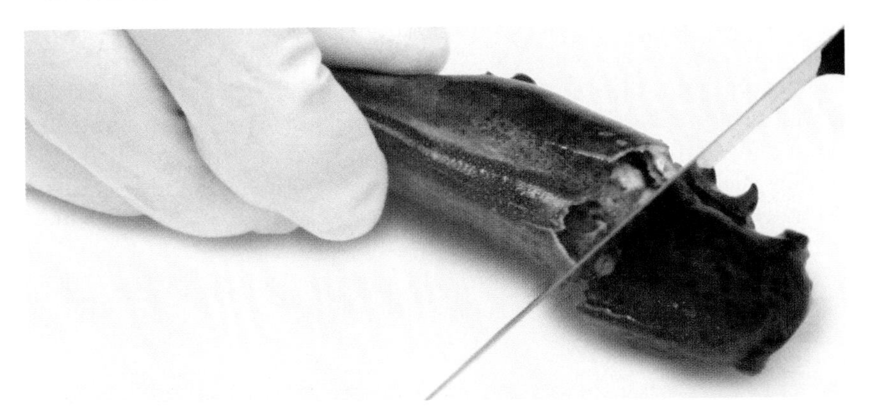

9. After removing the broken shell from the end of the claw, pull the meat out from inside the claw shell. Repeat Steps 7–9 on the other claw.

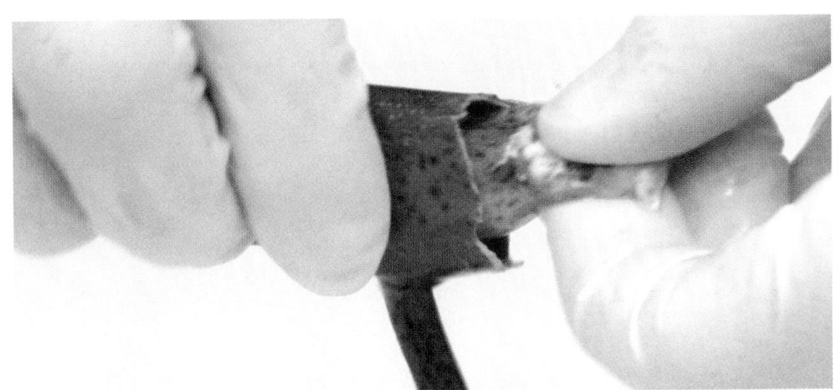

The majority of lobster meat comes from the knuckles, claws, and tail.

TECHNIQUE 47

CARVING MEATS ACROSS THE GRAIN ON THE BIAS

VIDEO 47

The meat from foods such as beef, pork, lamb, and poultry is mainly comprised of muscle fibers that run in a specific direction. The direction of these muscle fibers is referred to as the grain. The direction that the knife blade is held in relation to the grain impacts the tenderness and appearance of an item. For example, if beef is cut in the same direction as the grain, the end result is typically a tough, chewy piece of meat with a stringy appearance. In contrast, cutting beef across the grain breaks down and shortens the muscle fibers, resulting in a tender and visually appealing slice of meat. Cutting meat on the bias (at an angle) also shortens the muscle fibers and increases the likelihood of serving a tender product.

Note: Disposable gloves are typically worn in the foodservice industry when preparing ready-to-eat food items.

CARVING FLANK STEAKS ACROSS THE GRAIN ON THE BIAS

1. Locate the direction of the grain (muscle fibers) on a cooked and rested piece of flank steak.

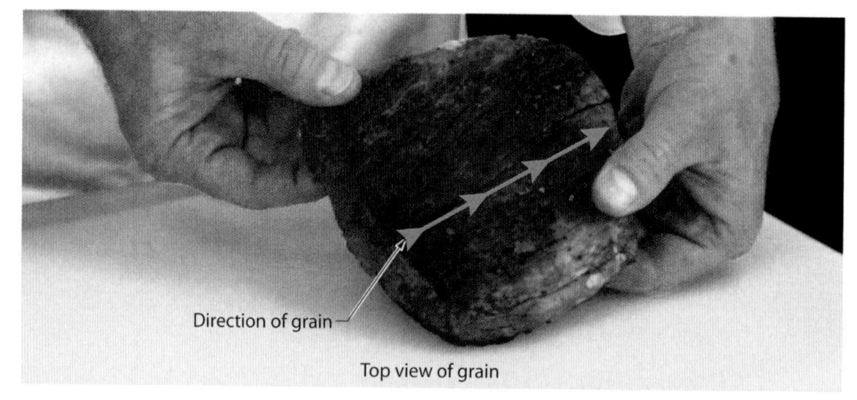

QUICK TIP

Flank steak is a boneless cut of lean beef that is long, narrow, and fibrous. For maximum tenderness, flank steak is typically marinated, grilled or broiled, and rested before thinly slicing it across the grain.

CARVING FLANK STEAKS ACROSS THE GRAIN ON THE BIAS (CONTINUED)

2. Position the steak so that the direction of the grain runs parallel with the body and hold the top of the steak with the guiding hand. To make a bias cut, place the edge of the knife blade across the grain near the end of the steak on the knife hand side. Twist the knife hand so that the spine of the knife angles slightly toward the guiding hand and the knife blade forms a 45° angle with the meat.

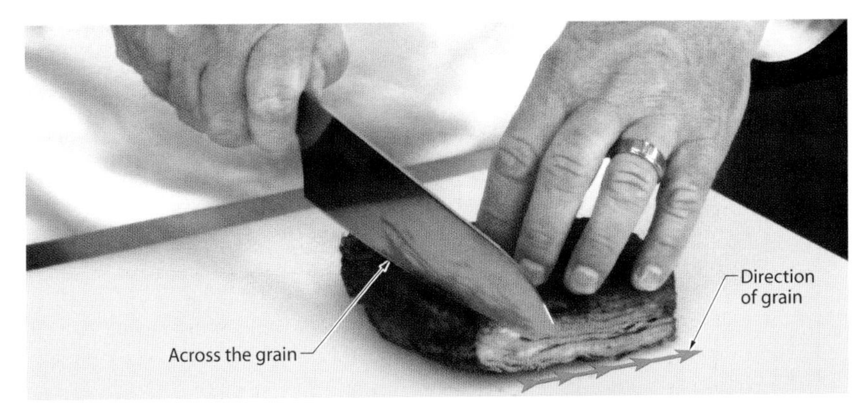

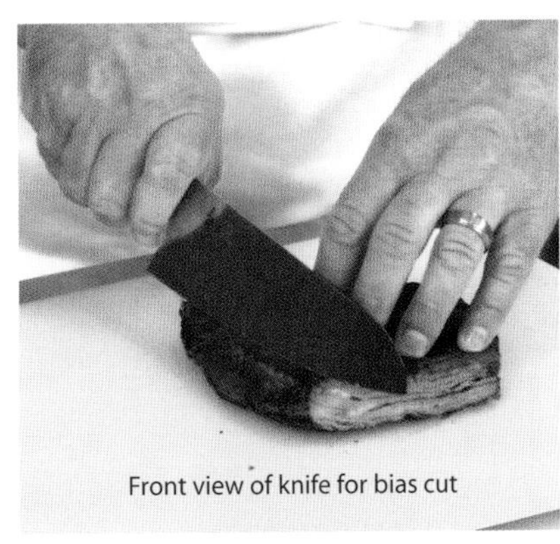

Front view of knife for bias cut

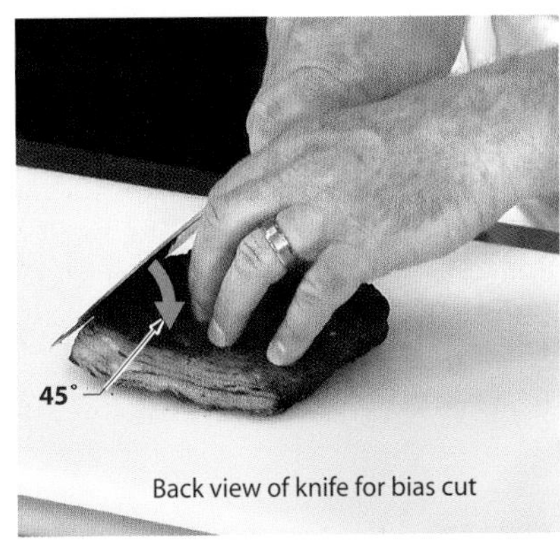

Back view of knife for bias cut

3. Maintain the angle of the knife blade and use a gentle sawing motion with smooth, long strokes to cut a slice of meat.

QUICK TIP

A 3 ounce serving of trimmed and cooked flank steak has 156 calories, 6 grams of total fat (3 grams of saturated fat), and 24 grams of protein.

4. Reposition the knife across the grain at a 45° angle near the end of the steak on the knife hand side. Cut another slice, again using a gentle sawing motion with smooth, long strokes.

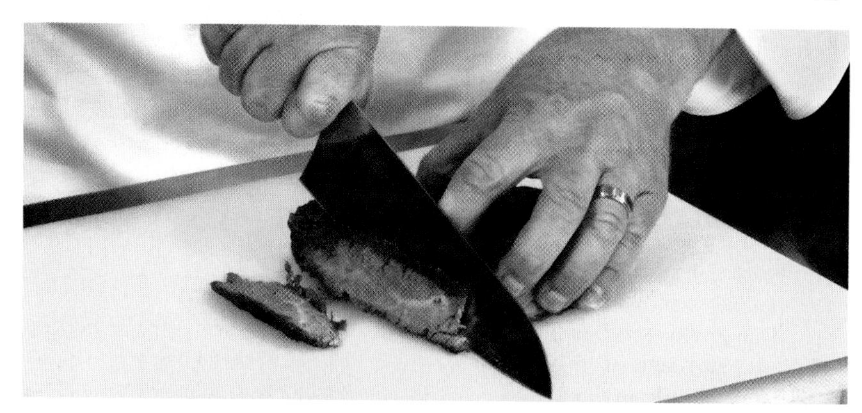

5. Repeat Step 4, taking care to cut slices of equal thickness.

Note: Cutting across the grain yields tender, visually appealing slices, whereas cutting with the grain yields tough fibrous slices.

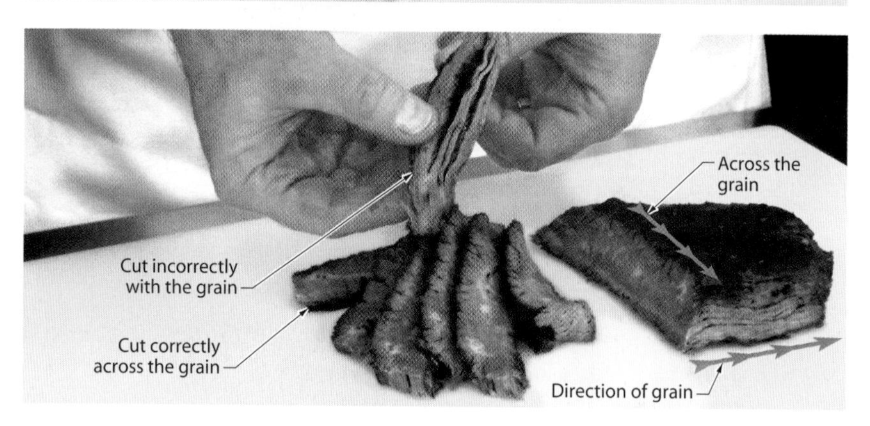

VIDEO 48

CARVING PRIME RIB ROASTS

TECHNIQUE 48

Prime rib is a common term used for a cut of beef that is actually a rib roast. In true culinary terms, "prime" refers to the highest beef grade given by the United States Department of Agriculture (USDA). The prime label is reserved for the finest beef, and what is labeled as "prime rib" often refers to the cut of beef, not the grade. "Prime rib" roast as a cut of beef usually includes seven rib bones from the upper rib section of a steer.

Before the roast is carved, it is rested to maintain juiciness. At this point, the fat cap and bones are often removed before further carving takes place. When the roast is carved with the rib bones intact, the first cut or end piece typically does not include the bone. Slices that contain a bone commonly alternate with slices that do not contain a bone. For example, the second slice would include a bone, the third slice would not include a bone, and so forth.

Messermeister

Note: Cooked and rested poultry and meats are typically carved while still hot and secured with a chef's fork. For demonstration purposes, the meat shown in this technique has been carved cold.

Scimitar

— or —

Messermeister

Stiff Boning Knife

Slicer

— or —

Carving Knife

CARVING BONE-IN PRIME RIB ROASTS

* This technique was executed by a left-handed chef.

1. Place a cooked and rested prime rib roast bone-side down on the cutting board with the ribs facing the knife hand side.

* This technique was executed by a left-handed chef.

2. Secure the roast with the guiding hand and use a scimitar to slice off the fat cap that runs along the rib bones.

   Note: Use caution to trim only the fat cap and not the meat.

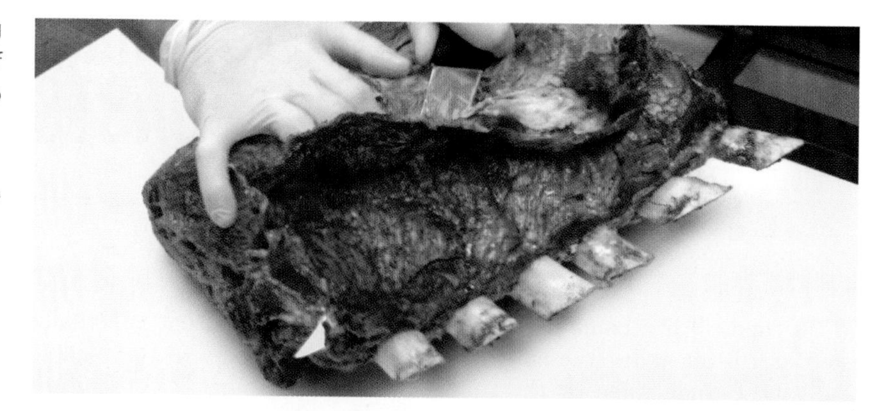

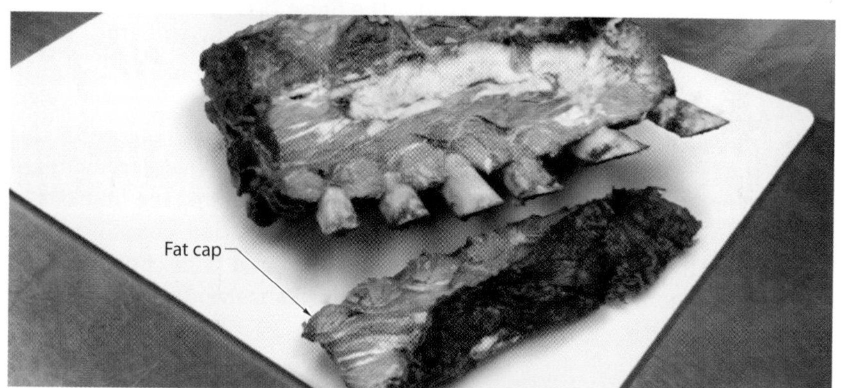

3. Rotate the roast 90° so that the end of the rib bones are facing the body. Place the guiding hand on top of the roast and use a slicer to cut off the end piece just before the first bone.

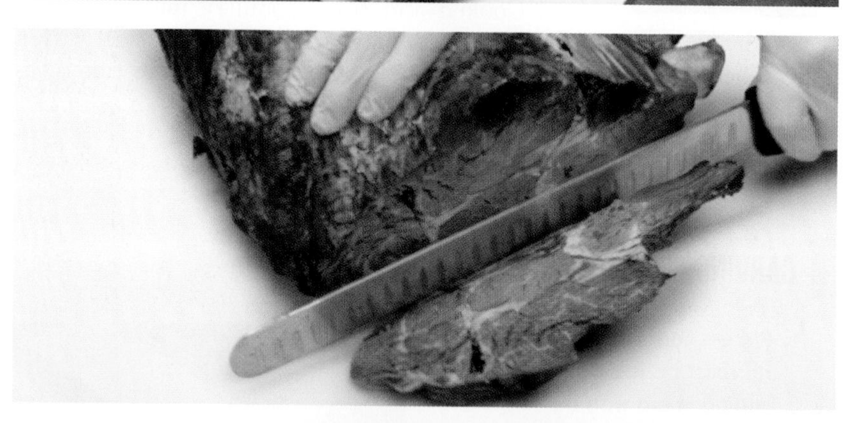

4. Continue cutting slices of prime rib to the desired thickness while avoiding the bones when necessary.

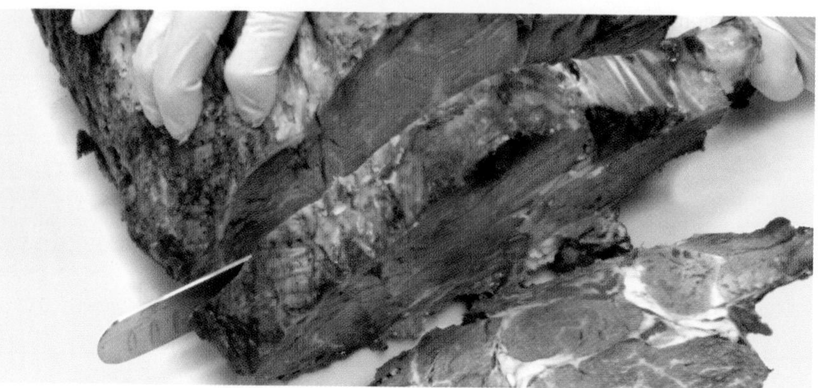

QUICK TIP

A bone-in slice of prime rib is often called a king cut, whereas a smaller, boneless slice may be referred to as a queen cut or an English cut.

DEBONING AND CARVING PRIME RIB ROASTS

* This technique was executed by a left-handed chef.

1. After removing the fat cap, stand a cooked and rested prime rib roast upright with the underside of the rib bones facing the guiding hand side. Hold the bones with the guiding hand and position a scimitar on top of the roast so that the side of the knife blade is flush against the ribs at the far end of the roast.

2. Use a sawing motion to cut along the natural curvature of the ribs. As the knife cuts deeper into the meat, use the guiding hand to help separate the meat from the ribs to ensure that the knife is cutting close to the bones.

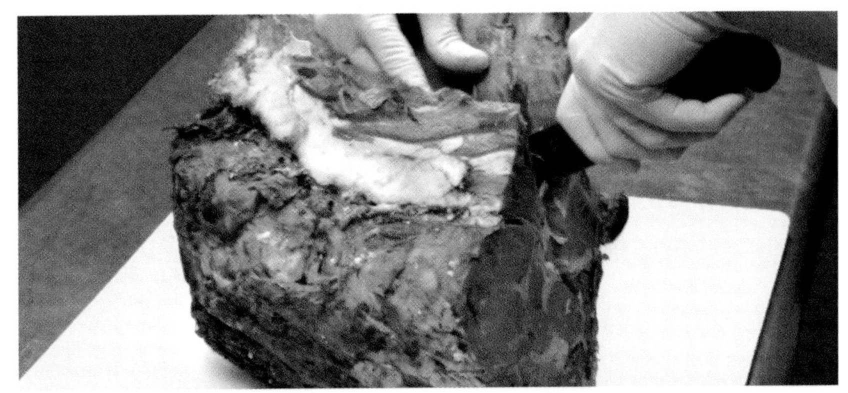

3. Continue cutting down the side of the rib bones until the ribs are separated from the rest of the roast.

> **QUICK TIP**
> *The beef ribs from a deboned prime rib roast are ideal for braising and serving with a barbeque-style sauce.*

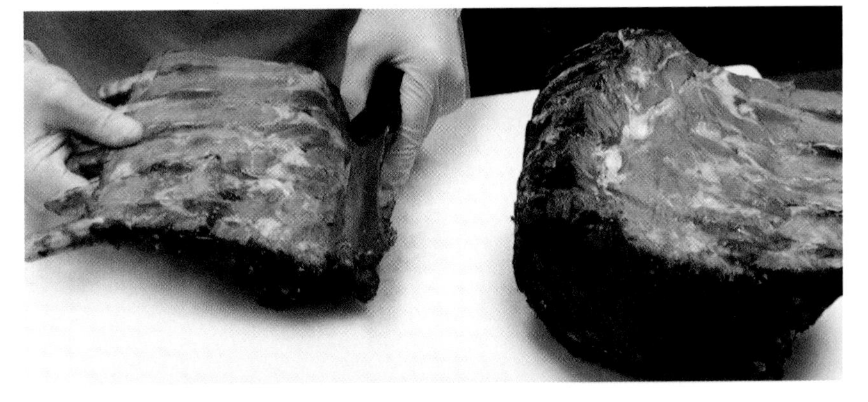

4. Position the roast rounded-side up with the tapered side facing the body. Secure the roast with the guiding hand and use a slicer to cut the roast into portions of the desired thickness.

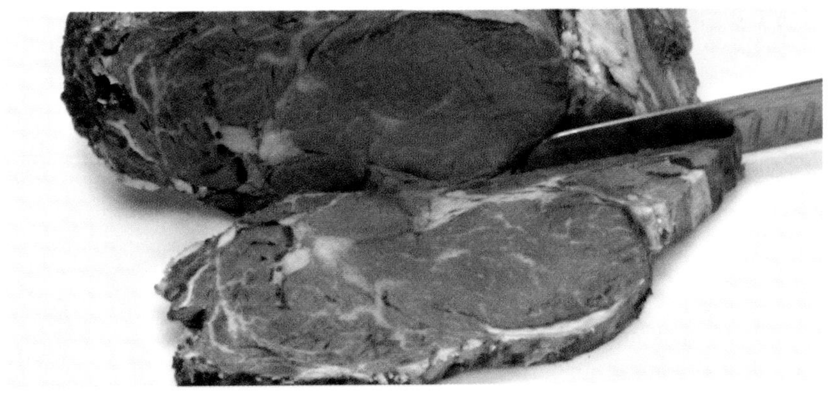

TECHNIQUE 49

CARVING HAM

Ham is a fabricated cut from the hind leg of a hog that is typically cut from the middle of the shank bone to the aitchbone, or hip bone. Hams are sold bone-in, partially deboned (semiboneless), and boneless. The advantage of bone-in hams is the flavor that the bones contribute during the cooking process.

After a cooked bone-in ham has rested, the aitchbone is removed. The meat can then remain on the shank and femur (leg bone) where it is cut into slices, or an entire side of ham can be cut away from the bones. A removed side of ham is boneless, making it convenient to slice, dice, julienne, or cut into thin strips.

Messermeister

Note: Cooked and rested poultry and meats are typically carved while still hot and secured with a chef's fork. For demonstration purposes, the meat shown in this technique has been carved cold.

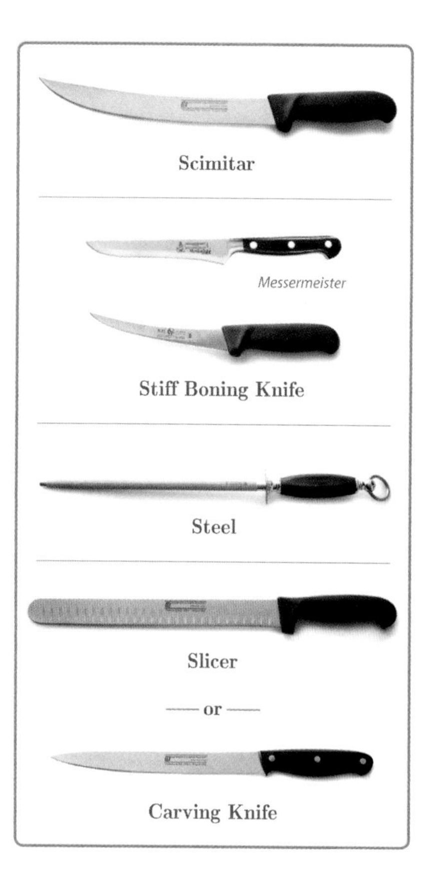

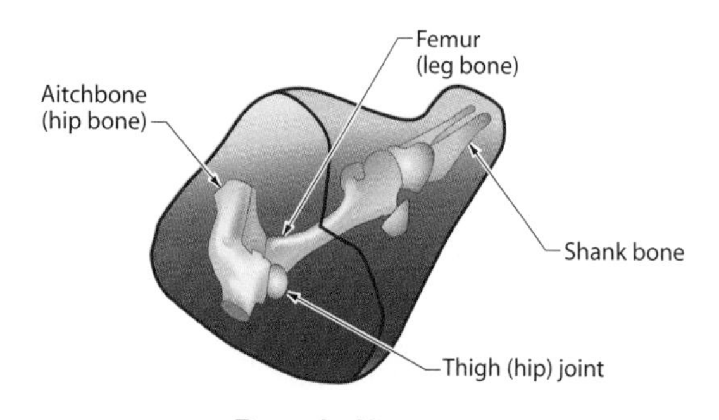

Bone-In Ham

CARVING WHOLE BONE-IN HAM

* This technique was executed by a left-handed chef.

1. Place a cooked and rested ham rounded-side (scored-side) up on the cutting board with the shank bone angled toward the guiding hand side. Secure the ham with the guiding hand and use a scimitar to trim most of the fat and skin from the sides and top of the ham.

   Note: Use caution to trim only the fat and skin and not the meat.

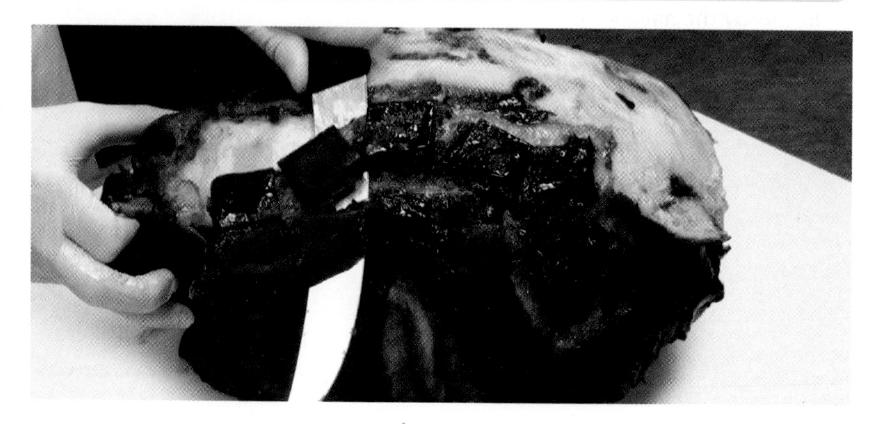

CARVING WHOLE BONE-IN HAM (CONTINUED)

* This technique was executed by a left-handed chef.

2. Flip the ham over so the side that has the skin removed is against the cutting board and the shank bone is facing away from the body. Use the tip of a stiff boning knife to cut away enough flesh to reveal the circular hole in the aitchbone. Continue cutting around the aitchbone to loosen it.

3. Insert the tip of a steel through the hole in the aitchbone. Secure the ham with the guiding hand and firmly bring the handle of the steel over the ham and down toward the shank bone to dislodge the aitchbone.

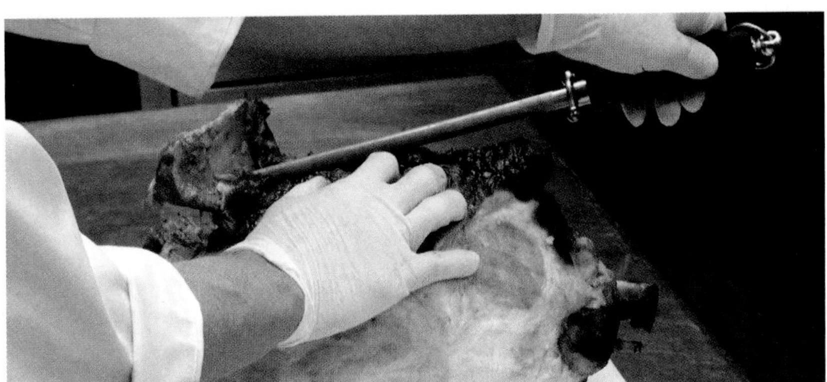

4. Use the stiff boning knife to cut away the flesh that remains around the aitchbone. Lift the aitchbone away from the ham.

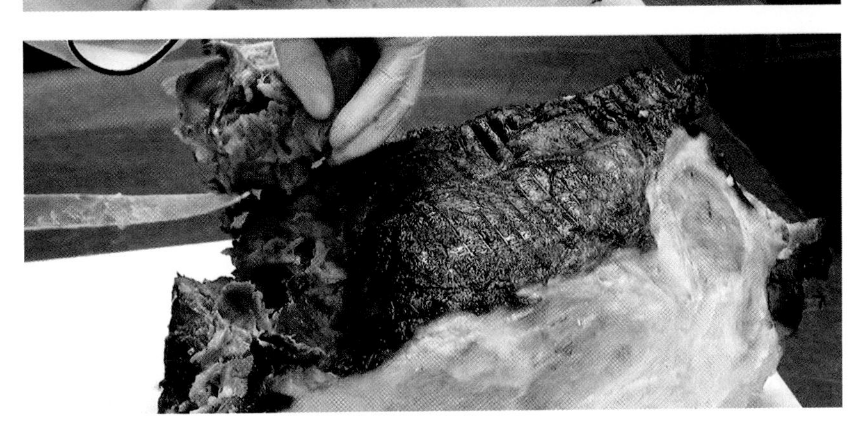

5. Steady the ham with the guiding hand and slice a small piece of ham from the thinner side of the leg to create a flat, steady base.

* This technique was executed by a left-handed chef.

6. Set the ham cut-side (base-side) down and secure the shank bone with the guiding hand. Hold the scimitar parallel to the cutting board and cut back along the top of the shank bone approximately 1 inch.

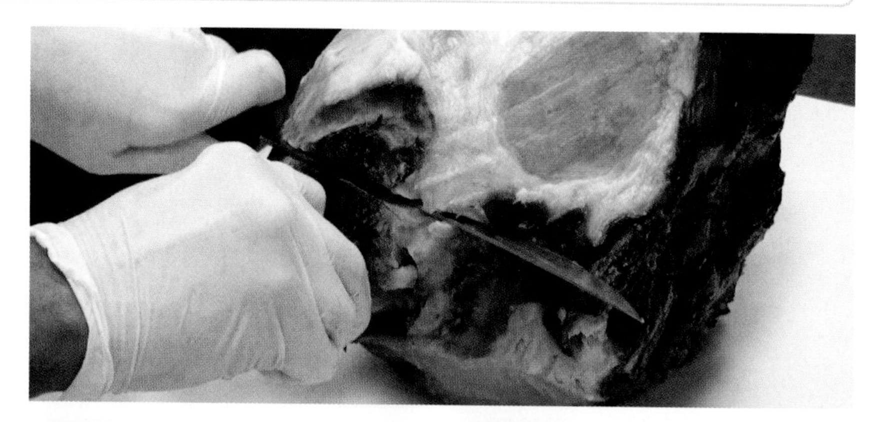

7. To remove a small wedge of ham and create a flat surface for slicing, reposition the knife perpendicular to the shank bone and cut down until reaching the end of the cut made along the shank bone in Step 6.

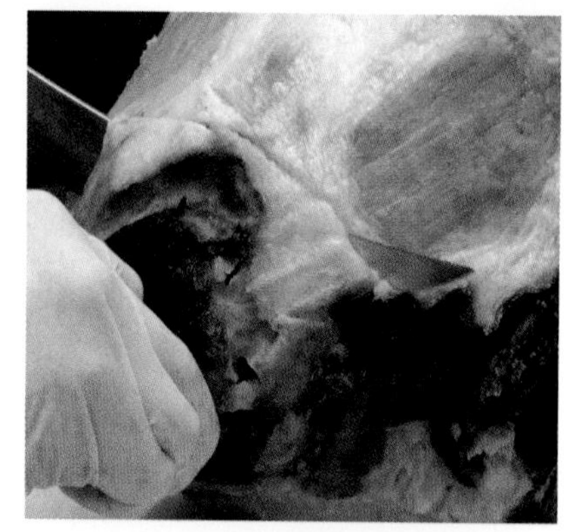

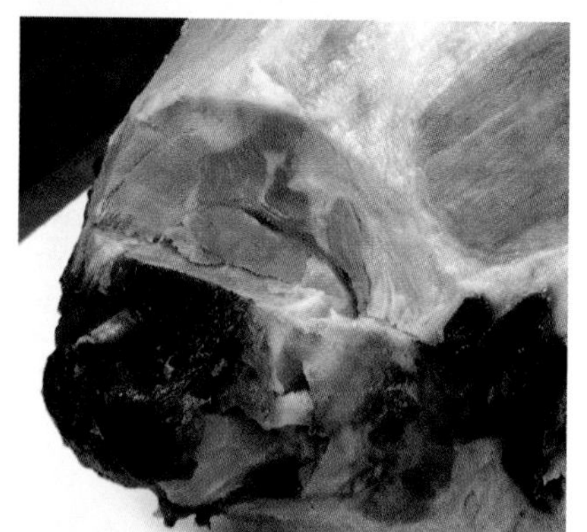

8. Use a slicer to cut the ham across the grain of the muscle fibers into consistently sized slices.

Note: Cutting across the grain yields tender, visually appealing slices, whereas cutting with the grain yields tough, fibrous slices.

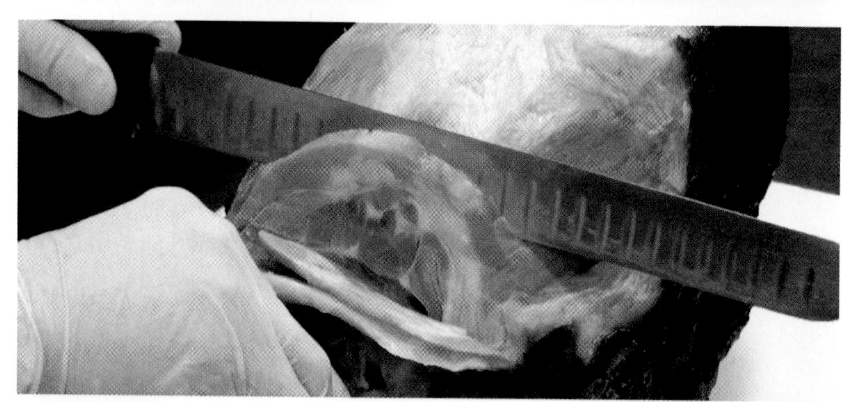

9. Cut along the leg bones to release the slices.

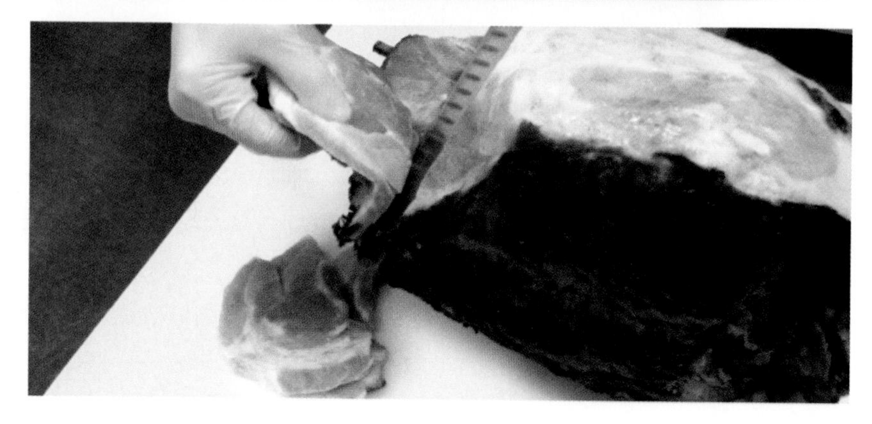

REMOVING AND CARVING SIDES OF HAM

* This technique was executed by a left-handed chef.

1. Start with a cooked and rested ham trimmed of fat and skin with the aitchbone removed. Place the ham rounded-side up on the cutting board with the shank bone facing the guiding hand side.

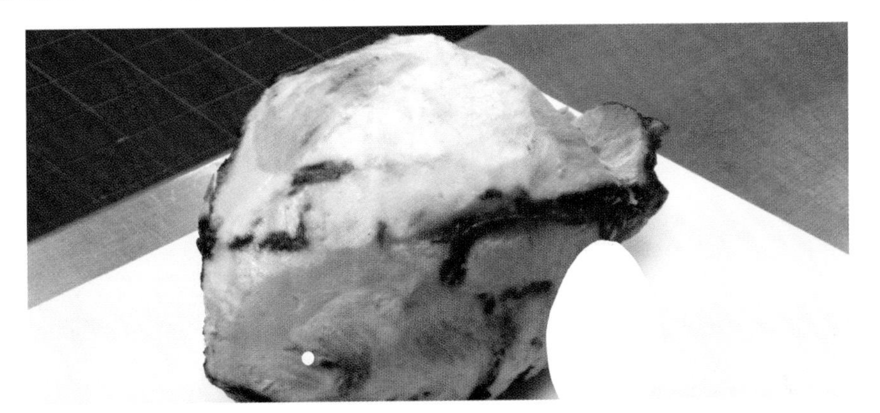

2. To remove a side of ham, hold a scimitar parallel with the cutting board and cut along the top of the shank and femur bones the entire length of the ham. Use the guiding hand to keep the ham secured in place.

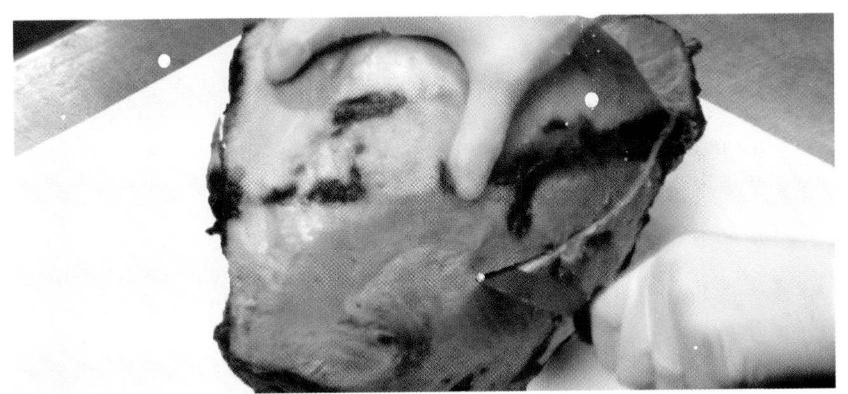

3. Place the removed side of ham cut-side (flat-side) down and secure with the guiding hand. Use a slicer to cut the ham across the grain of the muscle fibers into consistently sized slices.

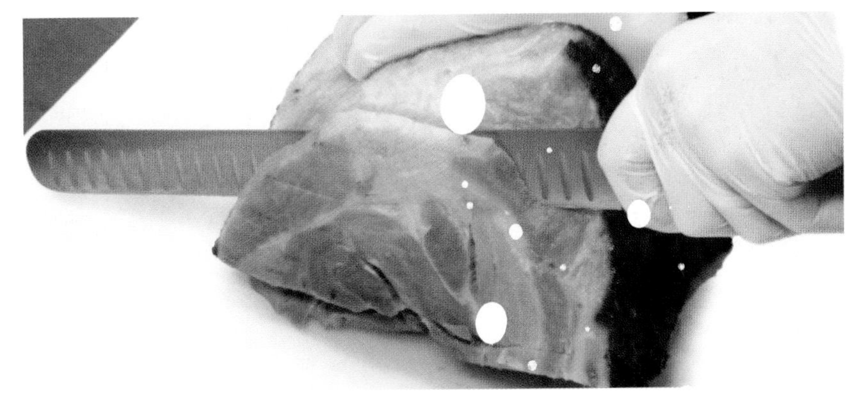

QUICK TIP

According to the United States Department of Agriculture (USDA), fresh (raw) ham should be cooked to a minimum internal temperature of 145°F and allowed to rest for at least 3 minutes. As the ham rests, the internal temperature will stay the same or rise slightly to help ensure that harmful germs are destroyed.

TECHNIQUE 50

CARVING LEGS OF LAMB

VIDEO 50

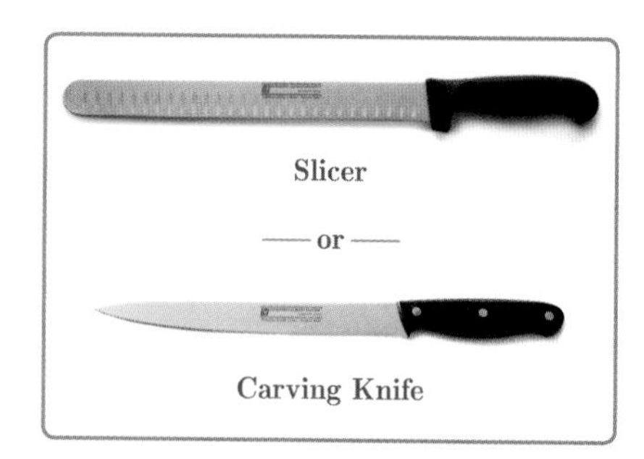

Legs of lamb can be prepared bone-in, partially deboned (semiboneless), or boneless. In a partially deboned leg of lamb, only the shank bone remains.

The advantage of cooking a partially deboned leg of lamb is that both the outside and inside of the leg can be seasoned with herbs and spices to layer and develop flavor. After seasoning the leg, it is often tied with butcher's twine in order to maintain its shape and cook evenly. Once the meat has been cooked and rested, the twine is removed and the leg of lamb is typically sliced across the grain and on the bias.

Messermeister

Note: Cooked and rested poultry and meats are typically carved while still hot and secured with a chef's fork. For demonstration purposes, the meat shown in this technique has been carved cold.

CARVING PARTIALLY DEBONED LEGS OF LAMB

* This technique was executed by a left-handed chef.

1. Start with a partially deboned leg of lamb that has been cooked and rested. Place the leg rounded-side up on the cutting board with the shank bone facing the guiding hand side and secure the leg with the guiding hand. Angle a slicer across the grain of the muscle fibers near the end of the leg on the knife hand side.

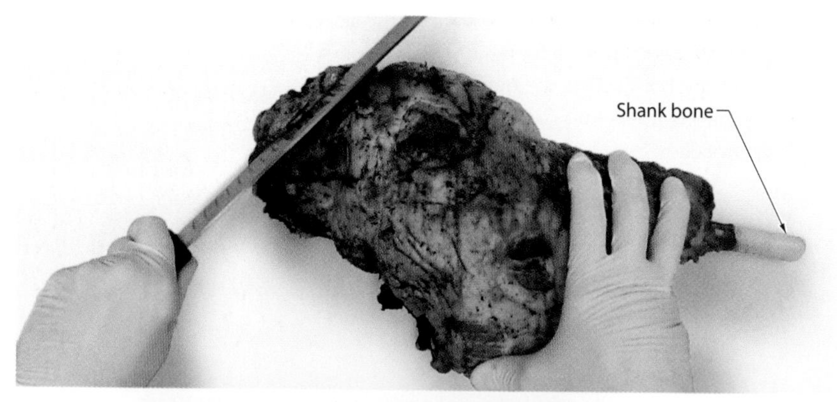

2. Use a smooth sawing motion to cut the leg across the grain and on the bias to yield two or three consistently sized slices.

   Note: Cutting across the grain yields tender, visually appealing slices, whereas cutting with the grain yields tough, fibrous slices.

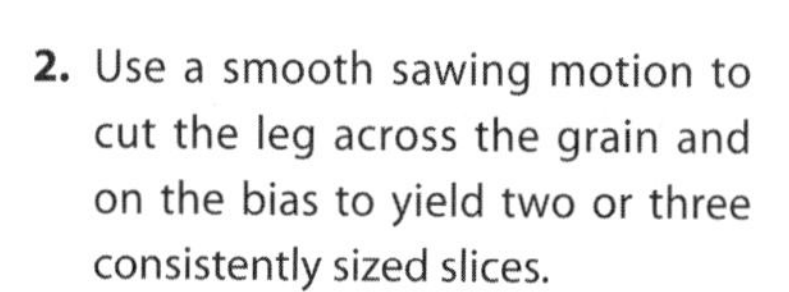

CARVING PARTIALLY DEBONED LEGS OF LAMB (CONTINUED)

* This technique was executed by a left-handed chef.

3. Rotate the shank bone toward the body and secure the leg with the guiding hand. Reposition the slicer so that it is again angled across the grain near the end of the leg on the knife hand side.

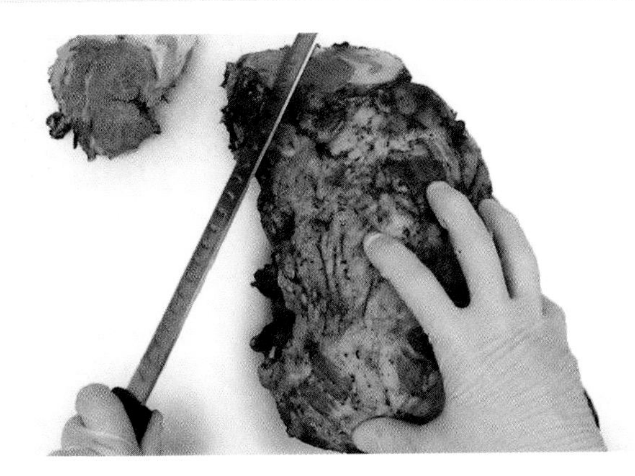

4. Repeat Step 2 by using a smooth sawing motion to cut the leg across the grain and on the bias to yield two or three consistently sized slices.

5. Continue to rotate and slice the lamb across the grain and on the bias to yield slices of equal thickness.

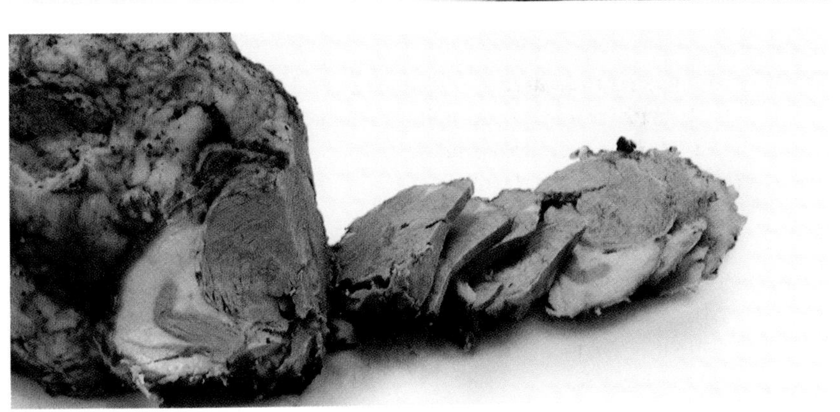

QUICK TIP

When carving a leg of lamb, cut only the desired amount. This helps ensure that the lamb stays juicy, tender, and flavorful.

GLOSSARY

A

airline poultry breast: A skin-on, semiboneless poultry breast with the first bone of the wing (drumette) still attached.

B

batonnet cut: A knife technique that produces a stick-shaped cut ¼ inch wide × ¼ inch high × 2 inches long.

bevel: The angled region of a knife blade that has been ground to form the cutting edge.

bivalve: A type of mollusk that has a top and a bottom shell connected by a central hinge such as an oyster, clam, mussel, or scallop.

blade: The sharp, flat portion of a knife that is used for cutting food items.

bolster: A thick band of metal located where a knife blade joins the handle.

boning knife: A knife with a thin, pointed blade that is 5–6 inches long and is used to separate flesh from bones with minimal waste.

bread knife: A knife with a serrated blade that is 8–12 inches long and used to cut through bread without crushing the soft interior.

brunoise cut: A knife technique that produces a dice-shaped cube with six equal sides measuring ⅛ inch each.

butcher's knife: A heavy knife with a curved tip and a blade that is 7–14 inches long.

butterfly: A knife technique in which a food item is cut almost completely through the center, resulting in two halves of flesh that can be spread apart to lay flat.

C

carving knife: A knife with a tapered blade and pointed tip that is 8–10 inches long and used to slice cooked meats and poultry.

cephalopod: A type of mollusk, such as squid, octopuses, and cuttlefish, that lacks an external shell, has arms around the head, has developed eyes, and often contains an ink sac.

channel knife: A cutting tool with a thin metal blade within a raised channel that is used to remove large strings from the surface of a food item.

chef's knife: A versatile knife typically with an 8-, 10-, or 12-inch tapering blade that is used to slice, dice, chop, and mince foods. Also known as a French knife.

chiffonade cut: A knife technique that produces long strips of herbs or leafy greens.

chisel-ground edge: An asymmetrical knife blade that is ground with one angled side that slants toward the cutting edge.

chopping: A knife technique that involves rough-cutting an item into small pieces that lack uniformity in shape and size.

clam knife: A short knife with a sharp edge and rounded tip that is used to open clams.

cleaver: A heavy knife with a rectangular blade that is typically used to cut through large bones.

compound-ground edge: A symmetrical knife blade that is ground to start forming a V-shape, but before reaching the edge, the blade tapers in again to form a smaller V-shape at the cutting edge.

concassé: A preparation method in which a tomato is peeled, seeded, and then diced or chopped.

convex-ground edge: A symmetrical knife blade that is ground with arced (convex) sides that angle smoothly downward to meet at the cutting edge.

D

diagonal cut: A knife technique that produces flat-sided, oval or semioval cuts.

dice cut: A knife technique that produces precise cubes cut from uniform stick-shaped cuts.

drawn fish: A fish that has had only the internal organs (viscera) removed.

E

edge: The sharpened part of a knife blade that extends from the heel to the tip.

F

fatback: The layer of fat that runs along the back of a hog and is generally trimmed during fabrication.

fat cap: The fat that surrounds a muscle.

fell: A thin, tough membrane that lies directly under the hide and over the fat layer.

fillet knife: A knife with a thin, flexible blade and a fine point that is 6–9 inches long and is used to cut delicate flesh.

fine brunoise cut: A knife technique that produces a dice-shaped cube with six equal sides measuring 1⁄16 inch each.

fine julienne cut: A knife technique that produces a stick-shaped cut 1⁄16 inch wide × 1⁄16 inch high × 2 inches long.

fish fillet: A lengthwise piece of flesh cut away from the backbone.

flatfish: Any thin, wide fish with both eyes located on one side of the head and a backbone that runs from head to tail through the midline of the body.

fluted cut: A knife technique that produces a decorative spiral pattern on the surface of an item by removing only a sliver of the item with each cut.

frenching: The process of removing the meat, fat, and membranes from the end of a bone.

fruit-vegetable: A botanical fruit that is sold, prepared, and served as a vegetable.

full tang: The tail of a knife blade that extends to the end of the handle and typically contains several rivets.

G

granton-edge blade: A knife blade with hollowed out grooves along both sides of the edge.

H

ham: A fabricated cut from the hind leg of a hog that is typically cut from the middle of the shank bone to the aitchbone, or hip bone.

handle: The area of a knife designed to be held in the hand.

heel: The rear portion of a knife blade that is most often used to cut thick food items in which more force is required.

hollow-ground edge: A symmetrical knife blade that is ground with sides that arch inward (concave) to meet at the cutting edge.

honing: The process of aligning a knife blade's edge and removing any burrs or rough spots on the blade. Also known as truing.

I

included angle: The sum of the angles on both sides of a knife blade.

J

julienne cut: A knife technique that produces a stick-shaped cut ⅛ inch wide × ⅛ inch high × 2 inches long. Also known as matchstick.

K

kitchen shears: A cutting tool that operates like heavy-duty scissors and is used to cut through the skin, bones, joints, and ligaments of poultry, the fins of fish, and the shells of some crustaceans.

L

leg of lamb: A primal cut of lamb that contains the last portion of the backbone, hip bone, aitchbone, femur (leg bone), hindshank, and tailbone.

M

mandolin: A cutting tool with adjustable steel blades that is used to cut food into consistently sized pieces.

medallion: A pounded disc-shaped cut of meat that weighs approximately 2–4 ounces.

mincing: A knife technique that involves finely chopping an item to yield very small pieces that are not entirely uniform in shape.

O

oblique cut: A knife technique that produces a wedge-shaped cut with two angled sides. Also known as a rolled cut.

oyster knife: A short knife with a dull edge and tapered tip that is used to open oysters.

P

paring knife: A short knife with a stiff blade that is 2–5 inches long and typically used to trim or peel fruits and vegetables.

parisienne scoop: A cutting tool with either one or two sharp-edged scoops attached to a handle that is used to cut fruits and vegetables into uniform spheres. Also known as a melon baller.

partial tang: A shorter tail of a knife blade that has fewer rivets than a full tang.

paysanne cut: A knife technique that produces a flat, tile-shaped square, circular, or triangular cut ½ inch wide × ½ inch high × ⅛ inch thick.

peeler: A cutting tool with a swiveling, double-edged blade attached to a handle that is used to remove the skin or peel from fruits and vegetables.

point: The foremost section of a knife tip that can be used as a piercing tool.

pome: A fleshy fruit that contains a core of seeds and has an edible skin.

pork loin: A primal cut that extends along the greater part of the backbone, from about the second rib, through the rib and loin area of a hog.

poultry: A term used to describe chickens, turkeys, ducks, geese, guinea fowls, and squabs.

poultry tenderloin: A thin strip of muscle that runs along the inside, lower section of the breast and is situated close to the bone. Also known as a poultry tender.

primal cut: A large cut from a whole or a partial carcass.

R

rat-tail tang: A narrow rod of metal that runs the length of a knife handle but is narrower than the handle.

rivet: A metal fastener used to securely attach the tang of a knife to the handle.

rondelle cut: A knife technique that produces flat-sided, circular cuts. Also known as a round cut.

roulade: A thin piece of meat or poultry that is stuffed (filled), rolled, and cooked.

roundfish: Any fish with a cylindrical body, an eye located on each side of the head, and a backbone that runs from head to tail in the center of the body.

S

santoku knife: A knife with a broad blade and a razor-sharp edge that is less tapered than a chef's knife.

scimitar: A long knife with an upward-curved tip that is used to cut steaks and primal cuts of meat.

serrated-edge blade: A knife blade with a scallop-shaped or sawlike edge.

shucking: The process of opening and removing the shell of a bivalve.

silverskin: Tough, rubbery, silver-white connective tissue that does not break down when heated.

slicer: A knife with a rounded tip and a 10–14 inch long blade that is used to slice cooked meats, poultry, and fish.

spine: The unsharpened top part of a knife blade that is opposite the edge.

steel: A steel rod approximately 18 inches long that is attached to a handle and used to align the edge of knife blades. Also known as a butcher's steel.

straight-edge blade: A knife blade with a smooth edge from the tip of the knife to the heel.

supreme: The flesh from a segment of citrus fruit that has been cut away from the membrane.

T

tang: The unsharpened tail of a knife blade that extends into the handle.

tenderloin: A tapered strip of muscle in animals such as cows and pigs that extends into both the short loin and sirloin and runs lengthwise with the backbone below the ribs.

tip: The front quarter of a knife blade.

tourné cut: A knife technique that produces a seven-sided, football-shaped item with small flat ends.

tourné knife: A short knife with a curved blade that is primarily used to carve vegetables into a seven-sided football shape with flat ends called a tourné. Also known as a bird's beak knife.

U

utility knife: A multipurpose knife with a stiff blade that is 6–10 inches long and is similar in shape to a chef's knife but much narrower at the heel.

V

V-ground edge: A symmetrical knife blade that is ground to form a V-shaped cutting edge. Also known as a taper-ground edge.

W

whetstone: A stone that is used to grind the edge of a blade to the proper angle for sharpness.

Z

zester: A cutting tool with tiny blades inside small holes that are attached to a handle.

INDEX

E

F

G

H

I

J

K

L

T

U

V

W

Z